Explanation and its Limits

ROYAL INSTITUTE OF PHILOSOPHY LECTURE SERIES: 27
SUPPLEMENT TO *PHILOSOPHY* 1990

EDITED BY

Dudley Knowles

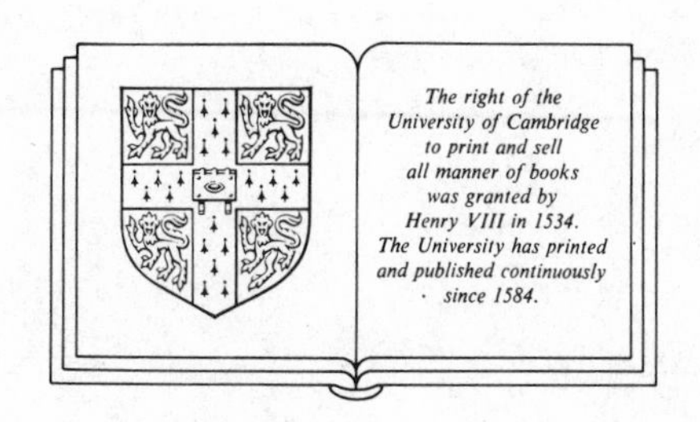

CAMBRIDGE UNIVERSITY PRESS

CAMBRIDGE
NEW YORK PORT CHESTER MELBOURNE SYDNEY

Published by the Press Syndicate of the University of Cambridge
The Pitt Building, Trumpington Street, Cambridge, CB2 1RP
40 West 20th Street, New York, NY 10011, USA
10 Stamford Road, Oakleigh, Melbourne 3166, Australia

British Library Cataloguing in Publication Data

Explanation and its limits.—(Royal Institute of Philosophy lecture series).
1. Explanation
I. Knowles, Dudley II. Philosophy III. Series 160

ISBN 0 521 39598 4

Library of Congress Cataloguing in Publication Data

Explanation and its Limits/edited by Dudley Knowles,
p. cm.—(Royal Institute of Philosophy lectures: 27)
'Supplement to Philosophy 1990'
Includes bibliographical references and index
ISBN 0 521 39598 4
1. Science—Philosophy—Congresses.
2. Explanation (Philosophy)—Congresses.
I. Knowles, Dudley. II. Philosophy (London, England).
1990 (Supplement). III. Series: Royal Institute of Philosophy lectures: v. 27.
Q175.E96 1990 90-37622
501—dc20 CIP

Origination by Precise Printing Company Ltd, Reigate, Surrey
Printed in Great Britain by the University Press, Cambridge

Contents

Preface

The papers collected in this volume were delivered at the Royal Institute of Philosophy Conference on 'Explanation and its Limits' in September 1989 at the University of Glasgow.

Dr Flint Schier, one of the original organizers of the conference, died in May 1988. He worked hard to prepare a list of speakers. It is sad that he did not see his ideas bear fruit. Readers may judge of the quality of the proceedings from the essays that follow. It is worth reporting in addition that all the discussions were spirited and probing, the company (of scientists and laymen, as well as academic philosophers) was friendly and congenial and the staff of Queen Margaret Hall made sure everyone was welcome and well-fed!

Special thanks are due to the Royal Institute of Philosophy, the Royal Society of Edinburgh and the University of Glasgow for their financial support. Without this generosity we would not have been able to assemble such an impressive, international group of contributors.

Finally, thanks are due to the many people who helped me to organize the conference and prepare this volume, especially Robin Downie, Pat Shaw, Anne Southall, Anne Valentine and Helen Knowles. I should like to thank, too, in advance of publication, the staff of Cambridge University Press on whose good offices I shall continue to presume.

Department of Philosophy
University of Glasgow

Dudley Knowles

List of Contributors

Peter Clark
Department of Logic and Metaphysics, University of St. Andrews

Stephen R. L. Clark
Professor of Philosophy, University of Liverpool

Jim Edwards
Department of Philosophy, University of Glasgow

Philip Gasper
Assistant Professor of Philosophy, Middlebury College, Vermont

Joyce Kinoshita
Assistant Professor of Philosophy, College of the Holy Cross, Worcester, Mass.

Peter Lipton
Assistant Professor of Philosophy, Williams College, Williamstown, Mass.

John Maynard Smith
Professor of Biology, University of Sussex

David Papineau
Department of History and Philosophy of Science, University of Cambridge

Michael Redhead
Professor of History and Philosophy of Science, University of Cambridge

David-Hillel Ruben
Department of Philosophy, Logic and Scientific Method, The London School of Economics and Political Science

Matti Sintonen
Research Fellow, The Academy of Finland; acting Professor of Philosophy, University of Turku, Finland

John Skorupski
Professor of Moral Philosophy, University of St Andrews

J. J. C. Smart
Formerly Professor of Philosophy, University of Adelaide and the Australian National University; Hon. Research Fellow, Automated Reasoning Project, Research School of Social Sciences, Australian National University

Elliott Sober
Hans Reichenbach Professor of Philosophy, University of Wisconsin – Madison

Richard Swinburne
Nolloth Professor of the Philosophy of the Christian Religion, University of Oxford

James Woodward
Associate Professor of Philosophy, California Institute of Technology

Explanation—Opening Address

J. J. C. SMART

It is a pleasure for me to give this opening address to the Royal Institute of Philosophy Conference on 'Explanation' for two reasons. The first is that it is succeeded by exciting symposia and other papers concerned with various special aspects of the topic of explanation. The second is that the conference is being held in my old *alma mater*, the University of Glasgow, where I did my first degree. Especially due to C. A. Campbell and George Brown there was in the Logic Department a big emphasis on absolute idealism, especially F. H. Bradley. My inclinations were to oppose this line of thought and to espouse the empiricism and realism of Russell, Broad and the like. Empiricism was represented in the department by D. R. Cousin, a modest man who published relatively little, but who was of quite extraordinary philosophical acumen and lucidity, and by Miss M. J. Levett, whose translation of Plato's *Theaetetus*[1] formed an important part of the philosophy syllabus.

Despite my leaning towards empiricism and realism, I remember having been impressed by Bernard Bosanquet's *Implication and Linear Inference*.[2] In the main, one must regard this book as a book not on logic but on epistemology. Linear inference is the sort of inference we get in a mathematical proof, but Bosanquet's notion of implication is much closer to the contemporary notion of inference to the best explanation, which Gilbert Harman, who introduced the term, has argued to be the core notion of epistemology.[3] I shall in general be agreeing with Harman here, though I shall not take a stand on the question of whether *all* inductive reasoning can be so described. Bosanquet's notion of implication stresses the notion of fitting into a system. In this way, also, it is not too far from Quine's and Ullian's metaphor of the web of belief.[4] The notion of argument to the best explanation indeed goes back to the idea

[1] *The Theatetus of Plato*. Later reissued with a preface by D. R. Cousin (University of Glasgow Press, 1977). (Originally published by Jackson, Wylie and Co.: Glasgow, 1928.)

[2] London: Macmillan, 1920.

[3] Gilbert Harman, 'The Inference to the Best Explanation', *Philosophical Review*, **74**, 88–95, and *Thought* (Princeton University Press, 1975).

[4] W. V. Quine and J. S. Ullian, *The Web of Belief*, revised edition (New York: Random House, 1978).

of coherence as we find it in the writings of writers such as Bradley, Bosanquet and Joachim, though it is important to stress that coherence should be seen as an epistemological notion, not a semantical one. In other words, coherence provides a bad theory of truth, but a good theory of warranted assertability.[5]

This brings us to the main theme of this conference. I want to characterize explanation of some fact as a matter of fitting belief in this fact into a system of beliefs. The affinities to the coherence theory and to the notion of the web of belief are obvious. In fact I think it just *is* the coherence theory, so long as it is remembered that I take the coherence theory to be a theory of warranted assertability, not a theory of truth, an epistemological theory, not a semantical or ontological theory. Of course the notion of fitting into a system or of coherence is not at all a precise one. This is, I think, because the notion of explanation itself is not a very precise one. Nevertheless though the notion of coherence may not be a very precise one, it seems to me to be a philosophically important one, and I think that, as I shall shortly try to explain, it has gained in scientific respectability, with all its imprecision, by being put into relation with concepts and investigations of artificial intelligence, in particular by some work by Paul Thagard,[6] and also by recent work in the dynamics of belief systems.[7]

The concept of explanation is a somewhat polymorphous one, though that is not the source of the sort of imprecision that I had in mind in the last paragraph. Mainly I shall be concerned with scientific and historical explanation i.e. the explanation of particular facts (and in the case of the exact sciences) of laws. There are peripheral uses of the terms 'explain' and 'explanation'. For example, in the previous paragraph I said that I would shortly try to explain something to do with the notion of coherence. This was not a promise to try to give a scientific or historical explanation of the origins of the notion of coherence. Similarly, a teacher might explain to a pupil the meaning of a word. This is also a peripheral sense. Nevertheless even these peripheral cases can perhaps be said to fall under the general notion of fitting something into a web of belief. I class mathematical explanation with scientific

[5] The term 'warranted assertability' apparently goes back to John Dewey, though Dewey did not make the sharp distinction which I hold that we ought to make, between warranted assertability and truth.

[6] See Paul Thagard, *Computational Philosophy of Science* (Cambridge, Mass.: M.I.T. Press, 1988), and 'Explanatory Coherence', *Behavioral and Brain Sciences* **12** (1989), 435–67.

[7] See Brian Ellis, *Rational Belief Systems* (Oxford: Blackwell, 1979), Peter Forrest, *The Dynamics of Belief* (Oxford: Blackwell, 1986), Peter Gärdenfors, *Knowledge in Flux: Modeling the Dynamics of Epistemic States* (Cambridge, Mass.: M.I.T. Press, 1988).

explanation. This would be resisted by those who, unlike me, regard the notion of causation as essential to scientific explanation. Thus one might ask for the explanation of the fact that $e^{i\pi}=-1$. The explanation would come from the proposition that $\cos\theta+i\sin\theta=e^{i\theta}$, and the derivation of this last formula. All these cases whether central or peripheral, come under the hospitable notion of fitting into a web of belief.

What explains what? We sometimes say that at a certain time a person explains something to another person. This suggests that an explanation (an explaining) is a speech act. This is how Peter Achinstein has suggested that we think of an explanation.[8] On the other hand one may say that the theory of natural selection explains the fact of evolution as discovered from the palaeontological record, or that Newtonian celestial mechanics does not explain the advance of the perihelion of Mercury, while Einstein's general theory of relativity does. This is surely a legitimate notion of explanation, but it could be subsumed under the speech act one by saying that someone's belief in the fact of evolution coheres with his or her belief in Darwin's theory, and so on. C. G. Hempel, with his Deductive-Nomological and Inductive-Statistical patterns of explanation was concerned to explicate our rough notion of explanation in purely syntactical terms. The notion of explanation as a speech act obviously allows us to bring in pragmatic considerations. For McTavish to explain the tides to Jones there must be some considerations about the state of knowledge of Jones. Suppose that Jones already knows the explanation of tides. Does McTavish then explain the tides to Jones? Again suppose that Jones is not competent to follow McTavish's proof that the tidal phenomena follow from laws of nature and statements of initial conditions that Jones accepts or is prepared to accept on McTavish's authority. It would sound a bid odd to say that in this case McTavish succeeds in explaining the tides to Jones.

We might talk of the sentences or propositions asserted by McTavish in his attempt to explain the tides to Jones as an explanation in some sense, akin to Hempel's perhaps, in which an explanation was a linguistic or abstract entity. It is an explanation in the sense that in typical circumstances it would furnish an explanation in the speech act sense.

Talking of x explaining y to z may concentrate attention too much on the didactic type of explanation. Crucially important in the development of science is the discovery of explanations, i.e. cases in which z is identical to x, as when, for example, Newton learned to explain the

[8] Peter Achinstein, *The Nature of Explanation* (Oxford: Clarendon Press, 1983). See also Sylvain Bromberger, 'An Approach to Explanation', in R. J. Butler (ed.), *Analytical Philosophy* Second Series (Oxford: Basil Blackwell, 1965).

tides to himself. Crucial to this was his already discovered theory of mechanics and gravitational attraction. The theory looms so large in our account of what happened that it is not surprising that we sometimes use 'explanation' in an abstract sense, not in that of a speech act. Nevertheless I think that we should follow Achinstein in taking the speech act sense as primary. After all a theory may function as an explanation to one person and yet fail so to function when addressed to another person. What fits into Jim's web of belief will not necessarily fit into George's. Thus certain links in a creation scientist's web may be so strong that they cannot easily be broken by the discussion of an orthodox biologist. This sort of thing does not only occur in the conflict between science and pseudo-science. It of course also occurs within orthodox science when scientists passionately, but reasonably, differ over the merits of some new theory.

As I have indicated earlier Paul Thagard's work on artificial intelligence modelling of explanation may help to give some scientific respectability to the general notion of argument to the best explanation (and of explanation itself) and of the notion of coherence or fitting into a system.[9]

Thagard has laid down various principles of explanatory coherence and incoherence.[10] The first principle is that coherence and incoherence are symmetrical as between pairs of propositions. The second principle is that if a set of propositions explains a further proposition, then each proposition in the set coheres with that further proposition, and that propositions in the set cohere pairwise with one another. He takes the degree of coherence to be inversely proportional to the number of propositions in the set. The third principle is one of *analogy*: if two analogous propositions separately explain different ones of a further pair of analogous propositions, then the first pair cohere with one another, and so do the second (explananda) pair. The fourth principle says that observation reports have a degree of acceptability on their own. Fifthly, contradictory propositions incohere. Sixthly, acceptability of a proposition in a system depends on its coherence with the propositions in that system.

There are features of this account that do prevent it from replacing more detailed notions of explanation in a full conceptual analysis of the concept. Thagard takes 'explain' as primitive and gets acceptability as a kind of coherence out of that. Explanation is a matter of fitting into a system, but there are a number of ways, deductive, statistical, causal

[9] Nicholas Rescher in his book *The Coherence Theory of Truth* (Oxford: Clarendon Press, 1973), has worked on revising the coherence theory, though I do not agree that it should be seen as a theory of *truth*.

[10] 'Explanatory Coherence', op. cit.

and so on, of doing this. However to explain a fact *successfully* requires having hypotheses that are acceptable because they fit into one's whole system.

These principles do not of course give us an AI model. Such a model has to be developed by hypothesis and experiment. Thagard has a connectionist model called ECHO. Being connectionist it contrasts with Bayesian models. The model can be thought of as consisting of various neuron like entities connected by various excitatory and inhibitory fibres, but in fact it is represented by a computer programme. Subsystems of the network can be explanatory, but full explanation comes from the whole network. Now if a hypothesis is in the subordinate way explanatory of observation facts from diverse fields then it possesses *consilience*. This term was introduced by William Whewell.[11] It can be modelled in the network with an inhibitory link from hypothesis 1 to hypothesis 2 if hypothesis 1 explains more pieces of evidence than does hypothesis 2. Simplicity is modelled as follows. Suppose that n hypothe ses together explain a fact E and m other hypotheses together explain E, and $m<n$. Then the excitatory links from E to the first n hypotheses will be weaker than the excitatory links from E to the other m. If a hypotheses H in the first set is inconsistent with a hypothesis H' in the second set then H' will be preferred because it will have more activation from E than its rival does.

Numbers, or weights, are given to the excitatory and inhibitory links in the network. Typically in Thagard's programme ECHO the excitatory links are given the weight 0.05 and the inhibitory ones are given -0.2.

The idea is that the states of the network will evolve in such a way that with luck it will settle down so that only one hypothesis of any pair of contradictory hypotheses is accepted. So the network will have settled down to accepting the best explanation of the evidence (represented by excited evidence elements).

Thagard tried his programme ECHO[12] successfully on simplified versions of various episodes in the history of science, for example Darwin's theory of natural selection (as contrasted with the theory of special creation of species) and Lavoisier's theory of combustion (as contrasted with the phlogiston theory).[13] The computer programmer sees to it that the right answers come out, and so the AI processes model the inductive reasoning (argument to the best explanation) of these

[11] William Whewell, *Novum Organon Renovatum* (London: John W. Parker, 1858).

[12] As reported in his 'Explanatory Coherence', op. cit.

[13] For earlier work see Thagard's *Computational Philosophy of Science*, op. cit.

historic figures. Thagard gives values to the hypotheses and evidence (in fact giving competing hypotheses and different pieces of evidence equal values) and to the excitatory and inhibitory links, and to the effect of simplicity, consilience and analogy. These values are not entirely arbitrary, since they were arrived at by experimenting with what values give good performance. The empirical success of the model therefore suggests that these values correspond to something objective (and the same could be said about the values corresponding to the so-called intuitions of human scientists), even though it is not clear what. This empirical and connectionist approach thus has advantages over the usual Bayesian approach, where probabilities of hypotheses cannot be assigned except in special cases. As Thagard remarks, Bayesian methods work well in diagnostic AI because probabilities can be assigned on the basis of medical statistics. The Ramsey method of determining subjective probabilities in terms of betting quotients cannot be applied: as Thagard says, he would not know how to bet on the theory of evolution. I take the point, but I am not sure how to understand it. I would bet any amount on the theory of evolution (its details apart). How would we collect the bet from a creation scientist? Is the idea that someone *before* learning about and studying the theory of evolution would bet on the conclusion he or she would come to after learning about it and studying it, thus giving it an *initial* subjective probability?

I have hopes that this quasi-empirical support will perhaps give some scientific respectability to the more intuitive notions of the coherence theory of warranted assertability. Thus encouraged I shall make use of the admittedly imprecise notions of consilience, simplicity, analogy and fitting into a web of belief, or in short of coherence.[14]

It was noted that Thagard gives observation reports an independent degree of acceptability. This is necessary, I suppose, for his simplified models, but the full coherence theory need not make this concession to

[14] This is not of course to decry in any way the efforts of those who have tried (or who do or will try) to precisify these notions. Eminent among them is one of the participants in this conference, Elliott Sober. In his book *Simplicity* (Oxford: Clarendon Press, 1975) Sober defines simplicity as informativeness with respect to a question: a hypothesis is simpler than another to the extent that it requires less extra information for us to answer the question. Sober's approach is syntactic which in effect is to assume that all atomic propositions are equally informative. Thagard (*Computational Philosophy of Science*, p. 85) remarks that scientists typically are not concerned about this sort of syntactic simplicity, but regard the number of statements of initial conditions required to supplement a theory to derive an observational statement as irrelevant to the simplicity of a theory. A semantic notion of simplicity seems as fraught with as many difficulties as the notion of subjective probability. But all the same, we need both notions.

an atomistic empiricism. It is part of our web of belief that observation is usually reliable. We know about the way vision works, how light rays enter the eyes, are refracted to produce images (in the sense of geometrical optics) on the retina, about the visual cortex, and so on. We also know about illusions and how to try to avoid them. Furthermore the theory of evolution by natural selection leads us to another understanding of why we should expect our eyes, ears and other senses to be on the whole reliable. So the belief that observation statements cohere with the rest of our web of belief means that giving extra weight to observation statements is not really to go outside the coherence theory of warranted assertability.

Similar reflections lead us to understand why the coherence theory should lead us to make fresh observations in order to test our theories. Furthermore one explanation will be a better explanation than another if it also explains a set of phenomena from a different field (consilience). So the requirement that scientists and others consider total evidence available to them and also seek more evidence is taken care of in a way in which it is not taken care of by the simple requirement that the hypothesis should be the most probable (relative to the evidence). That is, not mere self-coherence but *comprehensiveness* belongs to the notion of coherence. The web of belief has its own tendency to expand. We want our beliefs to be true, and truth differs from coherence. We may be warranted in asserting something that is false. Indeed, I think that even a Peircean ideal science might be false in important respects. The universe might trick us. Note that I elucidate this 'might' in terms of consistency with core science. There would indeed be a pragmatic paradox (but not a semantic one) in asserting that core science itself might be false. It is part of our web of belief that we are able to weed out false theories by testing as severely as possible the theories we already have. Hence it is part of our web of belief that we can increase the chance of our theories being true, or at least possessing verisimilitude (approximation to truth) if we make further observations and test our theories with respect to them.

In Thagard's models the observation statements, or what serve as such, are taken to be unproblematic. In real life this is not so, and an AI model to take account of this fact will require new features. There is no bedrock, and explanation is holistic. So sometimes we do not explain facts but explain them away. For example, I am extremely resistant to considering or trying to think up hypotheses that might explain putative paranormal phenomena. Since I cannot fit telepathy or clairvoyance into my web of belief I prefer to put alleged reports of such things down to guillibility, wishful thinking, bad handling of statistics, sheer coincidence, or other ways of resisting belief in paranormal phenomena. The simplest way of fitting the observed phenomena, which are not the supposed paranormal ones, but the putative reports, into my web of belief is to refuse to take them at face value. We may compare the case of conjurors who have done the most magical looking things in front of our eyes, and in ways of which we do not have the slightest conception. Still, it is part of my web of belief that my eyes can be made to deceive me.

Continually, parts of the web of belief are of course being reconstructed. Generally, though not always, observations that cannot be fitted into the web lead to replacement of one hypothesis within the web by another. Sometimes this can come about by considerations of simplicity and analogy. Considerations of simplicity and analogy may lead us to replace one hypothesis within the web by another. In Thagard's simulations the set of alternative hypotheses is given, but in real life new hypotheses get thought of by the not very well understood process that C. S.Peirce called 'abduction'. The special theory of relativity has received good confirmation from the Michelson–Morley experiment and with observations of fast particles, so that one could not design a cyclotron without using the dynamics of special relativity. Nevertheless Einstein was motivated to a large extent by considerations of simplicity and analogy. Until relatively recently the *general* theory of relativity had very little, if any, hard observational or experimental support. Einstein and others accepted it largely on the sorts of *a priori* considerations that I have been mentioning.

Thus when I talk of fitting something into a web of belief, the web must not be thought of as something static into which the new observation reports or sub-theories must fit. Sometimes the fit comes by ripping apart large chunks of what had seemed to be a strong and secure region of the web. How beliefs should be accepted, rejected, strengthened or weakened is a matter for the dynamics of belief. It is an open question how precise and how objective such a study may be. We do need such a study to improve on the metaphor of 'fitting in' to a web of belief or the turn of the century talk of 'coherence'. I have referred to the work of Thagard on AI modellings of scientific controversies as pointing to one way of getting objectivity into a study of argument to the best explanation. Another line of research which may go in the same general direction is the work on the statics and dynamics of belief systems by Brian Ellis, Peter Gärdenfors and Peter Forrest. As with the case of the AI modelling, I am not very knowledgeable about this line of research and I shall have to remain on a more imprecise and elementary level.

It is not surprising, perhaps, that such a vague theory of explanation as that of coherence can incorporate the insights of various well known accounts of explanation.

Explanations, like screwdrivers, can be good and bad. Indeed the notion of 'argument to the best explanation implies that some explanations can be better than others.'[15] Note here that 'best explanation' here really means 'best hypothesis'. (It can hardly mean 'best explanatory speech act'.) However the term 'argument to the best explanation' indicates in what way the hypothesis is best, namely that it is the hypothesis belief in which fits best into our web of belief. This contextuality is taken care of by the conception of explanation as a speech act. Not only can explanations be better than others, but we can surely talk of explanations as being positively (if that is the right word here) *bad*. An explanation may be bad if it fits only into a bad web of belief. It can also be bad if it fits into a (possibly good) web of belief in a bad sort of way.

Thus a creation scientist's web of belief is a bad one. Essential to his or her web is a belief that *Genesis* and other books of the Bible are the literal word of an infallible Deity. Grant this proposition and no doubt the rest follows easily. However this proposition is a highly unplausible one and the best hypothesis surely is that *Genesis*, etc. is based on even earlier myths and on the ideas of ancient people with none of the advantages of modern cosmology who were doing the best they could. Modern cosmology, together with evolutionary biology, provides a more comprehensive web of belief. Moreover the creation scientist's total theory is far from simple: it requires a separate creation hypothesis for each species of plant and animal. It is no good just referring to the will of God, because it is still unexplained why God made the various decisions in the different particular cases.

An explanatory hypothesis may be bad if it fits into a good web of belief in a bad sort of way, e.g. to a wrong analogy, a mathematical mistake, a wrong assessment of plausibility, and so on.

One way in which an explanatory hypothesis may be best is of course that it be true. As critics of the correspondence theory of truth have pointed out, we cannot directly compare a hypothesis with facts in the world. This is a poor criticism of the correspondence theory of truth, for why should an adherent of such a theory (or similar theory) wish to say that we could make such a direct comparison? Though a poor criticism of the correspondence theory of truth, it does make a point

[15] Peter Forrest prefers to speak of 'argument to the best theory' instead of 'argument to the best explanation'. I partly agree, but still prefer 'argument to the best explanation', because it is the best theory that provides the best explanation, and it is best *because* it provides the best explanation for the person who wants the explanation. People with different presuppositions, whose puzzles or 'why?' questions are different, might require different hypotheses to provide an explanation for them. The hypothesis is not a speech act, but the explanation is, even when it is talking to oneself.

about warranted assertability. We cannot directly compare hypothesis with fact. The best we can do is to choose the best warranted hypothesis (in this sense the best explanation) in the hope that warranted assertability is a pointer towards truth. The pointer may lead us astray, since the universe may always trick us. Still, we do the best we can, which is pretty well. At least it is pretty central to my web of belief that I do pretty well. (For example evolutionary theory is part of my web of belief, and this makes it most unplausible that my senses are very unreliable.)

So when we choose the best explanation, the criteria for 'best' do not include 'true'. In the AI model the best hypothesis is the one that settles down as best on the operation of the network. In a wider conception, allowing for possible rejection of observation reports, it is the one which survives in the modified web. Moreover it is the best of the alternative explanatory hypotheses that occur to us, or that the network is able to add to its store by an AI process analogous to Peircean abduction. It is too much to ask that it be the best of all possible alternative hypotheses, laid up in some platonic or set theoretic heaven. We do the best we can, and we cannot consider hypotheses of which we cannot think.

The history of science suggests that theoretical assertions often get replaced by others with greater verisimilitude. Not always. I think that there are many hypotheses in science that will never be improved on. Will it ever be denied that the lightest isotope of hydrogen contains one proton and one electron? However much we change our conception of electrons and protons still this proposition about hydrogen will hold up. Generalizing, there is a vast reservoir of accumulated scientific fact and theory that will never be overturned. We can recognize Kuhnian revolution without jettisoning much of the cumulative or 'Whig' conception of scientific advance. Still, in fundamental matters we hope for verisimilitude rather than truth. Now a better model of the web of belief would have to take account of the fact that in the face of new evidence or of new serendipitous abductions a good web settles down to a state that does not so much probably maximize verisimilitude, as (in terminology analogous to that of expected utility of economists and utilitarian moralists) maximizes the expected verisimilitude. (In saying this, I am assuming that good sense can be made of the concept of verisimilitude. That is a big 'if' and yet the concept of verisimilitude seems to me to be indispensible to much philosophy of science and metaphysics.[16])

[16] For work on verisimilitude and for references to other recent work on the subject see Graham Oddie, *Likeness to Truth* (Dordrecht: D. Reidel, 1986). The subject is a highly technical and contentious one, and yet if it can be brought to fruition the work on it is surely of the greatest importance for the philosophy of science.

Vague and imprecise as I must admit that the characterization of explanation as fitting the explanandum proposition into a system is, this very vagueness and imprecision allows me to subsume many of the well known theories of explanation as special cases.

We may begin with the notion that explanation is a matter of reduction to the familiar. As a general characterization of explanation this is clearly quite wrong. When the facts of black body radiation did not fit in into the webs of belief of classical physicists, Planck introduced the notion of quanta, which affected parts of the web which then had to be repaired in quite novel ways. The history of science suggests that most often explanation is reduction to the unfamiliar. Of course good popularization of novel scientific theory, while stressing the novelties and unfamiliarities, does try, so far as possible, to make the theories acceptable to the intended readers by making copious use of common sense analogies. Even in technical science reduction to the familiar has at least played some part in explanation. Consider, for example, the early kinetic theory of gases, in which we find the gas laws explained in terms of the collisions of small elastic particles. Something like the notion of explanation as reduction to the familiar was prominent in N. R. Campbell's philosophy of physics.[17] Campbell held that good theories could be distinguished from absurdly trivial ones by the presence of analogy, and the analogy of gas particles with such things as billiard balls, provided an example. In other cases, such as that of the Fourier theory of heat conduction, Campbell saw a mathetical analogy between the form of the laws of heat conduction and the empirical laws that the theory is intended to explain. Campbell remarked that Fourier's equation is developed by considering the flow of heat through an infinite plane parallel slab. In quantum mechanics we could point to frequent analogies between Hamilton equations with operators and Hamilton equations in classical mechanics. (In opposition to this, it could be held that the use of such analogies is heuristic only and that the analogies do not add to the strength of an explanation.)

The presence of analogy is not quite the same as that of reduction to the familiar. Campbell could allow analogy even in the case of very unfamiliar looking and esoteric theories, there being analogical resemblance between successive theories but little or none between the first and last theories in the chain. On the other hand, in modern physics we may find little or no analogy between successive theories and I suppose that there is no analogy in the case of the replacement of Aristotle's mechanics by Galileo's. Looking at the matter from the

[17] N. R. Campbell, *Foundations of Physics* (New York: Dover Inc., 1957), originally published by Cambridge University Press, under the title *Physics: The Elements* (1920).

point of view of Thagard's AI investigations we can say that Campbell's theory works well when the excitatory analogical links are set high.

Indeed in fact Campbell does not see all explanation as reduction of the unfamiliar to the familiar.[18] He distinguishes two sorts of explanation, one being substitution of the familiar for the unfamiliar and the other being substitution of the simple for the complex. A given explanation can be of both sorts, but Campbell agrees that the latter plays the more important part and is characteristic of scientific explanations. As has been pointed out, simplicity plays a part in Thagard's models.

Indeed it is interesting that there are things that are difficult to explain and yet which are very familiar to common sense and infant physics. We all know that glass transmits light whereas wood does not, but not everyone knows why. Ohm's law is familiar to those beginning to study physics, but its explanation requires the unfamiliar and sophisticated ideas of solid state physics.

In mythology, in the traditional legends of the Australian aborigines, we come across what seems in part to be meant as explanations, though there may be other purposes, in that the stories of a mythical past may have expressed wishes and desires that were unable to be realized in the people's own contemporary desert environment.[19] Consider the mythical explanations of land forms, which are put down to the activities of totemic ancestors. Thus in Aranda tradition a gap in a range of hills is the remains of a broken fish weir, and cracks and water gutters were supposed to have been scoured out by the body of a great and hideous water serpent.[20] This represents a sort of fitting into a web of belief: beliefs about present land forms are fitted into beliefs about the activities of water serpents and of people who construct fish weirs. To the scientific eye the explanations are of course extremely unsatisfactory. The web of belief is extended at one point at the cost of irreparable damage and loss at other points. The activities of humans and other animals were so familiar that it was assumed that they did not need explanation. This assumption was mistaken. Indeed the human or animal brain is a horrendously complex entity, whose workings are still very imperfectly known, even though modern neuroscience is taking its first tentative steps along the way. On rare occasions also the myths may speak of mountains or formless plains arising by their own volition.

Of course the supposed mythical events are larger than life and so are unfamiliar, but the concepts used, 'water snake', 'fish weir' and so on, are familiar enough. Even the notion of 'turning into' is a common sense

[18] Campbell, *Foundations of Physics*, 114.

[19] See T. G. H. Strehlow, *Aranda Traditions* (Melbourne University Press, 1947), 26.

[20] Ibid.

one. Caterpillars turn into butterflies. Children find no difficulty with stories in which a prince turns into a frog. Perhaps such ideas are reinforced by dreams. So the idea of a totemic ancestor turning into the totemic animal, which then solidifies into a mountain range, does not use unfamiliar concepts, even though the supposed events are of unfamiliar sorts. In modern times we have come to see that the messy patterns of common sense events can be simplified only by the replacing of familiar concepts by unfamiliar ones. Thus nowadays an unfamiliar concept such as that of 'parity' does not present itself as puzzling, whereas the familiar concept of volition may raise various sorts of difficulties.

It is no wonder, therefore, that reduction to the familiar does not figure much in contemporary philosophizing about explanation. Some good explanations are of this sort and more are not. Bad and unscientific explanations are quite commonly of this sort.

A very influential theory of explanation has been C. G. Hempel's Deductive-Nomological and Inductive-Statistical patterns of explanation. The deductive-nomological pattern is that of an argument in which we deduce a statement concerning a particular event or else a statement of law. I shall concentrate on the former. The Laplacean ideal was that of an infinite intelligence which was possessed of knowledge of the whole state of the universe at a given instant and which could predict or retrodict any future or past event. In practice of course we have to concern ourselves with finite systems that we can treat as closed from outside influences, which in Newtonian mechanics would queer the pitch if they entered the system from outside with action at a distance (perhaps from infinity), between the time of prediction or retrodiction and the event to be explained. We must also neglect Gödelian possibilities that the statement of the event might be true but not deducible from the premises.[21]

The Hempelian deductive-nomological model of explanation clearly fits in well with the notion of explanation in terms of coherence. One way of fitting a belief into a system is to show that it is deducible from other beliefs. However, as is well known, Hempel's model suffers from counter-examples, examples for which we need to consider also whether the coherence account does or does not avoid them. Hempel has of course reacted to these counter-examples in his own way, but I shall here simply assume that they are damaging to the deductive-nomological account of explanation as a general definition of explanation (apart from Hempel's Inductive-Statistical pattern which I shall discuss separately). Thus consider Wesley Salmon's example 'No men

[21] This is one reason why the Laplacean syntactical definition of determinism is better replaced by a model theoretic one.

who take the birth control pill become pregnant, Jones is a man who regularly takes his wife's birth control pills, therefore Jones does not become pregnant'.[22] The major and minor premises are true and the conclusion follows from it. Yet the putative explanation of Jones' non-pregnancy is at least misleading, because it suggests that taking the birth control pill had something to do with Jones not becoming pregnant. It is worse than misleading if it leads Jones to think that if he stopped taking the birth control pills he would very likely become pregnant. The putative explanation does not conduce to coherence of Jones's set of beliefs because it invites adding to his web of belief the proposition that at least some men who do not take the pill become pregnant, which is contrary to common experience, as well as to anatomy and physiology. (I shall ignore the case of the man with the XXY chromosome who had a baby and later became officially a woman. To avoid such counter-examples we would have to define 'man' more narrowly, perhaps as persons with XY chromosomes.)

A clearer problem arises over Sylvain Bromberger's early and well known example of the flag pole.[23] Consider a flagpole 20 metres high distant from an observer with his eye at ground level, distant 40 metres from the base of the flagpole. By the laws of Euclidean geometry and simple trigonometry we deduce that the angle subtended at the observer's eye is 26° 31′. Now it is intuitive that the height of the flagpole explains the fact of its subtending an angle of 26° 31′ at ground level 40 metres away but that the converse does not hold. Nevertheless, each fits Hempel's pattern. I think that the coherence theory suggests an answer to the asymmetry. The proposition that the flagpole is 20 metres high fits naturally into a system of beliefs about the purposes of flagpoles, how they are made from trees and made a certain length in order to subserve such purposes. Also beliefs about how you would make a flagpole to be 20 metres high. Of course this would *ipso facto* make a flagpole that subtends 26° 31′ 40 metres away at ground level, but the measurements that would need to be made would, except in bizarre circumstances, be of the length of the flagpole. Of course you could imagine a person measuring off 40 metres and then with a theodolite finding a suitable spot on the flagpole to cut it, in which case it would be natural to say that the flagpole was 20 metres high because it subtended the appropriate angle 40 metres away at ground level. Since such circumstances would be very unusual we naturally say that the flagpole

[22] Wesley C. Salmon, *Scientific Explanation and the Causal Structure of the World* (Princeton University Press, 1984), 47.

[23] Cf. Sylvain Bromberger, 'Why-Questions', in Robert G. Colodny (ed.), *Mind and Cosmos* (University of Pittsburgh Press, 1966), 86–111, especially 92.

subtends the angle because of its height and not vice versa. See also Bas van Fraassen's amusing story[24] explaining why a tower was a certain height by the fact of the shadow being a certain length. It is an ingenious story, and in less out of the way cases we of course expect to explain the length of the shadow by the height of the tower, not the other way around.[25]

Why is the light from distant galaxies shifted to the red? Answer: the recession of the galaxies as the universe expands. We are puzzled by the red shift and the hypothesis of the expansion of the universe is what best enables us to fit the fact of the red shift into a web of belief about the wave nature of light, the Doppler shift and so on. The theory of the expansion of the universe renders the red shift no longer puzzling, whereas the expansion of the universe is hardly rendered less puzzling by facts about red shifts. The pragmatic aspect of explanation renders the asymmetry of 'fitting in' or coherence understandable.

Some philosophers may well say here that I could save myself a lot of trouble by defining explanation in terms of causality. For reasons of my own, I am inclined (unfashionably these days though it was fashionable enough fifty or more years ago) to argue that the notion of causality is unimportant in science, or at least in theoretical physics. Certainly I do not want the notion of causality if it has to be defined in terms of counterfactual conditionals. Counterfactual conditionals are sometimes elucidated, notably by David Lewis, as about possible words other than the actual one.[26] I hold that science and metaphysics should concern themselves only with the actual. I would give a metalinguistic account of counterfactuals, and I hold that science and metaphysics should not concern themselves with the metalinguistic (save as a linguistic phenomenon).[27]

Frank Jackson regards causality as an unanalysable theoretical notion, which arises postulationally by the usual hypothetico-deductive method. This certainly seems to me to be an improvement on defining causality in terms of counterfactuals. Salmon's notion of causality is an

[24] Bas C. van Fraassen, *The Scientific Image* (Oxford: Clarendon Press, 1980), 132–33.

[25] Philip Kitcher and Wesley Salmon have critically discussed van Fraassen's application of the story in their paper 'Van Fraassen on Explanation', *Journal of Philosophy* (1987), 315–30, especially 317. They think that van Fraassen brings in extraneous considerations. I am inclined to think that the coherence account, for which no considerations need be extraneous, may avoid these criticism.

[26] David Lewis, *Counterfactuals* (Oxford: Basil Blackwell, 1973).

[27] Some of the ways of constructing what Lewis calls 'ersatz possible worlds' are essentially metalinguistic. I have not space here to discuss the relevance of ersatz worlds that are supposed to exist in the world of set theory.

eliminible one, I think. He defines causality in terms of statistical relevance relations, and what he defines are causal processes, which consist in transfer of matter or energy. For Salmon, causality is a process, not a relation between events.[28]

Salmon does not avoid counterfactuals altogether. Thus he treats space-time joined by time-like geodesics and yet by means of a causal theory, in the manner of A. A. Robb.[29] One objection to this is that there are surely points of space-time not connected with one another causally. So we need to talk of *possible* processes. Now surely geometry is more clearly understood than counterfactuals, and the postulation of absolute space-time seems preferable to postulation of possible worlds other than the actual world (which is the basis of David Lewis's treatment of counterfactuals) or going metalinguistic when one should remain within the object language of science. The explanation of gravity in the general theory of relativity seems geometrical rather than causal (though I suppose that it could be held that the variable curvature of space-time which accounts for the geodesics of freely falling bodies—such as planets of a star—is caused by the presence of matter). Salmon, of course, acknowledges problems in connection with the experiments that show violation of Bell's inequality, in which causality seems to be ruled out.

Also there have been attempts to geometrize the whole of physics, with unified theories of the four forces of nature by means of an eleven dimensioned space-time. Philosophically I was a bit disappointed that J. A. Wheeler had to abandon his 'worm hole' theory in which particles are the ends of worm holes in a multiply connected space-time. (Spinoza would have loved it!) Geometrization has been an ideal of physical explanation that should not be ruled out philosophically by too much emphasis on causality.

However, I shall give an example from special relativity of a geometrical explanation that is not causal. (I could have given an even simpler example, e.g. of Putnam's example of why square pegs fit into square holes and not into round ones.) Consider the situation of the twin paradox. Twin Peter stays at home while his twin brother Paul is shot off to the vicinity of a distant star where he is decelerated and reaccelerated so as to return home to join his twin. Except for small curved bits corresponding to the periods of acceleration and deceleration at start and finish of his journey and in the vicinity of that distant

[28] Salmon, ibid., 170ff.

[29] And extended to general relativity by subsequent writers. See Adolf Grünbaum, *Philosophical Problems of Space and Time*, 2nd edn (Dordrecht: D Reidel, 1973), 735–50. For A. A. Robb, see his *Geometry of Time and Space* (Cambridge University Press, 1936).

star we can regard the relevant part of Peter's world line as consisting of the two equal sides *AB* and *BC* of an isosceles triangle, while the relevant part of Paul's world line is *AC*. And since *AB*, *BC* and *AC* are all time-like lines, the geometry of Minkowski space-time ensures that *AB* +*BC* is *less* than *AC*. (Analogous to Euclidean geometry, in which two sides of a triangle are greater than the third side. The difference is in the minus signs in the metric.) Now this geometrical explanation gives far greater insight than does an operationalist treatment in the manner of Einstein's (pre-Minkowski) 1905 paper. (Just as Minkowski's exhibition of the Lorentz transformations of special relativity as simply a rotation of axes in space-time gives more insight than does an operationalist treatment of special relativity.) If there are worries about Paul's periods of acceleration and deceleration it can be remarked that the calculation could have taken these into account with the aid of a little differential calculus. If it be said that the real explanation is causal, because the accelerations and decelerations were caused by rocket motors, I reply that this does not give the main insight. Let *AB* be part of the world line of a clock that passes a clock part of whose world line is *AC* and let *BC* be a clock which passes the *AB* clock at *B* and the *AC* clock at *C*. Then we explain why the sum of the time differences registered by the *AB* clock and by the *BC* clock is less than the time difference registered by the *AC* clock.[30] No acceleration, and no causes here.

Salmon's philosophical approach seems to fit rather well with quantum field theory in which forces (other than gravitational ones—though there have been speculations about gravitons) are seen as exchange forces. Two particles are attracted to one another as another particle is exchanged between them. (This is a little counter-intuitive. If you are standing on smooth ice and I throw a cricket ball for you to catch, we slide apart, not towards one another!) So leaving Bell inequality or Einstein–Podolsky–Rosen considerations out of it, most of fundamental physics (outside the general theory of relativity) does fit Salmon's picture of the world as a network of causal processes (in his own harmless sense of 'causal').

Salmon sees explanation not as argument but as the giving of relevant information. This information can be statistical in nature as indeed it must be in quantum mechanics. Salmon is thus able to deal with statistical explanation of particular events in a better way than Hempel does. In Hempel's deductive-nomological pattern of explanation we have argument, but there is some uncertainty in his presentation of the 'inductive-statistical pattern'. Thus in the deductive-nomological pat-

[30] Adolf Grünbaum used the case of passing clocks in his paper 'The Clock Paradox in the Special Theory of Relativity', *Philosophy of Science* **21** (1954), 249–53 and replies to discussions of this, ibid. **22** (1955), 53 and 233.

tern, Hempel writes the conclusion of the argument under a single line, whereas in the inductive-statistical case he uses a double line and a 'probably' in brackets. This makes the so called argument look like a metalinguistic statement of a relation of conclusion to premises, which is of course consonant with the fact that the validity of inductive arguments is affected by addition of premises, which is not the case with deductive arguments.

Hempel says that a statistical explanation of a particular event is a good one only if the occurrence of the event is given a high probability. Salmon, denying that an explanation is an argument, is able to say that if there is a 1/10 possibility of a radio-active atom emitting an alpha particle in a certain period of time, and the alpha particle is emitted, the probability statement gives a perfectly good explanation of the event. Quantum mechanics assures us that there are no ways of describing the atom which would give us a better probability. Salmon's denial that explanations are arguments works well in this sort of context. What would we say if the probability predicted by quantum mechanics was not 1/10 but $1/10^{1,000,000}$? In this case if I observed the event I would not try to fit the observation statement into my web of belief: I would suppose that my eyes had deceived me or that there was something funny about the experiment. To speak a bit in an Irish manner, I would fit the observation into my web of belief by *not* believing it. But if the probability were 1/10 I would be perfectly happy.

Salmon describes his account of explanation as objectivist and ontological rather than epistemological. Explanation describes how events fit into an objective causal network. My own account is epistemological, we explain *x* to *y* by means of fitting *x* into *y*'s web of belief. Nevertheless, in our explanation we aspire to truth—we want a true web of belief, and in this respect there is not too much difference between the epistemological and the ontological concepts of explanation. Where I can talk of bad explanations as consisting of the bad fitting of beliefs into a pre-existing web, or of good fittings into a bad pre-existing web, the ontological conception can allow us to talk in terms of descriptions and misdescriptions of the causal network.

The main theories of explanation, then, seem at least to fit into the general coherence theory of explanation as special cases. The lack of general agreement among philosophers of science despite the brilliant, and often painstaking work of leading workers in the field suggests to me that the concept of explanation may be what Wittgenstein called a 'family resemblance' one. This would make it at least highly disjunctive, and I suspect also that even if we confine ourselves to good scientific explanation, it may also exhibit what Waismann called 'open texture', in that patterns of discourse may be proposed in science which we would immediately recognize as explanatory, even though we could not have predicted them in advance. The coherence account of explanation would surely fit anything of the sort that will arise, but I am only too aware that it may protect itself in the future as it may have done in the past only by its own woolliness, as a sheep does on a frosty night. (However, I have suggested that Paul Thagard's AI approach may show promise of acting a bit like a sheep shearer and reducing the woolliness.)

Finally, why do philosophers need a theory of explanation? As we all know there are some who will analyse any concept at the drop of a hat. Then there are those who see explanation as a key concept in epistemology, and so need a theory of it. For me, epistemology is the handmaiden of metaphysics, and since argument to the best explanation is important in the defence of realism, I would like to have a good account of what explanation is. In the face of the brilliant, lucid, and often painstaking work of Hempel, Ernest Nagel, Salmon, Achinstein, van Fraassen and others, all I have had to offer are some imprecise thoughts.

Fortunately most of the symposia to follow in this conference are concerned with explanation in the special sciences, Physics, Biology, Psychology, and the Social Sciences. Here I think that the issues concern the structure of these sciences as much as the nature of explanation itself. From these discussions I hope to learn much, as well as from the appropriately final symposium on the Limits of Explanation.[31]

[31] I wish to thank Paul Thagard for kindly commenting in correspondence on a draft of this paper, in particular on suggesting corrections to those pages in which I expound his views. He is of course not responsible for defects that remain, and I hope that readers may be encouraged to read his 'Explanatory Coherence' and his *Computational Philosophy of Science*, already cited.

Truth and Teleology

DAVID PAPINEAU

1 Introduction

A number of recent writers have argued that we should explain mental representation teleologically, in terms of the biological purposes of beliefs and other mental states.[1]

A rather older idea is that the truth condition of a belief is that condition which guarantees that actions based on that belief will succeed.[2]

What I want to show in this paper is that these two ideas complement each other. The teleological theory is inadequate unless it incorporates the thesis that truth is the guarantee of successful action. Conversely, the success-guaranteeing account of truth conditions is incomplete until it is placed in a teleological context.

I shall proceed as follows. In the next section I shall explain why mental representation is philosophically problematic. Then, in Section 3, I shall show how representational notions play a role in action explanation. This will lead to a version of the success-guaranteeing account of truth conditions, and in Section 4 I shall elaborate and defend this account. Section 5 will then show why the success-guaranteeing account of truth conditions needs to be incorporated within a more general teleological theory of mental representation. In the final section I show how a standard objection to the teleological theory can be answered.

2 Mental Representation

Why should we find mental representation philosophically problematic? A short answer is that representation of any kind is philosophically problematic. No sensible metaphysics ought to posit semantic relation-

[1] See Dennett (1969, Ch. 9; 1987, Ch. 8); Fodor (1984); Millikan (1984, 1986); Papineau (1984, 1986, 1987); McGinn (1989, Ch. 2).

[2] See Ramsey (1927, p. 159); Putnam (1978); Appiah (1986); Mellor (1988). Also see Stalnaker (1984, esp. pp. 15–18) which argues, as I shall, that the success-guaranteeing account is part, but only part, of the truth about truth.

ships as primitive features of the world. Semantic relationships need somehow to be explained in non-semantic terms.[3]

So mental representation is initially problematic as an instance of a *general* problem: how is it possible for anything, mental states included, to stand for things other than themselves? But there is also a more *specific* puzzle about mental representation. The most obvious role played by beliefs, desires and other representational mental states in our everyday thinking is in the explanation of action. But what exactly does action explanation have to do with representation? Why is it that action explainers represent things other than themselves? Acceleration explainers—forces—do not represent things other than themselves. So why do action explainers?

A natural strategy is to try to solve the general puzzle about mental representation by answering this more specific question about the connection between mental representation and action explanation. After all, when we invoke beliefs and desires in action explanations, we always identify them representationally, as the belief *that such-and-such*, the desire *for such-and-such*. So it seems likely that the representational powers of mental states are somehow important for their explanatory significance. Perhaps a detailed grasp of how this works will enable us to understand how mental representation can find a place in an intrinsically non-representational world.

However, there is a *prima facie* reason for doubting that this strategy can work. Let us focus on beliefs. The representational aspect of beliefs can be thought of in terms of truth conditions: any given belief will be true if the world is one way, and false otherwise. But the trouble now is that this difference, between truth and falsity, seems irrelevant to the explanatory significance of beliefs. For surely, when we explain some action in terms of some belief, all that matters is that the agent *has* the belief, not whether it is true or false: once you have a belief, it will still have the same influence on your actions, even if it is quite mistaken.

Let me articulate the principle behind this line of thought:

(A) The truth or falsity of a belief is irrelevant to its explanatory role.

This principle suggests that the specific problem of representation is ill-posed, in that there is not any real link between the representational and explanatory aspects of belief after all. Maybe the way we identify beliefs, in terms of 'that p' content clauses, makes it look as if the explanatory role of beliefs depends somehow on their representational powers. But principle (A) suggests that this is an illusion, and that content clauses are really doing two quite independent jobs—on the one

[3] Field (1972) argues that physicalism cannot allow primitive semantic relationships. The point, however, is not peculiar to physicalism.

hand, specifying when beliefs are true and when false, and, on the other, indicating the explanatory significance of beliefs.

A common response to this argument is to object that, while principle (A) might show that truth *values* are irrelevant to action explanations, it does not show that representational features in the form of truth *conditions* are so irrelevant. But this misses the point. It is certainly true that a belief's explanatory potential depends on what belief it is—on whether it is, as we say, the belief *that p* or the belief *that q*. And in this sense its explanatory potential certainly depends on its 'truth condition'. But we still need to be told what having a 'truth condition' in *this* sense has to do with *truth*. It is one thing to use the phrase 'truth condition' to describe how we identify beliefs for explanatory purposes. But, if principle (A) is right, such identifications would serve their explanatory purposes just as well even if they did not specify when beliefs were true and when false, that is, even if they were not representational identifications at all.

So principle (A) casts doubt on the proposal that we should solve the general problem of mental representation by considering the specific question of the relationship between mental representation and action explanation—for principle (A) suggests there is not any such relationship to start with. However, I do not think we should be too quick to abandon this proposal and start looking for some alternative explanation of mental representation instead. Despite the persuasiveness of the argument from thesis (A) (and it has persuaded a number of philosophers, myself included[4]) the conclusion ought surely to be viewed with suspicion: it surely cannot just be a *coincidence* that the same way of identifying beliefs, in terms of 'that p' clauses, should be suitable for both representational and explanatory purposes. If thesis (A) implies this conclusion, then perhaps we ought to re-examine thesis (A).

In the next section I shall show that, contrary to thesis (A), there is a species of action explanation to which the difference between truth and falsity does matter. By focusing on this species of action explanation we will be able to resolve the specific problem of how mental representation relates to action explanation. And this will then lead us round to an answer to the more general question of how mental representation is possible at all.

But before proceeding I would like to comment briefly on a related topic. Much recent philosophical debate has focused on beliefs with 'broad contents', beliefs whose correct ascription to believers answers not just to what is going on inside the believers' heads, but also to what is going on around the believers. Thus Hilary Putnam has argued that

[4] See Papineau (1984), especially Section XI. Also see Field (1978), McGinn (1982).

the identity of beliefs about natural kinds depends on what kinds are actually present in the believer's world (1975); Tyler Burge has argued that the content of theoretical beliefs can depend on features of the social context (1979, 1982); and Gareth Evans has maintained that the possession of singular beliefs demands the existence of the objects those beliefs are about (1982). All these stories have the upshot that two believers can be molecule-for-molecule identical, and yet one believes that p and the other not, because of differences in their surroundings.

Some philosophers seem to think that once broad beliefs are admitted, then philosophical problems about representation disappear (cf. McDowell, 1986, esp. Section 5). The suggestion here is that problems of representation only arise as long as we think of beliefs as things inside people's heads. Once we realize that the very possession of a belief can involve extra-cranial facts, we ought no longer to be puzzled about how things inside the head can be about things outside.

However, this thought does not really address the problems I have raised in this section. Recognizing that some beliefs have *broad* contents does nothing to account for their having *representational* contents as such. Let us accept that the possession of beliefs can in some cases require certain entities outside believers. However, so does the possession of such states as financial solvency, or popularity, or being married. Yet my financial solvency does not therefore represent the world as being a certain way. So why do beliefs represent the world as being a certain way? And what does this have to do with their explanatory role? These questions remain just as much for broad beliefs as for any others.

A related point applies to those philosophers who argue *against* the psychological importance of broad contents, on the grounds that it is difficult to see how facts outside the head can matter to the explanatory significance of beliefs (cf. Fodor, 1987). If this is a legitimate worry about broad contents, then surely there is a far more important prior worry, namely, that of understanding how representational contents of *any* kind can matter to the explanatory significance of beliefs. For, special cases aside, the difference between true and false beliefs is just as much outside the head as the entities that make broad beliefs broad.

As it happens, I think that a proper appreciation of broad contents emerges naturally from a satisfactory understanding of mental representation. In particular, once we understand how representational content *per se* can matter to action explanation, then we will see that it is quite unsurprising that some contents should be broad. I shall return to this point at the end of Section 4 below.

3 The Significance of Truth

Why is it initially plausible to accept (A), the principle that truth values are irrelevant to explanatory role? I think that the plausibility of (A) derives from a certain picture of the structure of action explanation:

(B) 1. X desires that t
2 X believe that s will bring about t
Therefore
3. X does s.

(B) obviously only gives us part of what is required for such means–end action explanations. There are plenty of cases where X does something, yet this is not adequately explained by X's belief that the action is an effective means to a desired end. The desire in question might be very weak, for instance, or X may also believe that the action s will preclude some more significant end. Fortunately, nothing in this paper requires me to be specific about what needs adding to (B). Some philosophers would hold that we only have an adequate action explanation if the beliefs and desires in question are related to the action by some natural law. Other philosophers would say that the beliefs and desires only have to show the action to be rational, in some non-natural sense of 'rational'. For present purposes, however, we can by-pass this issue, and simply take (B) to represent the agreed core of action explanation.

It is obvious enough why truth values do not matter in explanations like (B). What (B) explains is the *means*, s, adopted by X. But X's belief is about the *efficacy* of s. If the belief is true, then things will turn out as X expects, and, if it is false, they will not. But either way, X does s. So s is explained equally well, whether or not the belief behind the action is true.

So we can attribute the plausibility of principle (A) to explanatory structure (B). When we are explaining the means an agent adopts, the truth of the belief behind the action does not matter.

However, not all action explanations are explanations of the means adopted. This is why principle (A) is not unqualifiedly true. Sometimes we want to explain agents achieving their ends, rather than their adopting some means. And then truth values do matter.

Suppose my television set breaks down. I telephone my brother and he comes and fixes it. You are puzzled about his ability to do this. I tell you that he has read a good book on the subject, and so has information about which adjustments will cure which faults. I thereby explain his success at fixing the television.

In explaining his success I invoke his belief that a certain adjustment will fix the TV. But now, because I am explaining his achieving an end, and not his adopting a means, the truth of this belief is not irrelevant. If the belief he had acquired from his manual were not *true*, then his having that belief would not explain his fixing the fault, but merely his making the adjustment in question.

By appealing to the truth of my brother's belief, I am able to explain his achieving his desired end, and not just his adopting a certain means. That is, my explanation has the following structure, rather than structure (B):

(C) 1. X desires that t
2. X believes that a certain action will bring about t
3. This belief is true
Therefore
4. X brings about t.[5]

4 Truth as the Guarantee of Success

Explanations like (C) show why the difference between true and false beliefs matters to action explanation. If action explanations only explained the agent's means, as in (B), then representational powers would be irrelevant. But once we are concerned with explaining success, as in (C), then the explanatory significance of a belief depends precisely on whether or not it is true.

This now provides a solution to the specific problem of mental representation, the problem of the relation between the representational powers of beliefs and their explanatory role: explanatory schema (C) shows that to know the truth condition of a belief is precisely to know when it can explain success. But does this also amount to a solution to the general problem of understanding how mental representation is possible in an intrinsically non-representational world? I would like to suggest that it does. That is, I would like to suggest that, by understanding how representation relates to explanation, we can thereby understand what representation *is*: it is simply that aspect of beliefs which allows them to explain the success of actions.

[5] Is the mention of truth *essential* to such explanations of success? There is an extensive literature on whether or not explanations like (C) can be replaced by two-stage explanations which (a) explain the means s by the relevant belief and desire, and then (b) explain the success by the fact that s brings about t, and which therefore do not make any explicit mention of truth as such. (See Loar, 1981; Devitt, 1984; Field, 1986). However, this debate is of no immediate relevance to my present concerns. Even if the mention of truth is not *essential* to explaining success on any particular occasion, it can still be *illuminating* to view an explanation of success as an instance of the general pattern that true beliefs yield successful actions. The important question for my purposes is not whether we could somehow manage without truth, if we had to, but rather what role truth plays in our thinking, given that we do not have to manage without it.

Let me formulate this suggestion as an explicit analysis:

(D) The truth condition of any belief is that condition which guarantees that actions based on that belief will succeed. (A belief is then true if its truth condition obtains.)

Solutions to philosophical problems do not necessarily have to take the form of explicit analyses of concepts. But there is nothing wrong with explicit analyses, if you can get them. As it happens (and as I said in the Introduction) I do not think that the success-guaranteeing analysis of truth conditions is complete until it is placed in a teleological context. But I do think it is *part* of the truth about truth conditions. So at this stage it will be helpful to consider a number of initial objections to (D), and to make some preliminary observations. The need for teleology will be argued in the next section.

(1) Does not (D) apply only to beliefs of the form: s will bring about t? For these are the only beliefs which have so far been represented as relevant to the success of actions. Surely, however, an analysis of truth conditions ought to deal with beliefs of all forms, and not just with beliefs about means to ends.

It is not difficult, however, to see why (D) should be considered to hold for beliefs of all forms, as well as for means–ends beliefs. It is true that the relevance of beliefs to actions always depends in the last instance on what they imply about appropriate means. And in this sense it is only means–ends beliefs that are *directly* relevant to actions. But, still, such means–end beliefs, that s will bring about t, will as a rule be inferred by the agent from various other beliefs. And this then institutes the requisite general connection between truth and success. For if those other beliefs are true, and the inferences from them valid, then the belief that s will bring about t will be true too, and the resulting action will succeed. So it is a *general* principle that actions based on true beliefs will succeed, and not just a principle about means–ends beliefs as such.[6] Consequently, when we invert this principle into an analysis of truth conditions—analysis (D)—the analysis promises to apply to beliefs in general, and not just to beliefs of the means–end form.

(2) In general a *number* of beliefs will lie behind any given action. But this means that the truth of any *one* belief will be insufficient to guarantee the success of ensuing actions. For success will only be guaranteed if the *other* beliefs behind the action are also true. So strictly analysis (D) ought to be formulated:

> The truth condition of any belief is that condition which guarantees that actions based on that belief will succeed, assuming that any other beliefs it is acting in concert with are also true.

[6] I owe the argument for this principle to Horwich (1987).

But this then disqualifies (D) as analysis of truth-conditional representation, for it assumes the notion of truth in explaining it.

It might seem that we could deal with this difficulty by thinking of analysis (D) as applying specifically to cases where single beliefs generate actions on their own, without the assistance of other beliefs.[7] Truth conditions could then be identified as what guarantees success in such single-belief cases. But the trouble with this is that we then run into objection (1) again, since the only kind of belief that can generate actions on their own are means–ends beliefs.[8] If we want an analysis of truth that works for beliefs in general, and not just for means–ends beliefs, then we need a way of extending (D) beyond single-belief choices of action.

A better way to deal with the problem is to think of analysis (D) as being applied simultaneously to all the belief types in an agent's repertoire. That is, we should think of (D) as fixing the truth conditions for all those beliefs collectively by, as it were, solving a set of simultaneous equations. The 'equations' are the assumptions that the truth conditions of each belief will guarantee success, if other relevant beliefs are true; the overall 'solution' is then a collective assignment of truth conditions which satisfies all those equations.

(3) Another initial worry about (D) might be that it makes truth too easy. Surely we do not want to count beliefs as true whenever the actions they prompt have satisfactory results. Cannot an action achieve a desired result by accident, even though some of the beliefs behind it are false (as when they involve some self-correcting mistake)?

But (D) does not in fact rule out this possibility. The suggestion is not that it is enough, for the truth of a set of token beliefs, that a particular action, prompted by those particular tokens, should succeed. Rather (D) specifies a condition which *guarantees*, for *all* tokens of the relevant types, that ensuing actions will be successful.

(4) What about decisions made under uncertainty? In many cases an agent will act, not on full beliefs, but on partial beliefs. In such cases the agent's thinking will not pick out any action as certain to succeed, but rather select the action that is subjectively most likely to succeed. But then, if the action *does* succeed, that *will not* have been *guaranteed* by the agent's beliefs correctly representing the world.

It is an interesting question as to how far the well-foundedness of decisions made under uncertainty depends on objective features of the world, such as the existence of objective chances. But we can by-pass this issue here. For, once more, there is nothing in (D) which rules out

[7] Cf. Mellor (1988), 86.

[8] This point was put to me by Paul Horwich.

the possibility of actions whose success *is not* guaranteed by the truth of the beliefs behind them. The idea behind (D) is rather that we should focus on the kind of case where success *is* so guaranteed, and then analyse truth as what guarantees success in just those cases. So uncertain decisions issuing from partial beliefs are beside the point. To apply (D) to a given belief, we should stick to cases where that belief is held *fully*, and figures in decisions which *are not* uncertain: truth is what guarantees success in *those* cases.[9]

(5) Analysis (D) seems to imply that the virtue of truth is essentially pragmatic, that the reason for wanting truth is always so as to succeed in action. But surely truth can be pursued as an end in itself, and not just because of its pragmatic value. Indeed there are certain questions, about the farther reaches of the universe, say, or the distant past, where our interest in having true beliefs cannot possibly be practical, since such beliefs can make no difference to our actions.

But this complaint misses its target. (D) is not a theory about why we should *want* truth. It is a theory of what truth *is*: namely, for a belief, the obtaining of a condition which guarantees that, if an agent *were* to act on that belief, the ensuing action would succeed.[10] This does not presuppose that anybody will actually act on the belief. Nor does it presuppose that the only reason for wanting the truth in respect of that belief is to be able to act successfully on it. To be sure, if you do want to succeed in an action, then (D) does immediately imply that you have a motive for wanting the beliefs behind it to be true. But that leaves room for other motives for wanting truth, both in the case of practically significant beliefs and practically insignificant ones. In particular, it leaves room for truth to be valued as an end in itself. (Can't we now ask:

[9] This response to the objection about uncertainty was suggested to me by Hugh Mellor.

[10] It might still seem that there are some beliefs that could not, even counterfactually, be relevant to the success of an action. What about the belief *that there are no agents*, or the belief that *all my actions are doomed to failure*? (These examples were put to me by David Owens and David Sanford respectively.) At this point we need to appeal to the *compositionality* of beliefs. We should recognize that beliefs are made up of components ('concepts'), and we should seek to explain the representational significance of such components in terms of their systematic contribution to the truth conditions of the beliefs they enter into, that is, in terms of their systematic contribution to conditions which guarantee the success of actions based on those beliefs. In this way we can hope to pin down the representational significance of such concepts in terms of their contribution to beliefs which *can* be relevant to action, and then use those representational values to build up truth conditions for such special beliefs as cannot be relevant to action.

why should truth be valued as an end in itself? But I take it to be a virtue of (D) that it allows this as a significant question.)

(6) Analysis (D) talks of 'truth conditions', and of such conditions 'obtaining'. But does not this then commit us to dubious entities like propositions, or possible states of affairs, or sets of possible worlds?

Some philosophers would be untroubled by such commitments. They can skip ahead to the next objection. But, for those who are troubled, let me try to show that the reification of truth conditions is not essential to (D).

Consider the following schema:

(E) An action based on the belief$_n$ will be guaranteed to succeed just in case p.[11]

We can understand analysis (D) as asserting that claims of the form (E) specify the truth-conditional contents of beliefs. That is, analysis (D) can be understood as asserting that (E) is an equivalent substitute for:

(F) The belief$_n$ is true if and only if p.

Note now that neither (E) nor (F) refer to truth conditions as such. So if the import of analysis (D) is simply that (E) is equivalent to (F), then analysis (D) will be free of any substantial commitment to truth conditions too.[12] (I shall assume this understanding of analysis (D) henceforth: subsequent references to truth conditions should be viewed as expository conveniences.)

What now of truth itself? Well, for any particular belief, the relevant instance of (F) will tell us directly what is required for that belief to be true: for example, if it is the belief that snow is white, then it will be true that just in case snow *is* white; if it is the belief that grass is green, then it will be true just in case grass *is* green; and so on. But, more interestingly, since analysis (D) gives us a general account of the import of claims of form (F), an account which does not itself assume the notion of truth, we can generalize, and say that *any* belief will be true just in case actions based on that belief are guaranteed to succeed.[13]

[11] I speak of the 'belief$_n$', rather than the 'belief that p', so as to make it clear that I am not presupposing representational content in explaining it.

[12] What about the worry that the extensionality of (F) and (E) will allow instances that do not specify content? Thus: the belief that snow is white is true iff grass is green. Or, equivalently: the belief that snow is white will generate successful actions iff grass is green. This is not the place to pursue this issue, but I would hope that this worry will be answered once the analysis of representation has been extended to deal with the compositionality of belief (cf. footnote 10).

[13] If we were viewing the instances of (F) simply as truisms about truth, unexplained by a reductive analysis like (D), then generalizing would only yield the unilluminating, 'Any belief will be true just in case it is true'.

These last remarks can help us to understand what *kind* of analysis is offered by (D). Clearly (D) is not a *conceptual* analysis: the pre-theoretical concept of truth is not the same as the concept of the guarantee of successful action. Rather (D) must be offering some kind of *reductive* analysis of truth: it is explaining what truth really *is*, even if we do not think of it that way. But what then is the reductive basis? (D) certainly is not equating truth with any physical or other first-order natural property. I suggest that we see (D) as equating truth with a kind of *second-order* natural property: the property, in a belief, of *causing* actions that are guaranteed to *cause* successful results. This is analogous to functionalist definitions of second-order states in terms of causal roles. Just as a functionalist definition defines a state by invoking a theory which specifies the causal role of that state, so I am suggesting we should define truth by reference to our theory of the causal significance of truth, that is, by reference to the theory that actions caused by true beliefs cause success. The analogy can be carried further: just as a functionally defined state can be 'realized' by different first-order physical states in different systems, so will truth be differently 'realized' for different beliefs: in the case of the belief that snow is white, truth (that is, the property of causing actions which will cause successful results) is realized by snow being white; in the case of the belief that grass is green it is realized by grass being green; and so on.

There are similarities between the approach I am advocating here and the 'redundancy approach' to truth, in that I am arguing that we need not think of truth, for a belief, in terms of the belief's 'truth condition obtaining', but simply in terms of things being as the belief claims. But there are also similarities with the 'correspondence approach', in that when I explain what representational content is for a belief, that is, when I explain what it is for a belief to claim what it does, this involves specifying how the *world* must be in order for actions based on that belief to succeed. Even if this does not bring in truth conditions as such, it does make truth depend on how the world independently is.[14]

(7) Analysis (D) applies to only to beliefs whose truth is of potential causal relevance to the success of actions. Perhaps this will enable it to accommodate beliefs about the natural world. But what about moral, or

[14] This element of correspondence distinguishes the account offered here from 'disquotational' theories of truth of the kind defended in Quine (1970), Leeds (1978), Horwich (1982). Disquotational theories agree with the redundancy thesis that there is nothing more to a given belief being true than p (where 'p' is any sentence with the same content as that belief), but then deny that there is any remaining need for a correspondence notion of content. For a detailed discussion of the difficulties facing such theories see Field (1986).

modal, or mathematical judgments? In what sense, if any, can the truth of such non-natural judgments matter to the success of action?

I do not propose to pursue this complex topic here. Whether or not analysis (D) might apply to a given category of judgment depends on the details of the workings of such judgments, and such details are matters of active controversy for moral, modal, and mathematical judgments. My current thinking is that (D) can be made to apply to areas of discourse which are genuine expressions of *belief,* but that it is debatable how far moral, modal, and mathematical discourses fit this mould. But, as I said, this is too complex a topic to pursue here.

(8) In my answer to objection (1) I appealed to the notion of *validity*: I argued that analysis (D) could be extended from means–ends beliefs to other beliefs because valid inferences from true beliefs of any kind will lead to true conclusions about appropriate means. However, it might be argued that this appeal to validity is illegitimate, on the grounds that the notion of validity presupposes the notion of truth.

Analysis (D) certainly *needs* the notion of validity. Often agents will draw *invalid* inferences about means (imagine that they have to decide what to do quickly, or that their situation is very complicated) and then the truth of the beliefs on which those inferences are based *will not* guarantee the success of their actions. So if analysis (D) is to apply generally, and not just to means–end beliefs, it should strictly be formulated as:

> The truth condition of any belief is that condition which guarantees that actions *validly* based on that belief succeed.

But this now makes the problem clear: (D) can scarcely be held to constitute an analysis of truth, if it presupposes validity and validity presupposes truth.

One possible move here might be to deny that validity does depend on truth. Thus we might seek some purely syntactic notion of validity, defined in terms of a structure of rules of inference, rather than the semantic notion of a form of inference which is guaranteed to lead from truth to truths. However, this syntactic strategy seems unpromising. For a start, there are technical difficulties about the completeness of syntactic characterizations of non-first-order validity. And, in any case, given that syntactic characterizations are always answerable to the semantic conception of validity,[15] even for first-order validity, it is doubtful that the syntactic strategy will really dispose of the circularity, as opposed to just brushing it under the carpet.

For these reasons, I accept that the notion of validity raises a substantial difficulty for analysis (D). I think this difficulty can be solved, but

[15] Cf. Dummett (1974).

not until we have placed (D) in a teleological context. So let us shelve this difficulty until the next section.

This now completes my catalogue of initial objections to analysis (D). But before proceeding let me complete this section by returning briefly to the issue of broad contents.

If the only role for belief in our thinking was to explain the adoption of means, as in:

(B) 1. X desires that t
  2. X believes that s will bring about t
  Therefore,
  3. X does s,

then there would be no reason why any beliefs should be broad. For explanations of form (B) do not require beliefs to do anything except give a causal *push* to actions from the inside, as it were. And on this conception of beliefs it would indeed be puzzling that differences outside believers' heads can make any difference to what they believe.

However, from the perspective of analysis (D) we can see that this conception of belief leaves out the essential point of truth-conditional content. Truth-conditional content is not anything to do with internal pushes.[16] Rather such contents specify the *success* conditions for what is done: they specify a condition which will ensure that the means adopted will produce the desired end. Once we are clear about this, then it is easy to understand why some beliefs should have world-dependent contents. For this is simply the idea that, given two people who are physically identical, but in different worlds, or in different contexts, then what is required for one person's action to succeed can be different from what is required for the other's action to succeed.

The point is clearest for explicitly indexical beliefs. Suppose Bill and Ben are physically identical, and that they both have the desire and belief that they express by 'I want to be warm' and by 'Running around will make me warm'. Then they are both likely to start performing the same bodily movements, namely, running around. And to this extent their beliefs are the same: both beliefs 'push from the inside' in the same

[16] A corollary is that if we were not ever interested in explaining ends, but only means, then nothing would be lost by adopting a purely 'narrow' or even 'syntactical' notion of belief content. Note, though, that this point applies only if our explanatory interests are restricted exclusively to 'basic' means, to things the agent does at will, and do not extend to non-basic means, such as actions, t, which are means to further results, but which the agent does *by* doing something else, s. For even though such a non-basic t can be viewed as a means to its further result, we will still need to invoke the representational content of the belief that *s will bring about t* in order to explain how the agent succeeded in doing t in the first place.

way. But note now that the conditions that will satisfy their respective desires are *different*: Bill's desire will be satisfied by *Bill* getting warm, whereas Ben's desire will be satisfied by *Ben* getting warm. And because of this the condition required for Bill's and Ben's actions to succeed will be different: Bill's action will succeed just in case *Bill's* running around will make *Bill* warm, whereas the success of Ben's action requires the quite different condition than *Ben's* running around will make *Ben* warm. And that, according to (D), is why the truth conditions of Bill's and Ben's beliefs are different, despite their physical identity. It is simply due to the fact that Bill's and Ben's actions have different success conditions.

It is not hard to see how this story might be extended to apply to beliefs that are not explicitly indexical, such as beliefs about natural kinds. If we suppose that the condition that satisfies the desire expressed by 'water' depends on what kind of world the desirer is in, then the condition required for the success of actions aimed at that desire will similarly depend on the agent's world. And this, according to analysis (D), will then explain why beliefs about 'water' have world-dependent truth conditions.

Admittedly, this latter story takes it as given that *desires* involving natural kind concepts have broad *satisfaction* conditions. And perhaps this is contentious. With explicitly indexical desires, such as the desire expressed by 'I want to be warm', it is obvious enough that the satisfaction condition of the desire depends on who is doing the desiring. But it is not so clear why the satisfaction condition of a desire expressed by 'water' should depend on what kind of chemical stuff is present in the desirer's world, rather than on the trans-wordly availability of any liquid that is colourless, odourless and tasteless. But let me shelve this issue at this stage. I have at least shown that analysis (D) yields an explanation of broad beliefs, *if* we assume broad desires. I shall return to broad desires themselves at the end of the next section.

5 The Teleological Theory of Representation

We have just seen how questions about satisfaction conditions for desires raise a difficulty for the explanation of broad contents suggested by analysis (D). I now want to show how the issue of satisfaction conditions also raises a more general difficulty for analysis (D) as a whole.

Recall that, according to analysis (D), truth is what guarantees *success*. I have not said much about success as such so far, but, from what I have said, it clearly is not a primitive notion. Rather, it derives from the notion of a *desire* being *satisfied*: an action succeeds, in the relevant sense, just in case the satisfaction condition of the desire behind that action is fulfilled.

Now, satisfaction conditions, and satisfaction, for desires, are closely analogous to truth conditions, and truth, for beliefs. Just as each belief has a truth conditon, which, together with the world, determines a truth value for the belief, so each desire has a satisfaction condition, which, together with a world, determines whether or not the desire is satisfied.

I started off this paper by arguing that it is philosophically puzzling that beliefs should have truth conditions in virtue of which they represent the world as being a certain way. However, I could equally well have argued that it is philosophically puzzling that desires should have satisfaction conditions in virtue of which, they, so to speak, 'request' the world to be a certain way.

But then it is surely unsatisfactory to offer a solution to the problem of representation for beliefs which, as in (D), simply takes for granted the representational properties of desires. If that is all we can offer by way of philosophical explanation, the puzzlement we started with will remain with us.

Can we not simply analyse satisfaction along the same lines as we have analysed truth? To see how this might work, let me articulate the connection between truth and satisfaction in an explicit principle:

(G) Actions based on true beliefs will satisfy the desires they are aimed at.

In effect, we derived analysis (D) by focusing on the role of truth in this principle, and then 'inverting' the principle to get an analysis of what truth *is*. The suggestion now is that we should do the same with satisfaction, that is, that we should simply 'invert' (G) once more, but now with respect to desires rather than beliefs, to get:

(H) The satisfaction condition of a desire is that condition which is guaranteed to result from actions based on that desire, if the beliefs behind the action are true.

But this will not work. Trying to get an analysis of representation for both beliefs and desires out of (G) is like trying to solve a single equation with two unknowns. The trouble is that there will be any number of ways of attaching 'truth' and 'satisfaction' conditions to an agent's beliefs and desires, all of which satisfy the joint constraint that 'truth' guarantees 'satisfaction', but only one of which identifies the real representational contents of those beliefs and desires (cf. Stalnaker, 1984, pp. 17–18; Papineau, 1984, p. 555; Papineau, 1987, pp. 57). So to explain what makes the correct identification correct we need to add something to the explanation of representation offered by (G) alone.

A natural suggestion at this piont is to supplement the mutual coordination of truth and satisfaction required by (G) with the further thought, which has not figured in this paper so far, that is *normal* for beliefs to be present when their truth conditions obtain, and *normal* for desires to give rise to their satisfaction conditions.

How should 'normal' be read here? One possibility would be to understand it *statistically*. The truth conditions of beliefs would then come out as those circumstances that are *most frequently* present when those beliefs are held. And satisfaction conditions for desires would then come out as those circumstances that *most frequently* result when those desires are acted on.

It is now widely recognized, however, that this kind of statistical approach will not work (see Fodor, 1984; Papineau, 1984; Dretske, 1986). The basic difficulty is that many other conditions will be just as frequently co-present with beliefs and desires as their truth and satisfaction conditions. Thus consider the condition consisting of the *disjunction* of a given belief's truth condition with circumstances that characteristically deceive people into holding the belief even when it is false. This disjunctive condition will be co-present with that belief even *more* frequently than the belief's truth condition, since this disjunctive condition covers not only all the cases where the belief is true, but also some where it is false. Which shows that, whatever truth conditions are, they are not those circumstances that are most frequently co-present with beliefs. A similar point applies to desires. Given any desire, it is not difficult to construct conditions which follow the desire even more frequently than the fulfilment of its satisfaction condition, thereby disproving the thesis that the satisfaction conditions of desires are those conditions they are most frequently followed by.

This is where I think we need teleology. There seems little prospect of defining truth and satisfaction in purely statistical terms. If we introduce the notion of *purpose*, however, the problem becomes relatively simple. Then we can say that the truth conditions of a belief is that condition which it is the *purpose* of the belief to be co-present with. It is not the purpose of a belief to be present when deceptive circumstances obtain, and that is why those deceptive circumstances are not part of its truth condition. And similarly, it is not the *purpose* of desires to give rise to circumstances other than their satisfaction conditions.

This is not the place to offer a detailed defence of this teleological theory of representation. But a couple of general points will be in order.

Firstly, it is clear that the worth of this theory depends on what further account we can give of the teleological notion of 'purpose'. For without some such further account, purposes seem just as metaphysically dubious as representational powers. However, we can make purposes metaphysically acceptable by cashing them out in terms of selection processes. On this conception, it is appropriate to say that the purpose of A is to do B just in case A is now present because of the past operation of some selection process which selects items that do B (cf. Wright, 1973; Neander, 1984).

If we adopt this selectionist account of purposes, then claims about the purposes of beliefs and desires are no more metaphysically dubious than such biological claims as 'The purpose of the heart is to pump blood' or 'The purpose of shivering is to increase body heat'. The idea would be that beliefs have been selected because they are characteristically co-present with certain conditions, which therefore count as their truth conditions, and that desires have been selected because they characteristically give rise to certain results, which therefore count as their satisfaction conditions.

It is probably worth emphasizing that this selectionist-teleological approach to mental representation does *not* imply that all representational abilities must be genetically innate products of intergenerational selection. For selection-based teleology can also be a product of individual learning (cf. 'The pigeon is pressing the bar *in order to* get food'). And so, if some non-innate belief or desire is selected in the course of individual learning in virtue of the condition it is co-present with, or the result it gives rise to, then that belief or desire will have a genuine selection-based representational purpose, despite its non-innateness.[17]

The second general point I wish to make about the teleological theory of representation is that this theory does not *replace* the earlier success-guaranteeing account of truth, but rather *incorporates* it.

In order to see this, it will be helpful to think of beliefs and desires as components in an overall human decision-making system, and correspondingly to think of the purpose of beliefs and desires as deriving from their contribution to the overall purpose of this system.

The overall purpose of the human decision-making system can be thought of as generating actions that cause suitable results. Beliefs and desires both contribute to this purpose. However, they contribute in *different* ways. The role of desires is to do with the fact that different *results* are suitable at different times: our desires vary in order that our actions will produce different results at different times. The role of beliefs is to do with the fact that, given any result, different *means* are appropriate to that result at different times: our beliefs vary in order that we can choose the most effective means at any time to the results that we desire at that time.

[17] See Papineau (1987), Ch. 4.2. Also see Dretske (1988), who not only emphasizes the role of learning, but argues that that learning is essential to the kind of teleology involved in full-fledged mental representation.

In the end, all selection-based purposes depend on *results*: to have a purpose is to have been selected by a mechanism which favours certain *results*. However, these last remarks show that beliefs fit into this scheme in a slightly complicated way. For beliefs do not have any results of their own. Rather, their purposes are to produce whichever results will satisfy the desires they are acting in concert with. In effect, beliefs get selected at one remove, in virtue of being good at causing actions which cause desired results.

This implies that we need to think of the teleological approach to representation as applying first to desires and satisfaction. Any given desire will be present in order to produce a certain result r, which result is therefore its satisfaction condition. Given this explanation of satisfaction for desires, we can then explain the purposes of beliefs. Any given belief will be present in order to produce actions which will produce desired results if a certain condition p obtains, which condition is therefore that belief's truth condition.

But this is just to say that the truth condition of a belief is that condition which guarantees that the actions it produces will satisfy desires. And this is just our earlier success-guaranteeing analysis (D) of truth conditions once more. What the teleology has added is an independent account of desire satisfaction in terms of biological purposes. Given this account of desire satisfaction, beliefs can be understood as having biological purposes as well, namely that of enabling desires to fulfil their biological purposes. But beliefs have these biological purposes precisely because of the truth of analysis (D), and not for some other reason.

Let me now return to the two topics left hanging at the end of the last section, namely, validity and broad desires.

In order to deal with the difficulty about validity, consider the overall human decision-making system once more. So far I have presented this system as consisting solely of beliefs and desires. But this system clearly also needs as 'inference-procedure', so as to be able to infer conclusions about appropriate means from existing beliefs. Now, this inference-procedure might be more or less efficient at making such inferences. But, however efficient or inefficient it may be, its *purpose* will be to make such inferences *validly*, that is, to derive true conclusions about means from true premises. For in our biological history inference procedures will have been favoured precisely in so far as they are good at doing this, that is, good at telling us which actions will in fact produce successful results.

The objection considered in the last section was that analysis (D) was not entitled to define truth in terms of validity, since validity itself depends on truth. But once analysis (D) is incorporated within the teleological theory, this is no longer an objection. *Both* truth and validity have now been biologized, in the sense that both have now been explained as aspects of the successful functioning of the human decision-making system. Beliefs play their role in this system by being true, and inferences play their role by being valid.

Of course, an element of circularity is still present. Our inference-procedures can only have the purpose of validity in so far as they are usually supplied with true premises: valid inferences will not lead to successful actions if their premises are false. Conversely, the general run of beliefs can only have the purpose of being true in so far as the inferences drawn from them are generally valid: true beliefs will not lead to successful actions if the conclusions drawn from them are invalid. But such interlocking purposes are a familiar feature of biological systems. For example, our lungs only have the purpose of oxygenating the blood because the heart circulates the blood, but the heart only has the purpose of circulating the blood because the lungs oxygenate the blood. Natural selection is perfectly capable of designing systems which achieve results by the co-ordinated operation of different components. And in just this sense it has designed our decision-making system to choose successful actions by making valid inferences from true beliefs.

Now for broad desires. According to the teleological theory, the satisfaction conditions of desires are those results that desires have been selected to produce. This indicates various possible explanations for why some desires should be broad. To start with, there is a possible argument from genetic natural selection. Thus, for instance, it is arguable that human biological history has favoured organisms with states that led them to ingest H_2O when they needed it, whereas twin-earth organisms would have been favoured for having states which led them to seek XYZ when they lacked that. And this would then mean, given the teleological theory, that the desire we express by 'water' will be satisfied by H_2O, whereas the equivalent twin-earth desire would be satisfied by XYZ.[18]

One drawback to this biological argument, however, is that it is restricted to genetically-based desires. A more general account of broad desires might appeal to *learning*, and in particular to the learning processes involved in acquiring a language. It seems plausible that linguistic training is a process which selects cognitive states that will enable us to speak correctly: in this sense many of our cognitive abilities will be acquired in the first instance *so that* we can conform to community usage. It is also independently plausible that proper names and natural kind terms are used by human communities to refer to *the*

[18] This kind of argument for broad contents is explored in McGinn (1989), 157–9. For the origin of the H_2O-XYZ example, see Putnam (1975).

object, or stuff, which is the causal origin of the use of the name, or perhaps to *the object, or stuff, which experts will recognize as satisfying the name*. Putting these two points together, it follows that the purpose of beliefs and desires expressed by such proper names or natural kind terms will be to be co-present with, or give rise to, such objects and stuffs. And this, given the teleological theory, then means that these beliefs and desires will have broad truth and satisfaction conditions. (In a sense this sociolinguistic account of broadness assimilates proper names and natural kind terms to indexical terms, as the italicized phrases indicate. But note that this indexicality is located in the community, so to speak, rather than in the individual's head: what makes the concept expressed by 'aluminium' stand for *the stuff round here* is that it is the *purpose* of this concept to enable the individual to speak correctly by applying the term to the local stuff; and this purpose need not be reflected by anything inside the individual's head. This is why the references of proper name and natural kind concepts do not change as soon as humans take trips to twin earth.)

6 Functional Falsity

In this final section I want to show how one common objection to the teleological theory of representation falls away once we recognize that the teleological theory incorporates the earlier success-guaranteeing account (D) of truth conditions, rather than replacing it. This is the objection that certain beliefs have biological purposes which can require them to be present even when they are false, contrary to the claim that truth conditions can be analysed as circumstances in which beliefs are biologically supposed to be present.[19]

For example, consider the belief that you are not going to be injured in some unavoidable and imminent trial of violence. It is arguable that natural selection has bequeathed us an innate disposition to form this belief, even in cases where it runs counter to the evidence, in order to ensure that we will not flinch in battle. But it then seems to follow that, according to the teleological theory, the truth condition of this belief will include plenty of circumstances where we *will* be injured (since that is when we are biologically supposed to have the belief). But this then looks like a *reductio* of the teleological theory. For by hypothesis the truth condition of the belief is that we *will not* be injured.

Examples like this are interesting, but I do not think they suffice to discredit the teleological theory. In order to see why not, let me

[19] Ned Block has urged this objection on me. See Stich (1982), p. 53, for a related point.

distinguish two different kinds of biological purposes which beliefs can serve. I shall call these *normal* purposes and *special* purposes. Normal purposes involve the contribution of beliefs to the human decision-making system, as discussed in the last section. Special purposes involve the contribution of beliefs to other biological needs, in ways that will be explained shortly.

The teleological theory of representation needs to be understood as referring specifically to the normal purposes of beliefs. Recall that the contribution of beliefs to human decision-making is to indicate which means will achieve whichever results are currently *desired*. Now, in order to serve *this* purpose the belief in the above example will still need to be true, even if there are other special purposes which require it to be false. And so, provided we focus specifically on the normal purposes of beliefs, this example ceases to present a problem to the teleological theory.

If you are unconvinced the belief about invulnerability needs to be true in order to serve its normal function, consider the case, say, of Cuthbert Coward, who would far rather remain unscratched than win the battle. If Cuthbert believed that he will not be injured, when in fact he will, then this at least might induce even him to enter the fray, and thus to end up with the injury he was so keen to avoid. But then Cuthbert *will not* get what he desires, which was above all to remain unscathed. So, if Cuthbert is to be sure of satisfying his desire to remain unscathed, then his belief that he will not be injured needs to be true, just as the teleological theory as now proposed requires.

Of course, Cuthbert has somewhat unhappy desires. In particular, his desires are inappropriate from a biological point of view, in the sense that the satisfaction of his desires is unlikely to further his overall chances of survival and reproduction. This is where the *special* purposes of some beliefs come in. The point of these special purposes is in effect to by-pass the normal role of beliefs in satisfying desires, and to ensure instead that agents with biologically inappropriate desires do not end up performing biologically inappropriate actions. Cowards are a case in point. Their unfortunate desires mean that they are likely to end up running from battle, just in order to avoid a scratch. And so, in order to protect them against the biological danger of such consequences, natural selection predisposes them to believe that they are invulnerable, even when the evidence does not warrant this belief, so as to *stop* them performing those actions which would in fact satisfy their desires.

It might seem puzzling that natural selection should give some beliefs two different purposes. After all, natural selection presumably designs biological systems for one ultimate end, namely, the bequest of genes. So why do beliefs not simply have the single purpose of ensuring such gene bequests?

The answer relates once more to the nature of the human decision-making system. Note that this system does not work by always choosing that action which is most likely to ensure gene bequests. Rather it chooses that action which is most likely to satisfy existing desires. It is not impossible to imagine biological systems of the former kind, which always aimed directly for gene bequests. But it seems likely that the limitation of our cognitive capacities have prevented us from doing things in this way. Instead we aim for such relatively short-term goals as warmth and sex and chocolate ice-cream and listening to Chuck Berry records.

By and large such short-term goals correlate reasonably well with ultimate biological success, which is no doubt why our innate desires, and our ways of acquiring non-innate desires, have evolved as they have. But the satisfaction of our desires will not always coincide with biological success (not all sex leads, or even can lead, to reproduction). And this then means that there are certain biological risks consequent on our way of doing things. Now, it may be that some of these risks are *inevitable* by-products of our desire-based decision-making system: for example, it may be inevitable that humans will have extremely strong desires to avoid injuries, and so inevitable that in certain circumstances this will lead them to act against their biological interests. And this will lead to natural selection *interfering* with the normal operation of decision-making system, by giving us beliefs which lead us to act in ways that frustrate our desires, but satisfy our biological needs.

Let me sum up the argument of this last section. Certain beliefs do indeed have *some* biological purposes that require them to be false. However, this does not invalidate the teleological theory of representation. For we can understand the teleological theory as focusing specifically on the *normal* purposes of beliefs, namely, to guarantee the satisfaction of desires. And these normal purposes do not ever require beliefs to be false.[20]

References

Appiah, A. 1986. 'Truth Conditions: A Causal Theory', in J. Butterfield (ed.), *Language, Mind and Logic* (Cambridge University Press).

Burge, T. 1979. 'Individualism and the Mental', in *Midwest Studies in Philosophy*, **4**.

Burge, T. 1982. 'Other Bodies', in Woodfield, 1982.

Dennett, D. 1969. *Content and Consciousness* (London: Routledge and Kegan Paul).

[20] I would like to thank Tom Baldwin, Ned Block, Paul Horwich, Hugh Mellor, Philip Pettit, Huw Price and Jamie Whyte for discussing these topics with me.

Dennett D. 1987. *The Intentional Stance* (Cambridge, Mass: MIT Press).
Devitt, M. 1984. *Realism and Truth* (Oxford: Basil Blackwell).
Dretske, F. 1986. 'Misrepresentation', in R. Bogdan (ed.), *Belief* (Oxford: Clarendon Press).
Dretske, F. 1988. *Explaining Behaviour* (Cambridge, Mass: MIT Press).
Dummett, M. 1974. *The Justification of Deduction*, British Academy Lecture (Oxford University Press).
Evans, G. 1982. *The Varieties of Reference* (Oxford: Clarendon Press).
Field, H. 1972. 'Tarski's Theory of Truth', *Journal of Philosophy*, **69**.
Field, H. 1978. 'Mental Representation', *Erkenntnis*, **13**.
Field, H. 1986. 'The Deflationary Conception of Truth', in G. MacDonald and C. Wright (eds), *Fact, Science and Morality* (Oxford: Basil Blackwell).
Fodor, J. 1984: 'Semantics, Wisconsin Style', *Sythèse*, **59**.
Fodor, J. 1987. *Psychosemantics* (Cambridge, Mass: MIT Press).
Horwich, P. 1982. 'Three Forms of Realism', *Synthèse*, **51**.
Horwich, P. 1987. 'Pure Truth', unpublished draft.
Leeds, S. 1978. 'Theories of Reference and Truth', *Erkenntnis*, **13**.
Loar, B. 1982. *Mind and Meaning* (Cambridge, University Press).
McDowell, J. 1986. 'Singular Thought and the Extent of Inner Space', in P. Pettit and J. McDowell (eds), *Subject, Thought and Context* (Oxford: Clarendon Press).
McGinn, C. 1982. 'The Structure of Content', in Woodfield, 1982.
McGinn, C. 1989. *Mental Content* (Oxford: Basil Blackwell).
Mellor, D. H. 1988. 'I and Now', *Proceedings of the Aristotelian Society*, 89.
Millikan, R. 1984. *Language, Thought and Other Biological Categories* (Cambridge, Mass: MIT Press).
Millikan, R. 1986. 'Thought without Laws: Cognitive Science with Content', *Philosophical Review*, **95**.
Neander, K. 1984. 'Abnormal Psychobiology', PhD dissertation. La Trobe University.
Papineau, D. 1984. 'Representation and Explanation', *Philosophy of Science*, **51**.
Papineau, D. 1986. 'Semantic Reduction and Reference', in J. Butterfield (ed.) *Language, Mind and Logic*, (Cambridge University Press).
Papineau, D. 1987. *Reality and Representation* (Oxford: Basil Blackwell).
Putnam, H. 1975. 'The Meaning of "Meaning"', in his *Mind, Language and Reality* (Cambridge University Press).
Putnam, H. 1978. 'Reference and Understanding', in his *Meaning and the Moral Sciences* (London: Routledge and Kegan Paul).
Ramsey, F. 1927. 'Facts and Propositions', *Aristotelian Society Supplementary Volume*, **7**.
Stalnaker, R. 1984. *Inquiry* (Cambridge Mass: MIT Press).
Stich, S. 1982. 'Dennett on Intentional Systems', in J. Biro and R. Shahan (eds.), *Mind, Brain and Function* (Brighton: Harvester).
Quine, W. 1970. *Philosophy of Logic* (Englewood Cliffs, N.J.: Prentice-Hall).
Woodfield, A. (ed.) 1982. *Thought and Object* (Oxford: Clarendon Press).
Wright, L. 1973. 'Functions', *Philosophical Review*, **82**.

Functional Support for Anomalous Monism

JIM EDWARDS

1. Introduction

Donald Davidson finds folk-psychological explanations *anomalous* due to the open-ended and constitutive conception of rationality which they employ, and yet *monist* because they invoke an ontology of only physical events.[1] An eliminative materialist who thinks that the beliefs and desires of folk-psychology are mere pre-scientific fictions cannot accept these claims, but he could accept anomalous monism construed as an *analysis*, merely, of the ideological and ontological presumptions of folk-psychology. Of course, eliminative materialism is itself only a guess, a marker for material explanations we do not have, but it is made plausible by, *inter alia*, whatever difficulties we have in interpreting intentional folk-explanations realistically. And surely anomalous monism does require further explanation if it is to be accepted realistically and not dismissed as an analysis of a folk-idiom which is to be construed instrumentally at best. Some further explanation is needed of *how* beliefs, desires, etc. can form rational patterns which have 'no echo in physical theory'[2] and yet those beliefs, desires etc. be physical events.[3] To this end I propose to graft on to anomalous monism a modest version of functionalism.

At first sight functionalism might seem to support the monism but be antagonistic to the anomalism. For a basic idea of functionalism is that the normative rational patterns of folk–functional explanations are mirrored by purely causal inter-relations between the physical states which realize the beliefs, desires, etc. of a folk-explanation. Hence those rational patterns find not just an 'echo' but an exact image in physical theory. Thus Brian Loar comments on anomalism: 'the very possibility of a functional interpretation [of folk-psychology] shows that claim [i.e. anomalism] to be false' (Loar, 1981; 21). Loar has

[1] Donald Davidson, 'Mental Events', in Davidson (1980), 207–27.

[2] Donald Davidson, 'Psychology as Philosophy', in Davidson (1980), 231.

[3] Like Davidson I shall use the term 'physical' widely, to include all the natural sciences. So biology and physiology are in this sense physical. Where I use 'physics' to mean physics will be obvious from the text.

constructed a functional theory which, he claims, is a conservative explication of folk-psychology. However, as John McDowell has pointed out (McDowell, 1985), the conception of rationality employed in Loar's functional theory is a *minimal* conception, a conception included in but not co-extensive with the *wide* conception of rationality which, Davidson argues, makes psychological explanation anomalous. So Loar's functionalism, if otherwise acceptable, would not refute anomalism. Granted that the patterns required by Loar's minimal conception of rationality *are* mirrored in physical theory, it does not follow that the open-ended patterns required by Davidson's wide conception of rationality have an echo in physical theory. Accepting McDowell's point, the possibility emerges that folk-explanations are not co-extensive with but incorporate a functional theory, which functional theory employs a minimal conception of rationality, and that minimal conception of rationality does find an echo in, indeed is mirrored by the causal relations between the physical states which realize beliefs, desires, etc. Being rational in this minimal sense is only a necessary part of, and not sufficient for being rational in the open-ended, holistic, full-blown sense of rationality which makes the mental anomalous. If this can be made to work, the monism of anomalous monism is supported by the functionalism, since an explanation is given of *how* beliefs, desires, etc. can be causally active, but without imperilling the anomalism of the mental.

However, the philosophical presumptions of such a functionalism would be modest. The inventors of philosophical functionalism, David Lewis and Hiliary Putnam, made two claims: firstly, beliefs, desires, etc. are, or are realized by, internal states whose causal relations mirror the rational relations between the sentences which specify the contents of a subject's beliefs, desires, etc. And secondly, beliefs, desires, etc. are 'nothing but' states having such a nexus of relations, so that beliefs, desires, etc. are implicitly defined by the functional theory. Modesty confines the first claim to a minimal conception of rationality which falls short of the full-blown conception. It follows that the second claim will have to be modestly abandoned. To be a believer and desirer one must be rational in the wide sense. So realizing the functional component is not sufficient, and the functional theory alone cannot define belief etc. *Perhaps* beliefs, desires, etc. are implicitly defined by whatever full set of constraints on radical interpretation constitute the wide conception of rationality. However, if the wide conception of rationality is open-ended, it may be that these constraints cannot be stated in a precise but non-circular way which permits non-trivial definitions of beliefs, etc.[4]

[4] Functionalism has had a bad press recently—see Putnam (1988) and

The kind of functional theory which is intended wears its modesty on its sleeve. The functional component in folk-psychology will contain some input laws, perhaps

$$(z)(p)((O(p)\&R(z,p))\rightarrow B(z,p)) \quad (1)$$

where 'z' is a variable for persons, 'p' a variable for propositions, 'O(p)' says that the fact that p is observable, 'R(z,p)' says that z is suitably related to the fact that p, and 'B(z,p)' says that z believes p. And some internal-state-to-internal-state laws of transition, perhaps

$$(z)(p)(q)(((p\vdash q)\&B(z,p))\rightarrow B(z,q)) \quad (2)$$

And some output laws, perhaps

$$(z)(p)(I(z,p)\rightarrow D(z,p)) \quad (3)$$

Where 'I(z,p)' says that z intends to do p, and 'D(z,p)' says that z does p. Of course, more laws than just these are needed to form a functional theory. We need as well at least a law linking beliefs via desires to intentions. It matters not for the present that (1) to (3) are sketchy and simplified. In the above, and in what follows, I am using the term 'proposition' as a generalization of the notion of a statement, so that, for example, a conditional statement is a proposition, and it also contains at least two propositions which are not stated viz, its antecedent and its consequent. A proposition, as I am using the term, is thus a linguistic entity, and not a cut in the set of possible worlds.

We are to construe (1) to (3) functionally; that is to say, for any subject at any particular time, we are to take his beliefs etc. to be realized by certain physical states whose incoming causal relations, whose state-to-state causal relations, and whose outgoing causal relations mirror the rational relations prescribed by (1) to (3) when those internal states are indexed by the content-ascribing propositions of folk-psychological explanations. But if a subject's internal states causally mirror in this way the rational relations between attitudes specified by (1) to (3), they will *also* causally mirror the 'rational' relations specified by

$$(z)(p)((O(p)\&R(z,p))\rightarrow B(z,f(p))) \quad (1')$$

$$(z)(p)(q)(((p\vdash q)\&B(z,p))\rightarrow B(z,q)) \quad (2')$$

$$(z)(p)(I(z,f(p))\rightarrow D(z,p)) \quad (3')$$

where f(p) is any one-one function from propositions on to propositions of the same logical form: f(p) need not have the same sense and truth-

Schiffer (1987). But Putnam's and Schiffer's fire is directed at immodest functionalism, going over the head, I hope, of functionalism modestly construed.

value as p. All that we have done is re-index our subject's internal states with propositions of the same logical form, but not perhaps the same sense and truth value, and if our subject's internal states have internal and external causal relations which mirror the old set of indices as related by (1) to (3), then they equally mirror the new set of indices as related by (1′) to (3′). Construing folk-psychology as (in part) a functional theory is saying that we use propositional indices, as related by (1) to (3), as a *measure* of the causal relations of the physical states thus indexed. And in moving to (1′) to (3′) we have merely re-calibrated our measure, as when we move from the Fahrenheit to the Celcius temperature scale. So far as the facts which a functional theory records go, (1) to (3) and (1′) to (3′) record the same facts and are, in that sense, two versions of the same functional theory. So if we construe folk-psychological explanations as functional explanations, and nothing more, we can ascribe specific contents to a persons attitudes, but only relative to an understood indexing system. Tom's belief that snow is white (relative to our usual indexing system) *is* his belief that grass is green (relative to a different indexing system), *is* his belief that water is poisonous (relative to another indexing system), etc. The functional theory alone cannot show why one content rather than another is *the* content of Tom's attitude.

Moreover, our modest functionalism alone cannot explain why we ascribe intentional contents to beliefs, desires, etc. at all. If we arithmetically encode the syntax of our indices and the proof procedures involved in '$\vdash$' in the manner of Godel, then we can rewrite the *same* functional theory using numbers as indices of cognitive and volitional states. The same set of causal relations which were mapped by the logical relations between their propositional indices would then be equally mapped by arithmetic relations between their numerical indices. There is nothing about the causal relations between internal states and between those states and their environment, so far as those causal relations are recorded by our functional theory, to explain why those states have an intentional content. So far as our functional theory is concerned intentional contents are an adventitious feature of the indexing system, not an intrinsic feature of the states so indexed.

However, the anomalism of the mental and a modest construal of functionalism shifts the burden of explaining the intentional contents of folk-psychological explanations from the functional theory to the wider concept of rationality employed in radical interpretation. The new picture is that the intentional content is determined by whatever constrains radical interpretation. The wider, open-ended, holistic conception of rationality determines whether an object is a psychological subject and what the intentional contents of its psychological states are. Part of those constraints, but only part and not the whole, is that those contents should be related by the minimal concept of rationality employed by the functional part of total folk-psychological theory. But the functional component alone is not sufficient to determine intentional content, nor even that the states have intentional contents. Thus the functional theory is philosophically modest. What was a problem, when functionalism was immodestly taken to capture the whole explanatory force of folk-psychology, becomes a virtue when functionalism is modestly construed as part of an explanatory package which is, as a whole, anomalous.

Of course, Quine, Davidson and Putnam have all famously argued that radical interpretation itself fails to assign determinate contents, senses and references, to the sub-sentential parts of content-ascribing sentences. I do not find their arguments convincing, but that issue is not to the present point. The present point is that functional theories of which (1) to (3) are a prototype, *obviously* cannot determine a unique content, nor even that the functional states have any content. But that is not a fault of functional theories of the psychological, only a reason for giving them a modest philosophical role.

Putnam is a dualist: he distinguishes between mental states or events, on the one hand, and the physical states or events which realize them, on the other hand. No doubt this is because he takes the painfulness of a pain to be a *necessary* property of it, and takes some minimal intentional content of a belief to be a *necessary* property of it. A token physical state realizes a token pain state (or a token belief) because of its *contingent* causal relations to other states. So that token physical state in other possible words does not realize a pain (or a belief). Hence Putnam's need to distinguish mental states or events from the physical states or events which realize them. My aim is to support anomalous monism, so I shall follow Lewis by identifying token mental states or events with the token physical states or events which realize them, and bite the bullet: in other possible words this token pain is not painful, this belief not a belief, because in those possible worlds they lack the required causal connections. I do not think that the folk-psychologist-in-the-street has any worthwhile views as to whether or not the painfulness of a pain, or the intentional content of a belief, are necessary properties of them. The folk-psychologist-in-the-street is not required to take a stand on this issue when applying his folk-theory to his peers.

2. The Anomalism of Folk-explanations

My project is to show that a modest functional component in folk-psychology underpins monism whilst leaving room for anomaly. My object in this section is to find the sense in which folk-psychological explanations are claimed by Davidson to be anomalous, and to show that a functional component in folk-explanations is consistent with his best argument for such anomaly.

In one sense, functional explanations are themselves anomalous. The laws of the internal combustion engine, for example, apply to an object only if it is subject to certain boundary conditions—cars break down and cease to function. And it is a contingent matter whether or not the relevant boundary conditions continue to be met. The boundary conditions may differ from realization to realization—e.g. steel is stronger than cast iron, but more prone to rust, so a steel engine can stand higher stresses but less weather than a cast iron engine. Since we cannot list all the *possible* realizations of a given functional theory, we cannot incorporate a disjunctive list of all their relevant concrete boundary conditions into the functional theory. Functional theories and functional explanations generalize over different concrete realizations by ignoring any differences in the boundary conditions by which one concrete realization differs from another. Hence functional laws cannot be derived from the basic laws of physics. The most that can be derived are concrete particularizations of the functional 'laws' to a given particular realization via premisses specifying the boundary conditions peculiar to that concrete realization. And, of course, since the premisses include contingent specifications of boundary conditions, the conclusions are not really laws. Thus functional explanations enjoy a certain autonomy from basic physical explanations.

The same point applies to functional theories in biology. It is a functional 'law' that the heart causes the blood to circulate. Obviously obedience to this 'law' is subject to certain boundary conditions. The boundary conditions relevant to a frog will differ from the boundary conditions relevant to an elephant, despite their common remote evolutionary ancestors. And we cannot, in principle, list all the detailed boundary conditions for all possible species. Whilst it is certainly true that we give folk-psychological explanations in substantial ignorance of the relevant boundary conditions, this is an autonomy enjoyed by any functional theory—in this sense functional theories in the biological sciences and engineering are not reducible to explanations in physics. So if the anomaly of the mental amounted merely to the autonomy of a functional explanation, it would not be of much philosophic interest.

Without doubt, this autonomy of functional explanations does contribute to the phenomena which Davidson thinks make the mental anomalous.[5] Firstly, Davidson points out that the causal generalizations of folk psychology are not strictly true, but only true *ceteris*

[5] I am grateful to my colleague John Porter for an illuminating discussion of these points.

paribus, conditions which we cannot redeem with honest coin. And secondly, if we make a generalization of folk-psychology strictly true, by inserting an explicit *ceteris paribus* clause, then the result is not a falsifiable empirical generalization but is analytic.[6] Both these points apply to functional laws in general. Firstly, functional laws are not strictly true, but only true provided the relevant boundary conditions are maintained. Secondly, if we make a functional law strictly true by adding an extra condition to its antecedent 'and the relevant boundary conditions are maintained', then the resulting law is analytic, because the *relevant* boundary conditions are whatever conditions are sufficient to secure obedience to the law.

An argument for a distinctive brand of anomaly arises when we consider Davidson's views on radical interpretation. A radical interpreter ascribes semantic contents to his subject's utterances, and thereby to the subject's cognitive and volitional states. The radical interpreter produces, *inter alia*, a folk psychology of his subject. Of course, a radical interpreter taxomizes his subject's mental states by more than just their semantic contents. He taxonomizes them by an amalgam of semantic content and conceptual content. For example, he may ascribe to Tom the belief that Cicero denounced Cataline but not the belief that Tully denounced Cataline, despite the common semantic content of those two beliefs. It has become a philosophical commonplace to observe that the semantic contents ascribed to a subject's attitudes depend upon more than the facts about him which realize our modest functional laws. To adapt Putnam's famous example in 'The Meaning of "Meaning"' (Putnam, 1975; 215–71), Tom, on earth, believes that Archimedes leapt from a bath of water, but Twin-Tom on Twin-Earth believes that t-Archimedes (a different gentleman) leapt from a bath of t-water (a different substance). We need not suppose that Tom or Twin-Tom could identify or locate either Archimedes or t-Archimedes, nor that either knows that water is H_2O whilst t-water is XYZ. We suppose that Tom and his Twin are identical in sensory input, internal-state-to-internal-state causal transitions and behavioural output, for the whole of their lives. So there is nothing in the facts which realize our modest functional laws, viz the 'narrow' causal make-up of Tom and his Twin, to show why radical-interpretation applies the one set of contents to Tom's attitudes but not to Twin-Tom's, nor vice versa. So functional organization and history alone does not determine the contents of beliefs, desires, etc. assigned by radical interpretation.

We still need an argument to show that the conception of rationality employed in folk-explanations is anomalous in the sense of going

[6] See, for example, Donald Davidson, 'Freedom to Act', in Davidson (1980), 79–80.

beyond whatever minimal conception of rationality is captured by the laws of a folk-functional theory. Davidson does argue that there is no codification of the standard by which we judge a semantic theory of a subject's idiolect to be warranted, although that semantic theory is itself to be systematic in the Tarskian mould. And this is because judging the reasonablness of a semantic theory of an idiolect involves judging the reasonableness of whatever systems of beliefs that idiolect is used to express. And there is no codification of the standard by which beliefs are judged reasonable or otherwise. If we tried to codify the standard by which a system of beliefs is thought reasonable or not, our standard would have to apply to every possible theory that could be formulated in the idiolect in relation to every possible state of evidence.[7]

Now we have seen that folk-psychology individuates beliefs, and hence theories believed and evidence believed, by, in part, their semantic contents. And folk-psychology employs whatever conception of rationality is employed by a radical interpreter. So our codification of the full conception of rationality employed in folk-psychology would have to distinguish all possible theories by their distinct semantic contents, and all possible states of evidence by their various semantic contents, and relate the one to the other. I do not think that we have a proof that the task is hopeless, but it is plausible that it is, that the mental is, in this sense, anomalous. It is certain that the conception of rationality involved goes beyond the relatively weak conception of rationality required by a functional component in folk-psychology modelled on (1) to (3). And it is enough that the wide conception of rationality finds no echo in whatever physical theory specifies the causal relations which realize the minimal conception of rationality. So the

[7] Davidson tends to focus on *charity*: roughly, a semantic theory of a subject's idiolect is reasonable to the extent that it makes the beliefs of the subject, as expressed by his utterances in the idiolect, come out *true*. I am following Colin McGinn 'Radical Interpretation and Epistemology', in LePore (1986), 356–68, in supposing that it is charity about the *reasonableness* of a subject's belief-system rather than its *truth* which is crucial. Indeed, since I take the references and extensions of Tom's idiolect to be determined via the utterances of the language community to whom Tom holds his usage to be responsible, maximizing the *rationality* of an unfortunately placed member of that linguistic community need not maximize the *truth* of his beliefs. Perhaps even his perceptually based beliefs are massively false. Nonetheless, it may still be the case that the observational beliefs of the community as a whole could not be massively false. If so, a constraint on radical interpretation requiring individual rationality is consistent with a constraint requiring truth of commonplace observational beliefs. So having accepted McGinn's point, Davidson may still have an argument against some form of scepticism.

modest functional theory does not undermine the claim that the mental is anomalous.

Perhaps, if we could individuate belief-systems by the conceptual contents entertained by a believer, then we could codify the relations between any theory entertained and any evidence entertained. *Perhaps* we could codify the relations between conceptions entertained which make one conception evidence for another. Anomalism, construed as a doctrine about folk-psychological explanations, does not need to deny this possibility. For, as we have seen, folk-psychology individuates attitudes by, in part, their semantic contents and thus cross-cuts the conceptions entertained.

Ironically, in recent work Davidson[8] has defended 'what may as well be called a coherence theory of truth and knowledge'. A set of beliefs are coherent, it turns out, exactly if they are rational in the wide sense employed in radical interpretation. So if there were a *theory* of coherence, an answer to Davidson's question 'What exactly does coherence demand?' (LePore, 1986; 308), we might have a codification of wide rationality, and the 'anomalism' of the mental would shrink to the mere autonomy of functional explanations.

Some have argued that although radical interpretation and folk-psychology taxonomize mental states by an amalgam of semantic content and conceptual content, the semantic contents of propositional attitudes play no part in psychological explanations of behaviour (Field, 1978; Fodor, 1980 and McGinn, 1982, to name but three). Colin McGinn remarks: 'the explanatory force of the content ascription attaches only to the contribution the words in the content clause make in their capacity as specifiers of internal representations [i.e. of conceptual contents], their referential properties play no explanatory role' (Woodfield, 1982; 215). If this is so, then the fact, assuming it to be a fact, that the conception of rationality is anomalous when the rationality applies to attitudes taxonomized by an amalgam of their semantic and conceptual contents, this fact would *not* imply that folk-psychological *explanations* are themselves anomalous. For folk-psychological explanations take cognizance of only the conceptual contents of attitudes, ignoring their semantic contents, it is claimed.

However, whether or not psychological explanations *should* cite only conceptual contents, it is easy to show that folk-psychological attributions do *not* provide us with a route to such conceptual contents. Imagine that Tom and Dick are examining two pieces of jewellery which I will call 'A' and 'B'. Tom thinks that A is made of platinum and B of silver, because A is expensive and B relatively cheap. Tom knows

[8] Donald Davidson, 'A Coherence Theory of Truth and Knowledge', in LePore (1986), 301–19.

no essential or identifying properties of either of these metals. His only relevant beliefs are that A is called 'platinum' by the experts of his community and B 'silver', and that platinum is less common and more expensive than silver, beliefs he knows to be contingent. Now Dick is a visitor from Elsewhere, a place where they speak reverse-English, a language like English except that their experts call platinum 'silver' and silver 'platinum'. Suppose Dick has exactly the same *conceptions* regarding A as Tom, and exactly the same *conceptions* regarding B as Tom. E.g. Dick believes that the metal he and his experts call 'platinum' is less common and more expensive than the metal he and his experts call 'silver'—perhaps, platinum is common and silver rare in Elsewhere. The point is that Tom and Dick entertain exactly the same conceptions about A and exactly the same conceptions about B, and each entertains different conceptions about A from those he entertains about B, ex hypothesi. And yet folk-psychology, expressed in English, assigns to Tom the same attitude about A as it assigns to Dick about B, viz that it is platinum, and also assigns to Tom the same attitude about B as it assigns to Dick about A, viz that it is silver. (And folk-psychology expressed in reverse-English behaves similarly.) Hence, if folk-psychology gives exactly the same explanation of an action of Tom's as it does of an action of Dick's, it does not follow that Tom and Dick entertain the same conceptions. And if folk-psychology gives different explanations of actions of Tom and Dick, it does not follow that they were entertaining different conceptions. There is no route from folk-explanations to the concepts entertained, because folk-explanations specify attitudes by, in part, their semantic contents. Thus a folk-psychological explanation of a piece of Tom's behaviour may cite his belief that A is made of platinum, and no other beliefs of Tom's about A. And a folk-psychological explanation of a piece of Dick's behaviour may cite Dick's belief that A is silver, and no further belief of Dick's about A. There would be no knowing from these explanations that Tom and Dick entertained common conceptions about A. Also obviously, given an explanation which cites only Tom's belief that A is platinum and an explanation which cites only Dick's belief that B is platinum, we could not infer that Tom's conception of A differs from Dick's conception of B. The amalgam of semantic and conceptual contents by which folk-explanations taxonomize mental states cannot be separated into its components.[9]

[9] This is *not* the Fregean point that there is no route from the reference back to sense. Fregean sense determines reference. So if Tom and Dick's attitudes to A involve the same Fregean senses they would have the same semantic content and truth value (ditto B). But they do not: Tom and Dick's beliefs about A have opposite truth values (ditto B).

Perhaps, *some* belief tokens of the same folk-psychological type have the same conceptual contents. Perhaps beliefs about an *observable* property—e.g. beliefs that this is red or beliefs that that is water—have the same conceptual content. (Water, in our neck of the celestial wood, is readily recognizable. No matter that the judgment that this is water, based upon direct observation, may be corrected by theory, for the same is true of the judgment that that is red.) In contrast, whether something is platinum rather than silver is not readily recognizable. (I hope!) Nonetheless, it is not clear that we can infer from 'Tom believes that this is water' and 'Dick believes that this is water', both their beliefs being based upon direct observation, that they entertain the same conceptions. For Tom might be one of us and Dick a Martian with alien sensory modalities, but a creature of a kind that can recognize water.

If we retreat to a repertoire of more basic observable, sense-specific properties—colours for sight, pitch for sound, separating visual shapes from tactile shapes—can we say that if Tom and Dick both judge that this is red, then they entertain the same conception? It is still not clear, because Dick's alien physiology might exploit quite different causal mechanisms to gain the same information effect. Granted that all who *see* that this is red enjoy the same visual *experience*. But it does not follow that all who *believe* that this is red entertain the same conception, since the belief may be due to different experiences.

I have been assuming that folk-psychologists would judge on the basis of his performance (e.g. whether he can successfully sort things by colour etc.) whether or not alien Dick can observe that something is red (or round, or . . .) rather than on the basis of his sense modality (e.g. whether it is an optical receptor-at-a-distance or a contact probe). But because we folk-psychologists have not had to deal with talking Martians, it is not clear how we would extend our folk-explanatory practices to them. But if it is not clear, then it is not clear that sameness of folk-psychological attribution of even observational beliefs implies sameness of conception entertained. And the inference certainly does not hold for folk attributions in general, as the case of platinum and silver illustrates.

We have found a sense in which folk-psychology explanations can be plausibly claimed to be anomalous, an anomaly which goes beyond the mere autonomy of functional explanations. Suppose that folk-psychology taxonomizes attitudes by an amalgam of semantic and conceptual contents which cannot be unmixed. Grant that we are unable to give a general and illuminating codification of the conditions under which some beliefs so taxonomized justified other beliefs. Yet folk-psychological explanations employ this open-ended conception of rationality, since behaviour is explained to the extent that the attitudes associated with it are revealed to be rational. Hence we cannot hope to reduce folk-psychological explanations as a general class to any physical theory which we can grasp.

I am *not* supposing that we can give a general but illuminating codification of the conditions under which one set of beliefs about matters physical are evidence for a theory of physics—an inductive logic. Rather, the contrast is that given a physical theory it is in some sense determined what explanations of physical phenomena it sanctions—crudely, those phenomena whose descriptions are deductive consequences of the theory. Whereas the range of explanations of behaviour sanctioned by folk-psychology is open-ended. We cannot say in advance how attitudes must be related to explain behaviour, except to say that they must show the behaviour to be rational, employing our open-ended, uncodifiable conception of rationality.

So far I have supposed that the modest functional component of folk-psychology makes a sharp distinction between the functional law and the boundary conditions for the application of that law. Although this is so for functional theories in engineering and biology, where the boundary conditions vary from application to application, it is probably not true that the 'laws' of folk-psychology make a sharp distinction. Consider the following sketch of an input law:

> If there is a red object in Tom's line of sight and . . ., then Tom seems to see something red.

The gap '. . .' is to be filled by such conditions as the following: the object needs to be large enough in proportion to its distance from Tom; Tom needs to be normally sighted, alert and with his eyes open; the illumination must be good; there must be no opaque obstacles, etc. If this is to be an input law for a functional psychology, then its antecedent conditions must be specified without employing the consequent. We cannot specify 'being large enough' as being large enough to be seen, nor 'good illumination' as illumination sufficient to see colours, etc. Rather, we need to be able to determine whether the antecedent conditions have been fulfilled independently of whether Tom seems to see something red, since the latter will depend also upon Tom's other cognitive and volitional states and the other laws of the functional theory. Tom's seeming to see something red will need to be related in ways prescribed by all the functional laws to the rest of Tom's cognitive and volitional states. Furthermore, specification of the antecedent conditions cannot appeal to facts about visual perception known only to psychologists or physiologists but not to the man in the street. For if functional explanations are part of folk-explanations, then the relevant functional law must be known, implicitly if not explicitly, by the practitioners of folk-explanation.

It may be that folk-explanatory practices need some conservative explication to uncover a functional component. The problem arises because all the creatures we have met which are capable of seeming-to-see-something-red have been alike in the hardware by which they realize this functional state. And commonsense knows quite a lot about human (and animal) visual sense. So there is no clear line drawn in folk-explanations between the functional input-law and the boundary conditions of its realization by humans. But the philosophical explication of folk-explanations must draw such a line; must if there are to *be* perceptual input laws in the functional theory rather than merely *sketches* of input laws containing 'etc.' or '. . .' in their antecedents. Schiffer (1987; 31–2) rejects such sketches of putative functional input laws. He complains (a) that we cannot complete the list of conditions to be inserted in the gap '. . .', and (b) that some esoteric items of the list are unknown to practising folk-psychologists who are supposed to have an implicit grasp of the input law. The answer to Schiffer is that, because functional 'laws' are not really laws at all, the list is complete when we have included those conditions known to practicing folk-psychologists. The esoteric conditions known only to psychologists or physiologists are boundary conditions of the applicability of the functional law to humans.

3. Folk-explanations and Psychological Realism

Even with '$\vdash$' confined to topic-neutral inference rules, (4) leads to trouble. Assume that '$\vdash$' is the consequence relation of the lower predicate calculus. Church proved that LPC is undecidable. So *no* mechanism, not even an idealized mechanism with friction-free, perfectly rigid parts, could realize (4). For such an idealized mechanism would *be*, in effect, a decision procedure for LPC. One who had a blueprint for such an idealized machine could follow it through to determine whether or not $p \vdash q$. So we know from Church's theorem that there is no such idealized mechanism.

We cannot suppose that the functional relations prescribed by (4), with '$\vdash$' understood as the consequence relation of LPC, have even the lowly degree of reality which Daniel Dennett thinks beliefs have.

> The role of the concept of belief [in folk-psychology] is like the role of the concept of a centre of gravity, and the calculations that yield the predictions are more like the calculations one performs with a parallelogram of forces than like the calculations one performs with a blueprint of internal levers and cogs.
>
> . . . people really do have beliefs and desires, on my version of folk-psychology, just the way they really have centres of gravity and the earth has an equator. (Dennett, 1981; 46)

Unfortunately, even this degree of reality would violate Church's theorem. A mechanism which incorporated the consequence relation of LPC in (4), although only as a logical construct, would yield a proof-finding algorithm, provided only that the relevant logical constructs were constructively determinable from the primary physical properties of the mechanism.

Consider Anthony Appiah's conception of a computationally perfect agent (a CPA). He writes:

> Let us call the axioms and rules of inference which characterize the computational capacities of some agent the *proof theory* for that agent. (Appiah, 1986; 31–2)

And comments

> The proof theory *characterizes* the agent's capacities. There is no commitment to the agent's using its axioms and rules of inference; only to its being true that the agent would believe all the theorems of the proof theory, if computationally perfect.

However, Appiah commits himself, probably inadvertently, to an agent's proof theory being *decidable*, because he conceives of the CPA as being an idealized mechanism.

> computational errors have to be deviations from the normal functioning of the physical system that embodies the functional states. It follows that in calling something a computational error, we are committed to there being some explanation of it. A deviation from normal functioning is a deviation from what is prescribed by the functional theory . . . if people's behaviour deviates from what decision theory requires, this must be the result of some independently verifiable causal intervention with the functioning of minds. What's normal is what happens without such interventions. (Appiah, 1986; 35)

If the proof theory of an agent is not decidable, then the computationally perfect idealization of that agent's axioms and inference rules is not realizable, not even by an idealized mechanism which never malfunctions. So computational error, i.e. failure to be a CPA, cannot always be explained as the malfunctioning of a mechanism.

Even decidable consequence relations are problematic. For example, the propositional calculus is decidable. But Christopher Cherniak points out that a truth-table checker which computes each line in the time it takes a light ray to travel the diameter of a proton, and which has been running for the last 20 billion years (Big-Bang to now) would not yet have computed the table for a mere 138 independent variables. (Cherniak, 1986; 93). So even where we know an algorithm, that algorithm might not be feasible for realization in a problem-solving mechanism, because its demands on memory and real-time may be too expensive.

Yet we cannot abandon the conception of a CPA. The radical interpreter appeals to it, at least as an ideal of reason. The radical interpreter has to construct, *inter alia*, a semantic theory which can be used to show how the *truth* of Tom's beliefs is connected to the *success* of his actions. However, the connection between truth and success appeals to the concept of a CPA. Thus Appiah:

> TRUTH: the truth condition of a representation S of an agent is that condition whose holding is necessary and sufficient to guarantee that, for every action A, if he or she were computationally perfect and performed A on the belief that S, then the outcome would be that one of the two possible outcomes he or she preferred. (Appiah, 1986; 38)

Now, if Tom is anything like you or me, the consequence relation of his idiolect will be undecidable. But Appiah's guarantee is only good for the belief of a CPA. Belief systems put together by fast but dirty heuristics may be inconsistent, and action on them lead to an unwanted outcome. The radical interpreter is required to produce a systematic semantic theory of Tom's idiolect such that, when his beliefs are assigned semantic contents, and assuming Tom to be, *per impossible*, a CPA, then actions based on those beliefs will be successful if those beliefs are true. And furthermore, the systematic semantic theory must be such that, when Tom's beliefs are assigned semantic contents, Tom's actions based on those beliefs are more successful than not, at least in the context of his familiar environment. So the radical interpreter's semantic theory must make Tom out to *approximate*, in *some* sense, to a CPA, at least when operating in his home patch. Thus the conception of a CPA is an ideal of reason for a radical interpreter.

However, we cannot interpret a folk–functional theory which incorporates (4), the mark of a CPA, realistically. Rather, the folk theory would be an instrumental theory, and it would be the job of empirical psychology to trick out the fast but dirty heuristics which Tom actually uses to 'decide' whether or not $P_1, \ldots, P_n \vdash q$. But if the functional component of folk-theory is not to be interpreted realistically, then it cannot support the ideology and ontology of anomalous monism. If the beliefs, desires, etc. of folk-theory are merely instrumental fictions, then they cannot be the causes or effects of real events, and the monism of anomalous monism is unsupported.

We need a folk–functional theory which we *can* interpret realistically, yet one which shows that Tom approximates to a CPA—where approximating to a CPA does *not* mean that he would *be* a CPA if only the boundary conditions of the functional theory were never breached. Tom can approximate to a CPA by realizing, instead of (4), the following:

$$(p_1, \ldots, p_n, q)(\{B(\text{Tom}, p_1, \ldots, p_n \vdash q) \& B(\text{Tom}, p_1, \ldots, p_n)\} \rightarrow B(\text{Tom}, q)) \quad (5)$$

$$(p_1, \ldots, p_n, q)(\text{Tom is shown that } (p_1, \ldots, p_n \vdash q) \rightarrow B(\text{Tom}, p_1, \ldots, p_n \vdash q)) \quad (6)$$

When Tom discovers or is shown some proof involving the undecidable consequence relation, he adjusts his belief-system accordingly. The CPA is, in this sense, an ideal of reason for Tom too. Tom 'corrects' computational errors and oversights when these are pointed out to him. He adopts the attitudes which would characterize a CPA whenever he becomes aware that he is deviating from the attitudes of a CPA. The advantage of (5) and (6) over (4) is that the undecidability of $\vdash$ presents no bar to a mechanism realizing (5) and (6), subject to boundary conditions, as it did to a mechanism realizing (4).

Suppose Tom's belief-structures are really formed by some fast but dirty heuristic procedures, which, on this occasion, have dished him up dirt. His belief-set is $\vdash$inconsistent when given the semantic contents of the utterances which express them, as those utterances are themselves given semantic contents by the radical interpreter. If Tom will not acknowledge error when shown a proof of his $\vdash$inconsistency in his own idiolect, and if this cannot be dismissed as an abberation (failure of boundary conditions), nor given a special (anomalous) psychological explanation (obtuseness, stubborness, being publically committed and unable to back down, etc.), then that would be evidence that the semantic theory and its consequence relation were wrong. A systematic semantic theory of Tom's idiolect and its incorporated consequence relation are justified only if Tom conforms, in the sense of (5) and (6), to that consequence relation rather than to some other, or to none. An heuristic method of generating representational output from representational input is clean or dirty only relative to some semantic theory of those representations. And Tom's cleaving to (5) and (6) show to which consequence relation his heuristics are answerable. Because his real heuristics are thus answerable to that semantic theory, we are justified in ascribing the semantic contents prescribed by that semantic theory to Tom's beliefs, and interpreting those ascriptions realistically rather than merely instrumentally, despite Tom not really being a CPA.

A contrasting strategy, but one which also allows us to interpret the functional theory realistically, is to weaken the conception of rationality involved in the functional laws. With a sufficient weakened conception of rationality, we can suppose that the functional theory is realized, *modulo* breach of boundary conditions. This is, in effect, Brian Loar's strategy. Loar gives a set of 'rationality constraints', of which these are typical (Loar, 1981; 72):

1(a) Not B[−(p∨−p)]

1(b) B[p] ⇒ not B[−p]

2(a) B[p&q] ⇒ not B[−p], not B[−q]

2(b) B[p] and B[q] ⇒ not B[−(p&q)]

4(a) (B[p→q] and B[p]) ⇒ not B[−q]

where 'B' stands for belief and reference to the particular believer has been suppressed, '⇒' is a stronger conditional than the material conditional since it holds in counterfactual circumstances too, and I have added brackets '[. . .]' to indicate the scopes of the various beliefs. Apart from a couple of rules forbidding or attributing beliefs outright, Loar's rules all go from the attribution of a belief to a proscription on attributing certain related beliefs. Since this is so, I see no difficulty in building a machine which realizes Loar's rules. We do not require any proof-chains to be built up, because the output of one rule does not provide the input for any other rule—because the rules go from belief to non-belief. So all that is required is a machine which, given a finite set of beliefs as input, takes each of the finite set of rules in turn and goes through the beliefs deleting members as required. Loar is, of course, aware that his rationality constraint is very undemanding.

The weakness of Loar's strategy, it seems to me, is that the resulting functional theory fails to provide any substantial support for the semantic theory of the radical interpreter, because it fails to show how a creature which satisfies Loar's functional theory approximates to a CPA. A mechanism which realizes Loar's functional theory might, for example, hold the following triad of beliefs:

B[p&q], B[p→r], B[−r]

From rule 2(a) and B[p &q] we will be able to conclude only

not B[−p]

and from B[p→r] and B[−r] and rule 4(a) we can conclude only

not B[p]

which are mutually consistent. So Loar's functional theory finds nothing unsatisfactory about the above triad of beliefs. So the functional theory makes no contribution to explaining how or why a creature acting on its beliefs is likely to be successful—when its beliefs and desires have the semantic contents of their sentential indices, and those indices are given a normal semantic interpretation. Hence, the functional theory fails to connect with the standard semantic theory of the indices. There is then no warrant for supposing that the functional states indexed have the semantic contents of their indices, where those indices are given a standard interpretation. Hence, there is no warrant for giving Tom's utterances a standard interpretation, if Loar's functionalism captures the relevant mechanism in Tom. The moral of the tale is, I think, that the functional laws must incorporate the *full* syntactic consequence relation of the semantic theory of the subject's idiolect, but, if that consequence relation is undecidable, they must incorporate it in a way which does not block a realistic interpretation of the functional theory. This is what (5) and (6) seek to do. Only then can the functional component support the monism of anomalous monism—not a goal which Loar seeks, as we have already noted.

The functional theory which I have sketched applies only to language using animals who can learn from proofs, which is not surprising if that functional theory is seen as one constraint on the radical interpretation of a subject's idiolect. Furthermore, if the total set of constraints on radical interpretation implicitly 'define' in some loose sense the concepts of belief, desire, etc., then our dumb friends do not have beliefs or desires. I do not find this consequence repulsive. It is not to deny that other animals do have some kind of mental life, but merely to point out that our folk-psychology cannot be realistically applied to them. That our application of folk-psychology to other animals is instrumental, opportunistic and anthropomorphic is obvious. We say that a dog chases a rabbit into a burrow and then digs to find it. But do we really think the dog knows or cares whether the rabbit it finds is the same one as it saw run down the burrow? Would not any rabbit-feature do? And the dog just knows how one rabbit-feature leads to another. There is no reason why the folk-psychology we have devised and applied to ourselves should provide the means to realistically describe or explain the mentation of other forms of life. Nor should the anomalism of human mentation be assumed to be a necessary feature of mental life. If some animal's powers of reasoning are sufficiently humble, it may be that explanations of their 'attitudes', whether we can know them or not, are not anomalous. Perhaps their form of rationality *is* echoed in physical theory. We have no argument that cognition *per se* is anomalous.

4. Evolutionary Semantics

It seems to me that David Papineau's contribution to this volume may have uncovered a sense in which the semantic relations of mental states are anomalous.

Papineau gives a 'success-guaranteeing account of truth'—'the truth condition of any belief is that condition which guarantees that actions validly based on that belief succeed'. It is clear from the discussion above, and from Papineau's own discussion, that this guarantee holds only for a CPA. We are not CPAs, yet Papineau is commendably realist about our mental states actually having those truth conditions which are thus characterized in terms of a CPA. Papineau proceeds to a naturalistic account of the representational contents of our mental states, based upon natural selection, which 'incorporates' his earlier success-guaranteeing account. It is clear that the teleological account cannot incorporate the success-guaranteeing account by arguing that natural selection has produced a race of CPAs, since such creatures are impossible and therefore not available to be selected. Rather, natural selection favours an organism which approximates more to a CPA over one which approximates less. So the CPA is an impossible, unattainable ideal goal of the selection process.

Yet our real beliefs really do have the truth conditions which would guarantee success in the mind of a CPA. Those conditions do not guarantee success for us—for our beliefs, given those truth conditions, may well have hidden (from us) inconsistencies. To explain the actual truth conditions of our beliefs Papineau, in effect, invokes the naturalistically unattainable limit of the selection process. In this sense the explanation of the actual representational properties of our mental states is anomalous—it transcends any actual process of natural selection. In the last section we saw how the actual representational properties of a subject's mental states might involve that mythical creature the CPA. It is a necessary condition of an actual subject's mental states having those representational properties which would guarantee success to a CPA that the subject is actually disposed to bring his belief into line with any proof he discovers, stumbles upon or is shown. Natural selection has produced such creatures.

If instead we seek to explain the actual truth conditions of our beliefs without recourse to the myth of the CPA, then we must show that the process of natural selection has actually paired beliefs of type X with conditions of type Y because real actions based on beliefs of type X have usually been successful when conditions of type Y have obtained and usually not been successful when conditions of type Y have not obtained. I suspect that Papineau would seek to explain the semantic contents of mental states along such lines, but I do not know what evidence there is for such an explanation other than the appeal of an evolutionary account of representational content which avoids recourse to the transcendent notion of a CPA.

5. Conclusion

I have argued that a modest functional component in folk-psychology, incorporating a modest conception of rationality, underpins the monism of anomalous monism, and even contributes weakly via the autonomy of functional explanations to the claim of anomaly. More importantly, the modest functional component leaves room for the claim that, because folk-psychology taxonomizes mental states by an amalgam of the semantic and conceptual contents, folk-psychological explanations are anomalous in the sense that we cannot hope to codify the wide conception of rationality which governs folk-explanations. But we also saw that if the assignment of semantic contents depends upon a conception of rationality which *is* codifiable, but is *not* decidable, like the consequence relation of LPC, then the explanation of how our mental states get their semantic contents is anomalous in the sense that it makes essential appeal to an ideal which is not naturally attainable, viz the CPA.

References

Appiah, A. 1986. 'Truth Conditions: A Casual Theory', in Butterfield.

Jeremy Butterfield, 1986 (ed.), *Language, Mind and Logic* (Cambridge University Press).

Christopher Cherniak, 1986. *Minimal Rationality* (Cambridge, Mass.: M.I.T. Press).

Davidson, Donald 1980. *Essays on Actions and Events* (Oxford University Press).

Dennett, Daniel 1981. 'Three Types of Intentional Psychology' in Healey.

Hartry Field, 1978. 'Mental Representation', *Erkenntnis*, **13**, 9–61.

Fodor, Jerry 1980. 'Methodological Solipsism considered as a Research Strategy in Cognitive Psychology', *Brain and Behavioral Science*, **3**, 63–109.

Healey, Richard 1981. (ed.), *Reduction, Time and Reality* (Cambridge University Press).

LePore, Ernest and McLaughlin, Brian 1985. (eds), *Actions and Events* (Oxford: Blackwell).

LePore, Ernest 1986. (ed.), *Truth and Interpretation* (Oxford: Blackwell).

Loar, Brian 1981. *Mind and Meaning* (Cambridge University Press).

McDowell, John 1985. 'Functionalism and Anomalous Monism', in LePore and McLaughlin)

McGinn, Colin 1982. 'The Structure of Content', in Woodfield.

Putnam, Hiliary 1975. *Mind, Language and Reality* (Cambridge University Press).

Putnam, Hiliary 1988. *Representation and Reality* (Cambridge, Mass.: M.I.T. Press).

Schiffer, Stephen 1987. *Remnants of Meaning* (Cambridge, Mass.: M.I.T. Press).

Woodfield, Andrew 1982. (ed.), *Thought and Object* (Oxford University Press).

Explanation in Biology

JOHN MAYNARD SMITH

During the war, I worked in aircraft design. About a year after D-day, an exhibition was arranged at Farnborough of the mass of German equipment that had been captured, including the doodlebug and the V2 rocket. I and a friend spent a fascinating two days wandering round the exhibits. The questions that kept arising were 'Why did they make it like that?', or, equivalently 'I wonder what that is for?' We were particularly puzzled by a gyroscope in the control system of the V2. One's first assumption was that the gyroscope maintained the rocket on its course, but, instead of being connected to the steering vanes, it was connected to the fuel supply to the rocket. Ultimately my friend (who would have made a better biologist than I) remembered that the rate of precession of a gyro depends on acceleration, and saw that the Germans had used this fact to design an ingenious device for switching off the engines when the required velocity had been reached.

For an engineer, this was thinking backwards. Typically, an engineer wants to produce a result, and devises machinery to do it; on this occasion we were looking at machinery, and trying to deduce what result it was designed to produce. Many years later, I came to appreciate that biologists spend their lives thinking backwards in this way. We see a complicated structure—a wing joint, a ribosome, the cerebellum—and ask not only how it works, but what it is for. This sharply distinguishes biology from physics and chemistry. Of any particular phenomenon, we ask two questions, whereas they ask only one.

The existence of two questions—and hence of two answers—can be made clearer by an example. The question, 'Why does the heart beat?' can be interpreted in two ways. It may mean 'What makes the heart beat?' The answer would be in terms of the rhythmical properties of heart muscle, and of nervous control. This is the province of physiology: if the structure in question were a ribosome, and not a heart, it would be a question for a biochemist. But the original question might mean, 'Why do animals need a heart that beats?' If so, the answer would be along the lines, 'a heart pumps blood round the body, and the blood carries oxygen and nutrients to the tissues and removes waste'.

I can imagine that a physicist, confronted with the latter type of answer, might say, 'It may be true that the heart pumps blood, and that the blood carries oxygen, and so on. But that is merely a *consequence* of the fact that the heart beats. It is in no way an *explanation* of why the heart beats'. Yet the same physicist might well accept, as an explanation of the presence of a gyro in the V2, the answer that it was there to switch the engine off when a critical velocity was reached.

The reason for this apparent confusion over the word 'explanation' is that, when a biologist says 'The heart beats in order to pump blood round the body', this is a short-hand for 'Those animals which, in the past, had hearts that were efficient pumps survived, because oxygen reached their tissues, whereas animals whose hearts were less efficient as pumps died. Since offspring resemble their parents, this resulted in the fact that present-day animals have hearts that are efficient pumps'. In other words, he is giving an evolutionary explanation for the heart beat, and not a physiological one.

Before I go on to discuss what seem to me to be the real difficulties with this type of explanation, I want to discuss three semantic points. The first is the widespread use of teleological language, when an explanation in terms of natural selection is intended. I accept that the teleological usage does sometimes confuse people, but I suspect that it confuses philosophers more often than biologists. But I also regard the usage as unavoidable. You only have to compare the brevity of the teleological statement in the last paragraph with the long-winded nature of the (still very inadequate) evolutionary alternative. There is no way one could use the full version in everyday discourse. But it is important that, in critical cases (and, as I explain below, there are plenty of such cases) one should be prepared, on demand, to provide the necessary evolutionary exposition of any teleological statement one makes.

The second semantic point concerns the words 'causal' and 'functional'. The distinction between a short-term, physiological explanation and a long-term evolutionary one is usually obvious, and no special terms for them are needed. Difficulties do, however, arise in the field of animal behaviour. It has become customary to use the term 'causal' for a physiological or biochemical explanation, and 'functional' for an evolutionary one. In some ways this is unfortunate, because an explanation in terms of variation and selection is just as much (or little) causal as one in terms of nerve impulses or chemical reactions: it is only that the time scale is longer. Unfortunate or not, however, I fear we are stuck with the terms causal and functional. What we have to remember is that causal and functional explanations are not mutually exclusive alternatives: biological phenomena require both.

The third semantic point concerns the distinction between 'function' and 'consequence'. It is a *consequence* of the fact that a horse has a stiff backbone that people can ride it. But I would not want to say that the *function* of a stiff backbone is to enable people to ride, because I do not think that horses acquired a stiff backbone because that enables people to ride them. By 'function', I mean those consequences of a structure (or behaviour) that, through their effects on survival and reproduction, caused the evolution of that structure. It is a consequence of the fact that canaries sing that people keep them in cages, but the function of canary song is to defend a territory. Of course, one may not know the function of a particular structure. Indeed, I have spent most of my life trying to discover the function of various structures. But I do think it is helpful to use the words consequence and function in different senses. The usage is not universal among biologists, but it is spreading.

I now turn to what seem to me more difficult problems. There is nothing problematic about physiological ('causal') explanations—or, at least, nothing that is not also problematic for explanations in physics and chemistry. But 'functional' explanations depend on a particular theory of how evolution happened, and make sense only in so far as that theory is correct. In essence, the theory is as follows. Any population of entities with the properties of multiplication, variation and heredity (i.e. like begets like) will evolve in such a way that the individual entities acquire traits (often called adaptations) that ensure their survival and reproduction. Living organisms can be seen to have those three properties, and hence populations of organisms evolve. This process (i.e. natural selection) has been mainly responsible for past evolutionary change.

First, a few comments on this absurdly brief summary of Darwin's theory:

(i) One could spend a lifetime attempting to axiomatize the first sentence. Any such axiomatization would require additional assumptions, particularly about the nature of variation. However, the sentence is intended to be necessarily true: the falsifiable part of the theory comes in the final sentence.

(ii) There is a striking difference between this theory, and most theories in the physical sciences. For example, in the atomic theory of chemistry, there exists a body of data about chemical reactions, and the relative amounts of different substances that will combine (the law of constant proportions). These are 'explained' by hypothesizing entities (atoms) with properties that have been chosen because, if true, they will account for the data. It is my impression that most theories in physics have postulated entities in this way. In Darwin's theory, the entities (i.e. the organisms) are what we know about most directly: no one would seriously attempt to show that they do not multiply, or vary, or (in some sense) have heredity. It is the evolutionary consequences (analogous to the laws of chemical reactions) that we know least about. While on this topic, I should add that theories in genetics have, in this respect, resembled theories in physics: they have postulated entities that have only later been perceived directly.

(iii) The last sentence contains the words 'mainly' and 'past'. Few biologists regard natural selection as the only cause of change. Darwin certainly did not. Unhappily, this makes a strictly falsificationist attitude to Darwinism hard to maintain (although I have to confess to a liking for a kind of un-strict falsificationism that I would hate to have to defend before a jury of philosophers). The word 'past' was included because, if humans survive, future evolutionary changes are likely to occur by mechanisms different from those that were relevant in the past.

It is now possible to state what seems to me the source of most of the difficulties that arise when we try to say what we mean by an evolutionary explanation. What are the 'entities' that possess multiplication, variation and heredity, and that can therefore be expected to evolve adaptations to ensure their survival and reproduction? Are they genes, or chromosomes, or organisms, or populations, or species, or ecosystems, or maybe the whole biosphere? For Darwin the entities were individual organisms. But, as G. C. Williams and Richard Dawkins have emphasized, organisms do not replicate.[1] They die, and only their genes (or, more precisely, the information in their genes) are replicated and passed on to future generations. Should we not, then, treat the gene as the unit of evolution, which we expect to evolve traits that ensure its survival and reproduction, and regard organisms simply as temporary vehicles, programmed by the genes to ensure their survival?

From the other side, there have been biologists who have argued that entities larger than the individual organism are the relevant units of evolution. D. S. Wilson and E. Sober, in an unpublished manuscript, have argued that groups of organisms have 'heredity', because, if an organism that is a member of a group today contributes offspring to a group in the future, then there will be some resemblance, albeit slight, between the average properties of the future group and the present one. That is, groups have heredity, and can be treated as units of evolution. N. Eldredge and S. J. Gould, and S. M. Stanley have argued that the large-scale features of evolution are best explained by selection operating between species, rather than between individuals within species.[2]

[1] G. C. Williams, *Adaptation and Natural Selection* (Princeton University Press, 1966); Richard Dawkins, *The Selfish Gene* (Oxford University Press, 1976).

[2] N. Eldredge and S. J. Gould,'Punctuated Equilibria: An Alternative to Phyletic Gradualism', in T. M. Schopf (ed.), *Models in Paleobiology* (San Francisco: Freeman, Cooper and Co., 1972); S. M. Stanley, *Macroevolution: Pattern and Process* (San Francisco, Freeman, 1979).

Finally, there are those who have argued that whole ecosystems (or even, in the 'Gaia hypothesis', the earth itself) are the relevant entities.

How are we to cope with this plethora of levels? In particular, what effect do these arguments have on the validity of 'teleological shorthand'. If we are entitled to say 'the function of the blood is to pump blood round the body', are we entitled to say 'the function of territorial behaviour is to prevent the population from outrunning its food supply', or 'the function of earthworms is to speed up the re-cycling of dead leaves', or even, 'the function of green plants is to maintain the oxygen in the atmosphere'? My own view is that the statement about hearts is sensible, but those about territorial behaviour, earthworms, and green plants are not. Why?

It is useful to start with the apparently sensible statement. Why, if genes but not organisms replicate, can we treat organisms as units of evolution that evolve adaptations ensuring their survival? One answer would be that we have successfully treated organisms in this way ever since Darwin. Hearts, wings, eyes and ribosomes obviously *are* good for organisms. If our theories of genetics and evolution do not explain why this should be so, so much the worse for our theories. I have some sympathy with this approach, but I think we can do rather better for our theories than this. The essential point is that there is very little selection *between genes*, *within an organism*. If Mendel's laws are obeyed (which they are 99.9 per cent of the time), an individual that inherits two different alleles from its parents transmits them with exactly equal probability to its children. There are no properties of a gene that make it more likely to be transmitted than its allele. This is sometimes expressed by saying that meiosis is fair. Like most statements in biology, this is not a universal truth. There are cases, referred to as 'meiotic drive', in which particular alleles have an unfair advantage in meiosis. They are important, and I return to them briefly below. For the present, it is sufficient to say that they are rare—too rare to alter the fact that the only way a gene can increase its own chances of long-term survival is by ensuring the survival of the organisms in which it finds itself.

The conclusion is clear. Genes are indeed 'selfish', in the sense that they can be expected to evolve traits that ensure their own long-term survival. But, so long as mitosis and meiosis are fair, they can do this only by ensuring the survival and reproduction of their carriers. It is for this reason that their carriers—the individual organisms—evolve traits ensuring organism survival.

I now return to the three statements—about territorial behaviour, earthworms, and green plants—that I regard as not sensible. To be sensible, it is required that the following entities be units of evolution, with the properties of multiplication, variation and heredity: a species of birds displaying territorial behaviour; an ecosystem including earthworms; a biosphere with green plants. The ecosystem, and the biosphere, can at once be ruled out: there is no meaningful sense in which they could be said to have multiplication and heredity. But what about a bird species? Species do split into two (speciation, corresponding to multiplication); they go extinct (death); and when they speciate, the daughter species resemble the parent (heredity). Why, then, should a species not be a unit of evolution?

The answer is that a species is composed of components (the individual birds) which are themselves potential units of evolution, and between which (unlike the genes within an organism) selection does go on. There is between-bird, within-species selection. Whenever this is so, selection between units at the lower level (in this case, individual birds) is likely to swamp any selection between units at the higher level. This is a quantitative, not an absolute statement. But it does mean that, whenever there is selection at the lower level, one cannot assume that any trait (e.g. territorial behaviour) has evolved to ensure the survival of the higher level unit (species or population).

In general, I am unsympathetic to explanations in terms of species selection, for two reasons. First, selection between the individuals within a species will overwhelm any effects of between-species selection. Second, most of the traits studied by palaeontologists (who have tended to be the main proponents of species selection) are not properties of species at all, but of individuals. For example, the legs of antelopes evolved as they did because they enable antelopes to run fast. But species do not run—individual antelopes do. There is one possible exception to this generalization. Sexual reproduction may have evolved because it enables a species to evolve more rapidly to meet changing circumstances. Now species evolve, not individuals, so a species-selection explanation of sex cannot be ruled out on logical grounds. What of the quantitative argument, that individual selection in favour of parthenogenesis would overwhelm species selection in favour of sex? A possible reply is that individual selection in favour of parthenogenesis is indeed effective when a parthenogenetic mutant arises, but that such mutants are very rare (there are, for example, no parthenogenetic mammals). The matter is difficult, and still controversial: my own view is that species selection is part, but only part, of the explanation for sex. I have only raised the subject because this is the one context in which species selection may have been important.

Let me summarize the arguments so far. Explanations in biology are of two kinds, causal and functional. A functional explanation—'the function of the heart is to pump blood'—is a shorthand for an explanation in terms of natural selection. To say that the function of the heart is to pump blood is equivalent to saying that the heart evolved because natural selection favoured those organisms that had hearts which did pump bood. Thus functional explanations are also causal, but in terms of causes operating in populations over many generations. For a functional explanation to be valid, the behaviour to be explained (the pumping of the heart) must contribute to the survival and reproduction of an entity (in this case, the individual organism) that has the properties of multiplication, variation and heredity. For this reason, it is not valid to explain earthworms by saying that they contribute to the functioning of an ecosystem. To qualify as a 'unit of evolution', which can be expected to evolve traits ensuring its survival and reproduction (i.e. 'adaptations'), it is not sufficient that the entity has the properties of multiplication, variation and heredity. It is also necessary that it should not be composed of lower-level replicating entities, between which selection is occurring, because such lower-level selection would destroy integration at the higher level.

There are, I think, two major difficulties with this way of looking at things, one contemporary and one historical: both are active areas of current research. The contemporary difficulty is to decide how much selection between lower-level entities is compatible with the evolution of adaptations at a higher level. The simple answer, 'none', will not do, because organisms certainly evolve adaptations ensuring their survival and reproduction, and yet there is some within-individual between-gene selection occurring. A total absence of between-gene selection would require that each gene is replicated only when the cell divides, and that two alleles at a locus have exactly equal probabilities of entering a gamete. Unfortunately, neither of these requirements are met. The first is broken by 'transposable elements', which make up a large proportion of the genome in many organisms, including ourselves. The second is broken by the phenomenon of 'meiotic drive', which is rare but which does occur. We need to understand why transposition and meiotic drive do not prevent the evolution of organismic adaptation. We also need to think about higher-level entities—for example, insect colonies, and species. With the possible exception of sexual reproduction, I do not think there are any properties of species that look like species-level adaptations, but there are certainly features of insect colonies (for example, the heat-regulating architecture of a termite mound) that look like colony-level adaptations.

The historical difficulty is this. On a number of occasions during evolution, a higher-level entity has arisen that is composed of a number of lower-level entities that were, at least initially, capable of independent existence. Some examples will make this clearer. According to the most plausible hypothesis about the origin of life, the first units of evolution were replicating nucleic acid molecules. At some point, a set of different and competing molecules were linked together end-to-end, so that when one was replicated, all were: in this way, they became the chromosomes of a higher-level entity, a prokaryptic cell. In a later transition, several different prokaryotic cells (a host cell, an ancestral mitochondrion, and an ancestral chloroplast) joined to form a eukaryotic cell. Subsequently, and on a number of occasions, colonies of single cells have evolved into differentiated, multi-cellular organisms. In each of these transitions, a major problem is to understand how competition between the lower-level entities was supressed.

These transitions (the origin of life, of cells with chromosomes, of eukaryotes, of multicellular organisms, of animal societies) are the most important events in evolution. I cannot discuss them in any detail here, but I have two comments on how they should be approached. The first comment is that the different transitions have a good deal formally in common, and it may therefore be illuminating to study them simultaneously. The second is that one must start from the beginning, and not from the end. By this, I mean that a model of any one of these transitions must treat the lower-level entities as the basic units of evolution, and explain how natural selection brought about the emergence of the higher level. In doing so, of course, it is necessary to take into account both selection between the lower-level entities ('individual selection'), and selection between the emerging higher-level entities ('group selection'). Thus I am not recommending an 'individual selection' model on the grounds of parsimony: if there is differential survival of groups, this must be included in the model. My point is only that an adequate model must recognize explicitly the existence, at the start of the transition, of lower-level entities capable of independent replication, and explain how this capacity was lost.

Let's Razor Ockham's Razor

ELLIOTT SOBER

1. Introduction

When philosophers discuss the topic of explanation, they usually have in mind the following question: given the beliefs one has and some proposition that one wishes to explain, which subset of the beliefs constitutes an explanation of the target proposition? That is, the philosophical 'problem of explanation' typically has bracketed the issue of how one obtains the beliefs; they are taken as given. The problem of explanation has been the problem of understanding the relation 'x explains y'. Since Hempel (1965) did so much to canonize this way of thinking about explanation, it deserves to be called 'Hempel's problem'.

The broad heading for my paper departs from this Hempelian format. I am interested in how we might justify some of the explanatory propositions in our stock of beliefs. Of course, issues of theory confirmation and acceptance are really not so distant from the topic of explanation. After all, it is standard to describe theory evaluation as the procedure of 'inference to the best explanation'. Hypotheses are accepted, at least partly, in virtue of their ability to explain. If this is right, then the epistemology of explanations is closely related to Hempel's problem.

I should say at the outset that I take the philosopher's term 'inference to the best explanation' with a grain of salt. Lots of hypotheses are accepted on the testimony of evidence even though the hypotheses could not possibly be explanatory of the evidence. We infer the future from the present; we also infer one event from another simultaneously occurring event with which the first is correlated. Yet, the future does not explain the present; nor can one event explain another that occurs simultaneously with the first. Those who believe in inference to the best explanation may reply that they do not mean that inferring H from E requires that H explain E. They have in mind the looser requirement that H is inferrable from E only if adding H to one's total system of beliefs would maximize the overall explanatory coherence of that system. This global constraint, I think, is too vague to criticize; I suspect that 'explanatory coherence' is here used as a substitute for 'plausibility'. I doubt that plausibility can be reduced to the concept of explanatoriness in any meaningful way.

Another way in which philosophical talk of 'inference to the best explanation' is apt to mislead is that it suggests a gulf between the evaluation of explanatory hypotheses and the making of 'simple inductions'. Inductive inference, whether it concludes with a generalization or with a prediction about the 'next instance', often is assumed to markedly differ from postulating a hidden cause that explains one's observations. Again, I will merely note here my doubt that there are distinct rules for inductive and abductive inference.

Although I am not a card-carrying Bayesian, Bayes' theorem provides a useful vehicle for classifying the various considerations that might affect a hypothesis' plausibility. The theorem says that the probability that H has in the light of evidence E (P[H/E]) is a function of three quantities:

$$P(H/E) = P(E/H)P(H)/P(E).$$

This means that if one is comparing two hypotheses, H_1 and H_2, their overall plausibility (posterior probability) is influenced by two factors:

$$P(H_1/E)>P(H_2/E) \quad \text{iff} \quad P(E/H_1)P(H_1)>P(E/H_2)P(H_2).$$

P(H) is the prior probability of H—the probability it has before one obtains evidence E. P(E/H) is termed the *likelihood* of H; the likelihood of H is not H's probability, but the probability that H confers on E.

Likelihood is often a plausible measure of explanatory power. If some hypothesis were true, how good an explanation would it provide of the evidence (E)? Let us consider this as a comparative problem: we observe E and wish to know whether one hypothesis (H_1), if true, would explain E better than another hypothesis (H_2) would. Suppose that H_1 says that E was to be expected, while H_2 says that E was very improbable. Likelihood judges H_1 better supported than H_2; it is natural to see this judgment as reflecting one dimension of the concept of explanatory power.[1]

Hypotheses we have ample reason to believe untrue may nonetheless be explanatory. They may still have the property of being such that *IF*

[1] Although I agree with Salmon (1984) that a true explanation can be such that the *explanans* proposition says that the *explanandum* proposition had low probability, I nonetheless think that the explanatory power of a candidate hypothesis is influenced by how probable it says the *explanandum* is. See Sober (1987) for further discussion. It also is worth noting that philosophical discussion of explanation has paid little attention to the question of what makes one explanation a better explanation than another. Hempel's problem leads one to seek a yes/no criterion for being an explanation, or for being an ideally complete explanation. It is another matter to search for criteria by which one hypothesis is a better explanation than another.

they were true, they would account well for the observations. This judgment about antecedent plausibility the Bayesian tries to capture with the idea of prior probability.

There is little dispute about the relevance of likelihood to hypothesis evaluation; nor is there much dispute as to whether something besides the present observations can influence one's judgment about a hypothesis' overall plausibility. The main matter of contention over Bayesianism concerns whether hypotheses always have well-defined priors. The issue is whether prior probability is the right way to represent judgments about antecedent plausibility.

When the hypotheses in question describe possible outcomes of a chance process, assigning them prior probabilities is not controversial. Suppose a randomly selected human being has a red rash; we wish to say whether it is more probable that he has measles or mumps. In this case, it is meaningful to talk about the prior probabilities. The prior probability of a disease is just its population frequency. And the likelihoods also are clear; I can say how probable it would be for someone to have the red rash if he had measles and how probable the symptom would be if he had mumps. With these assignments of priors and likelihoods, I can calculate which posterior probability is greater.

Do not be misled by the terminology here. The prior probabilities in this example are not knowable *a priori*. The prior probability of the proposition that our subject has measles is the probability we assign to that disease when we do not know that he happens to have a red rash. The fact that he was randomly drawn from a population allows us to determine the prior probability by observing the population.

Matters change when the hypotheses in question do not describe the outcomes of chance processes. Examples include Newton's theory of gravity and Darwin's theory of evolution. A Bayesian will want to assign these prior probabilities and then describe how subsequent observations modify those initial assignments. Although likelihoods are often well-defined here, it is unclear what it would mean to talk about probabilities.

Bayesians sometimes go the subjective route and take prior probabilities to represent an agent's subjective degrees of belief in the hypotheses. Serious questions can be raised as to whether agents always have precise degrees of belief. But even if they did, the relevance of such prior probabilities to scientific inquiry would be questionable. If two agents have different priors, how are they to reach some agreement about which is more adequate? If they are to discuss the hypotheses under consideration, they must be able to anchor their probability assignments to something objective (or, at least, intersubjective).

Another Bayesian reaction to the problem of priors has been to argue that they are objectively determined by some *a priori* consideration.

Carnap (1950) looked to the structure of the scientist's language as a source of logically defined probabilities. But since scientists can expand or contract their languages at will, it seems implausible that this strategy will be successful. More recently, Rosenkrantz (1977), building on the work of Jaynes (1986), has argued that prior probabilities can be assigned *a priori* by appeal to the requirement that a correct prior should be invariant under certain transformations of how the variables are defined. I will not discuss this line of argument here, except to note that I do not think it works.[2] Prior probabilities, I will assume, are not assignable *a priori*.

I am not a Bayesian, in the sense that I do not think that prior probabilities are always available. But the Bayesian biconditional stated above is nonetheless something I find useful. It is a convenient reminder that hypothesis evaluation must take account of likelihoods and also of the hypotheses' antecedent plausibility. Only sometimes will the latter concept be interpretable as a probability.

Notice that the Bayesian biconditional does not use the word 'explanation'. Explanations have likelihoods; and sometimes they even have priors. This means that they can be evaluated for their overall plausibility. But there is no *sui generis* virtue called 'explanatoriness' that affects plausibility.[3] Likewise, Bayes's theorem enshrines no distinction between induction and abduction. The hypotheses may be inductive generalizations couched in the same vocabulary as the observations; or the hypotheses may exploit a theoretical vocabulary that denotes items not mentioned in the description of the observational evidence.[4] Bayesianism explains why the expression 'inference to the best explanation' can be doubly misleading.

Not only does the Bayesian biconditional make no mention of 'explanatoriness'; it also fails to mention the other epistemic virtues that philosophers like to cite. Parsimony, generality, fecundity, familiarity—all are virtues that do not speak their names. Just as Bayesianism suggests that explanatoriness is not a *sui generis* considera-

[2] See Seidenfeld's (1979) review of Rosenkrantz's book for some powerful objections to objective Bayesianism. Rosenkrantz (1979) is a reply.

[3] Here I find myself in agreement with Van Fraassen (1980), 22.

[4] This is why a Bayesian model of theory testing counts against Van Fraassen's (1980) constructive empiricism. According to Van Fraassen, the appropriate epistemic attitude to take towards a hypothesis depends on what the hypothesis is about. If it is strictly about observables, it is a legitimate scientific task to say whether the hypothesis is true or false. If it is at least partly about unobservables, science should not pronounce on this issue. These strictures find no expression in the Bayesian biconditional. I discuss the implications of the present view of confirmation for the realism/empiricism debate in Sober (forthcoming).

tion in hypothesis evaluation, it also suggests that parsimony is not a scientific end in itself. When parsimoniousness augments a hypothesis's likelihood or its prior probability, well and good. But parsimony, in and of itself, cannot make one hypothesis more plausible than another.

To this end it may be objected that scientists themselves frequently appeal to parsimony to justify their choice of hypotheses. Since science is a paradigm (perhaps *the* paradigm) of rationality, the objection continues, does not this mean that a theory's parsimoniousness must contribute to its plausibility? How much of twentieth-century discussion of simplicity and parsimony has been driven by Einstein's remark in his 1905 paper that his theory renders superfluous the introduction of a luminiferous aether? Removing the principle of parsimony from the organon of scientific method threatens to deprive science of its results.

This objection misunderstands my thesis. I do not claim that parsimony never counts. I claim that when it counts, it counts because it reflects something more fundamental. In particular, I believe that philosophers have hypostatized parsimony. When a scientist uses the idea, it has meaning only because it is embedded in a very specific context of inquiry. Only because of a set of background assumptions does parsimony connect with plausibility in a particular research problem. What makes parsimony reasonable in one context therefore may have nothing in common with why it matters in another. The philosopher's mistake is to think that there is a single global principle that spans diverse scientific subject matters.

My reasons for thinking this fall into two categories. First, there is the general framework I find useful for thinking about scientific inference. Probabilities are not obtainable *a priori*. If the importance of parsimony is to be reflected in a Bayesian framework, it must be linked either with the likelihoods or the priors of the competing hypotheses. The existence of this linkage is always a contingent matter that exists because some set of *a posteriori* propositions governs the context of inquiry. The second sort of reason has to do with how I understand the specific uses that scientists have made of the principle of parsimony. These case studies also suggest that there is no such thing as an *a priori* and subject matter invariant principle of parsimony.

The idea that parsimony is not a *sui generis* epistemic virtue is hardly new. Popper (1959) claims that simplicity reflects falsifiability. Jeffreys (1957) and Quine (1966) suggest that simplicity reflects high probability. Rosenkrantz (1977) seeks to explain the relevance of parsimony in a Bayesian framework. A timeslice of my former self argued that simplicity reduces to a kind of question-relative informativeness (Sober 1975).

What is perhaps more novel in my proposal is the idea that parsimony be understood locally, not globally. All the theories just mentioned attempt to define and justify the principle of parsimony by appeal to logical or mathematical features of the competing hypotheses. An exclusive focus on these features of hypotheses is inevitable, if one hopes to describe the principle of parsimony as applying across entirely disjoint subject matters. If the parsimoniousness of hypotheses in physics turns on the same features that determine the parsimoniousness of hypotheses in biology, what could determine parsimoniousness besides logic and mathematics? If a justification for this globally defined concept of parsimony is to be obtained, it will come from considerations in logic and mathematics. Understanding parsimony as a global constraint on inquiry thus leads naturally to the idea that it is *a priori* justified. My local approach entails that the legitimacy of parsimony stands or falls, in a particular research context, on subject matter specific (and *a posteriori*) considerations.[5]

In what follows I will discuss two examples of how appeals to parsimony have figured in recent evolutionary biology. The first is George C. Williams' use in his landmark book *Adaptation and Natural Selection* of a parsimony argument to criticize hypotheses of group selection. The second is the use made by cladists and many other systematic biologists of a parsimony criterion to reconstruct phylogenetic relationships among taxa from facts about their similarities and differences.

Williams' (1966) parsimony argument against group selection encountered almost no opposition in the evolution community. Group selection hypotheses were said to be less parsimonious than lower-level selection hypotheses, but no one seems to have asked why the greater parsimony of the latter was any reason to accept them as true.

Cladistic parsimony, on the other hand, has been criticized and debated intensively for the last twenty years. Many biologists have asserted that this inference principle assumes that evolution proceeds parsimoniously and have hastened to add that there is ample evidence that evolution does no such thing. Cladists have replied to these criticisms and the fires continue to blaze.

My own view is that it is perfectly legitimate, in both cases, to ask why parsimony is connected with plausibility. I will try to reconstruct the kind of answer that might be given in the case of the group selection issue. I also will discuss the way this question can be investigated in the case of phylogenetic inference.

[5] Miller (1987) also develops a local approach to confirmational issues, but within a framework less friendly to the usefulness of Bayesian ideas.

I noted earlier that the Bayesian biconditional suggests two avenues by which parsimony may impinge on plausibility. It may affect the prior probabilities and it may affect the likelihoods. The first biological example takes the first route, while the second takes the second.

2. Parsimony and the Units of Selection Controversy

Williams' 1966 book renewed contact between two disciplinary orientations in evolutionary biology that should have been communicating, but did not, at least not very much. Since the 1930s, population geneticists—pre-eminently Fisher (1930), Haldane (1932) and Wright (1945)—had been rather sceptical of the idea that there are group adaptations. A group adaptation is a characteristic that exists because it benefits the group in which it is found. Evolutionists have used the word 'altruism' to label characteristics that are disadvantageous for the organisms possessing them, though advantageous to the group. Population geneticists generally agreed that it is very difficult to get altruistic characteristics to evolve and be retained in a population. Field naturalists, on the other hand, often thought that characteristics observed in nature are good for the group though bad for the individuals. These field naturalists paid little attention to the quantitative models that the geneticists developed. These contradictory orientations coexisted for some thirty years.

Williams (1966) elaborated the reigning orthodoxy in population genetics; but he did so in English prose, without recourse to mathematical arguments. He argued that hypotheses of group adaptation and group selection are often products of sloppy thinking. A properly rigorous Darwinism should cast the concept of group adaptation on the same rubbish heap on to which Lamarckism had earlier been discarded.

Williams deployed a variety of arguments, some better than others. One prominent argument was that group selection hypotheses are less parsimonious than hypotheses that claim that the unit of selection is the individual or the gene.

This argument begins with the observation that 'adaptation is an onerous principle', one that a scientist should invoke only if driven to it. Flying fish return to the water after sailing over the waves. Why do they do this? Williams claims that there is no need to tell an adaptationist story. The mere fact that fish are heavier than air accounts for the fact that what goes up must come down. Thinking of the fish as having evolved a specific adaptation for returning to the water, Williams concludes, is unparsimonious and so should be rejected.

This idea—that it is more parsimonious to think of the fish's return to the water as a 'physical inevitability' rather than as an adaptation—is only part of the way the principle of parsimony applies to evolutionary explanations. Williams invokes Lloyd Morgan's rule that lower-level explanations are preferable to higher-level ones; Williams takes this to mean that it is better to think of a characteristic as having evolved for the good of the organism possessing it than to view it as having evolved for the good of the group. The principle of parsimony generates a hierarchy: purely physical explanations are preferable to adaptationist explanations, and hypotheses positing lower-level adaptations are preferable to ones that postulate adaptations at higher levels of organization.

Before explaining in more detail what Williams had in mind about competing units of selection, a comment on his flying fish is in order. I want to suggest that parsimony is entirely irrelvant to this example. If flying fish return to the water because they are *heavier than air*, then it is fairly clear why an adaptationist story will be implausible. Natural selection requires variation. If being heavier than air were an adaptation, then some ancestral population must have included organisms that were heavier than air and ones that were lighter. Since there is ample room to doubt that this was ever the case, we can safely discard the idea that being heavier than air is an adaptation. My point is that this reasoning is grounded in a fact about how natural selection proceeds and a plausible assumption about the character of ancestral populations. There is no need to invoke parsimony to make this point; Ockham's razor can safely be razored away.

Turning now to the difference between lower-level and higher-level hypotheses of adaptation, let me give an example of how Williams' argument proceeds. Musk oxen form a circle when attacked by wolves, with the adult males on the outside facing the attack and the females and young protected in the interior. Males therefore protect females and young to which they are not related. Apparently, this characteristic is good for the group, but deleterious for the individuals possessing it. A group selection explanation would maintain that this wagon-training behaviour evolved because groups competed against other groups. Groups that wagon train go extinct less often and found more daughter colonies than groups that do not.

Williams rejected this hypothesis of group adaptation. He proposes the following alternative. In general, when a predator attacks a prey organism, the prey can either fight or flee. Selection working for the good of the organism will equip an organism with optimal behaviours, given the nature of the threatening predator. If the threat comes from a predator that is relatively small and harmless, the prey is better off standing its ground. If the threat is posed by a large and dangerous predator, the prey is better off running away. A prediction of this idea is that there are some predators that cause large prey to fight and small prey to flee. Williams proposes that wolves fall in this size range; they make the male oxen stand their ground and the females and young flee to the interior. The group characteristic of wagon-training is just a statistical consequence of each organism's doing what is in its own self-interest. No need for group selection here; the more parsimonious individual-selection story suffices to explain.

Williams' book repeatedly deploys this pattern of reasoning. He describes some characteristic found in nature and the group selection explanation that some biologist has proposed for it. Williams then suggests that the characteristic can be explained purely in terms of a lower-level selection hypothesis. Rather than suspending judgment about which explanation is more plausible, Williams opts for the lower-level story, on the grounds that it is more parsimonious.

Why should the greater parsimony of a lower-level selection hypothesis make that hypothesis more plausible than an explanation in terms of group selection? Williams does not address this admittedly philosophical question. I propose the following reconstruction of Williams' argument. I believe that it is the best that can be done for it; in addition, I think that it is none too bad.

Williams suggests that the hypothesis of group selection, if true, would explain the observations, and that the same is true for the hypothesis of individual selection that he invents. This means, within the format provided by the Bayesian biconditional of the previous section, that the two hypotheses have identical likelihoods. If so, the hypotheses will differ in overall plausibility only if they have different priors. Why think that it is antecedently less probable that a characteristic has evolved by group selection than that it evolved by individual selection?

As noted earlier, an altruistic characteristic is one that is bad for the organism possessing it, but good for the group in which it occurs. Here good and bad are calculated in the currency of fitness—survival and

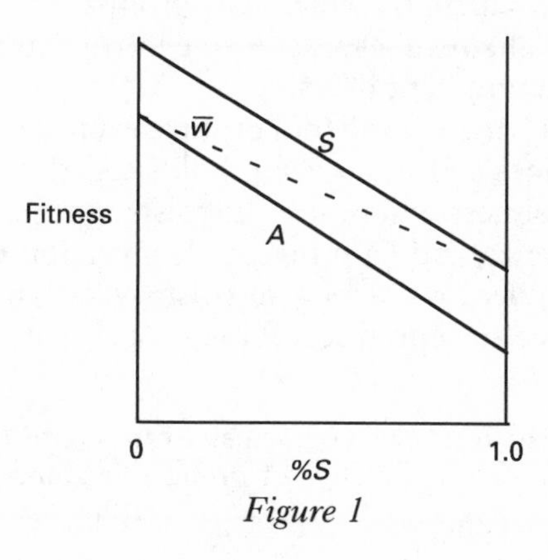

Figure 1

reproductive success.[6] This definition of altruism is illustrated in the fitness functions depicted in Figure 1. Notice that an organism is better off being selfish (S) than altruistic (A), no matter what sort of group it inhabits. Let us suppose that the fitness of a group is measured by the average fitness of the organisms in the group; this is represented in the figure by $\bar{w}$. If so, groups with higher concentrations of altruists are fitter than groups with lower concentrations.

What will happen if S and A evolve within the confines of a single population? With some modest further assumptions (e.g., that the traits are heritable), we may say that the population will evolve to eliminate altruism, no matter what initial composition the population happens to have.

For altruism to evolve and be maintained by group selection, there must be variation among group*s*. An ensemble of populations must be postulated, each with its own local frequency of altruism. Groups in which altruism is common must do better than groups in which altruism is rare.

To make this concrete, let us suppose that a group will fission into a number of daughter colonies once it reaches a certain census size. Suppose that this critical mass is 500 individuals and that the group will then divide into 50 offspring colonies containing 10 individuals each. Groups with higher values of $\bar{w}$ will reach this fission point more quickly, and so will have more offspring; they are fitter. In addition to this rule about colonization, suppose that groups run higher risks of extinction the more saturated they are with selfishness. These two assumptions about colonization and extinction ground the idea that altruistic groups are fitter than selfish ones—they are more reproductively successful (i.e. found more colonies) and they have better chances of surviving (avoiding extinction).

So far I have described how group and individual fitnesses are related, and the mechanism by which new groups are founded. Is that enough to allow the altruistic character to evolve? No it is not. I have omitted the crucial ingredient of *time*.

Suppose we begin with a number of groups, each with its local mix of altruism and selfishness. If each group holds together for a sufficient length of time, selfishness will replace altruism within it. Each group, as Dawkins (1976) once said, is subject to 'subversion from within'. If the groups hold together for too long, altruism will disappear before the groups have a chance to reproduce. This means that altruism cannot

[6] I will ignore the way the concept of inclusive fitness affects the appropriate definition of altruism and, indirectly, of group selection. I discuss this in Sober (1984, 1988d).

evolve if group reproduction happens much slower than individual reproduction.

I have provided a sketch of how altruism can evolve by group selection. One might say that it is a 'complicated' process, but this is not why such hypotheses are implausible. Meiosis and random genetic drift also may be 'complicated' in their way, but that is no basis for supposing that they rarely occur. The rational kernel of Williams' parsimony argument is that the evolution of altruism by group selection requires a number of restrictive assumptions about population structure. Not only must there be sufficient variation among groups, but rates of colonization and extinction must be sufficiently high. Other conditions are required as well. This coincidence of factors is not impossible; indeed, Williams concedes that at least one well documented case has been found (the evolution of the *t*-allele in the house mouse). Williams' parsimony argument is at bottom the thesis that natural systems rarely exemplify the suite of biological properties needed for altruism to evolve by group selection.[7]

Returning to the Bayesian biconditional, we may take Williams to be saying that the prior probability of a group selection hypothesis is lower than the prior probability of a hypothesis of individual selection. Think of the biologist as selecting at random a characteristic found in some natural population (like musk oxen wagon-training). Some of these characteristics may have evolved by group selection, others by lower-level varieties of selection. In assigning a lower prior probability to the group selection hypothesis, Williams is making a biological judgment about the relative frequency of certain population structures in nature.

In the ten years following Williams' book, a number of evolutionists investigated the question of group selection from a theoretical point of view. That is, they did not go to nature searching for altruistic characteristics; rather, they invented mathematical models for describing how altruism might evolve. The goal was to discover the range of parameter values within which an altruistic character can increase in frequency and then be maintained. These inquiries, critically reviewed in Wade (1978), uniformly concluded that altruism can evolve only within a rather narrow range of parameter values. The word 'parsimony' is not prominent in this series of investigations; but these biologists I believe, were fleshing out the parsimony argument that Williams had earlier constructed.

If one accepts Williams' picture of the relative frequency of conditions favourable for the evolution of altruism, it is quite reasonable to assign group selection explanations a low prior probability. But this

[7] I believe that this reconstruction of Williams' parsimony argument is more adequate than the ones I suggest in Sober (1981, 1984).

assignment cuts no ice, once a natural system is observed to exhibit the population structure required for altruism to evolve. Wilson (1980) and others have argued that such conditions are exhibited in numerous species of insects. Seeing Williams' parsimony argument as an argument about prior probabilities helps explain why the argument is relevant *prima facie*, though it does not prejudge the upshot of more detailed investigations of specific natural systems.

Almost no one any longer believes the principle of indifference (a.k.a. the principle of insufficient reason). This principle says that if $P_1, P_2, \ldots, P_n$ are exclusive and exhaustive propositions, and you have no more reason to think one of them true than you have for any of the others, you should assign them equal probabilities. The principle quickly leads to contradiction, since the space of alternatives can be partitioned in different ways. The familiar lesson is that probabilities cannot be extracted from ignorance alone, but require substantive assumptions about the world.

It is interesting to note how this standard philosophical idea conflicts with a common conception of how parsimony functions in hypothesis evaluation. The thought is that parsimony considerations allow us to assign prior probabilities and that the use of parsimony is 'purely methodological', presupposing nothing substantive about the way the world is. The resolution of this contradiction comes with realizing that whenever parsimony considerations generate prior probabilities for competing hypotheses,[8] the use of parsimony cannot be purely methodological.

3. Parsimony and Phylogenetic Inference

The usual philosophical picture of how parsimony impinges on hypothesis evaluation is of several hypotheses that are each consistent with

[8] Philosophers have sometimes discussed examples of 'competing hypotheses' that bear implication relations to each other. Popper (1959) talks about the relative simplicity of the hypothesis that the earth has an elliptical orbit and the hypothesis that it has a circular orbit, where the former is understood to entail the latter. Similarly, Quine (1966) discusses the relative simplicity of an estimate of a parameter that includes one significant digit and a second estimate consistent with the first that includes three. In such cases, saying that one hypothesis has a higher prior than another of course requires no specific assumptions about the empirical subject at hand. However, I would deny that these are properly treated as competing hypotheses; and even if they could be so treated, such purely logical and mathematical arguments leave wholly untouched the more standard case in which competing hypotheses do not bear implication relations to each other.

the evidence,[9] or explain it equally well, or are equally supported by it. Parsimony is then invoked as a further consideration. The example I will now discuss—the use of a parsimony criterion in phylogenetic inference—is a useful corrective to this limited view. In this instance, parsimony considerations are said to affect how well supported a hypothesis is by the data. In terms of the Bayesian biconditional, parsimony is relevant because of its impact on likelihoods, not because it affects priors.

Although parsimony considerations arguably have been implicit in much work that seeks to infer phylogenetic relationships from facts about similarity and differences, it was not until the 1960s that the principle was explicitly formulated. Edwards and Cavalli-Sforza (1963, 1964), two statistically minded evolutionists, put it this way: 'the most plausible estimate of the evolutionary tree is that which invokes the minimum net amount of evolution'. They claim that the principle has intuitive appeal, but concede that its presuppositions are none too clear. At about the same time, Willi Hennig's (1966) book appeared in English; this translated an expanded version of his German work of 1950. Although Hennig never used the word 'parsimony', his claims concerning how similarities and differences among taxa provide evidence about their phylogenetic relationships are basically equivalent to the parsimony idea. Hennig's followers, who came to be called 'cladists', used the term 'parsimony' and became that concept's principal champions in systematics.

Sparrows and pigeons have wings, whereas iguanas do not. That fact about similarity and difference seems to provide evidence that sparrows and pigeons are more closely related to each other than either is to iguanas. But why should this be so?

Figure 2 represents two phylogenetic hypotheses and the distribution of characters that needs to be explained. Each hypothesis says that two of the taxa are more closely related to each other than either is to the third. The inference problem I want to explore involves two evolutionary assumptions. Let us assume that the three taxa, if we trace them back far enough, share a common ancestor. In addition, let us assume that this ancestor did not have wings. That is, I am supposing that having wings is the derived (apomorphic) condition and lacking wings is the ancestral (plesiomorphic) state.[10]

[9] Although 'consistency with the data' is often how philosophers describe the way observations can influence a hypothesis' plausibility, it is a sorry explication of that concept. For one thing, consistency is an all or nothing relationship, whereas the support of hypotheses by data is presumably a matter of degree.

[10] I will not discuss here the various methods that systematists use to test

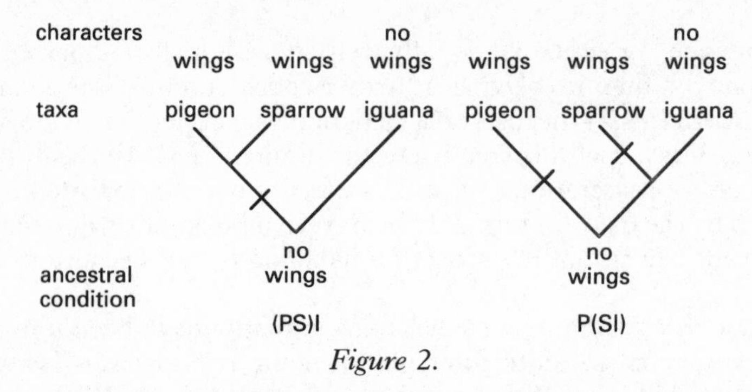

Figure 2.

According to each tree, the common ancestor lacked wings. Then, in the course of the branching process, the character must have changed to yield the distribution displayed at the tips. What is the minimum number of changes that would allow the (PS)I tree to generate this distribution? The answer is *one*. The (PS)I tree is consistent with the supposition that pigeons and sparrows obtained their wings from a common ancestor; the similarity might be a homology. The idea that there was a single evolutionary change is represented in the (PS)I tree by a single slash mark across the relevant branch.

Matters are different when we consider the P(SI) hypothesis. For this phylogenetic tree to generate the data found at the tips, at least two changes in character state are needed. I have drawn two slash marks in the P(SI) tree to indicate where these might have occurred. According to this tree, the similarity between pigeons and sparrows cannot be a homology, but must be the result of independent origination. The term for this is 'homoplasy'.

The principle of parsimony judges that the character distribution just mentioned supports (PS)I better than it supports P(SI). The reason is that the latter hypothesis requires at least two evolutionary changes to explain the data, whereas the former requires only one. The principle of parsimony says that we should minimize assumptions of homoplasy.

The similarity uniting pigeons and sparrows in this example is a *derived* similarity. Pigeons and sparrows are assumed to share a characteristic that was *not* present in the common ancestor of the three taxa under consideration. This fact about the example is important, because the principle of phylogenetic parsimony entails that some similarities

---

such assumptions about character polarity, on which see Sober (1988c). Also a fine point that will not affect my conclusions is worth mentioning. The two evolutionary assumptions just mentioned entail, not just that *a* common ancestor of the three taxa lacked wings, but that this was the character state of the three taxa's *most recent common ancestor*. This added assumption is useful for expository purposes, but is dispensable. I do without it in Sober (1988c).

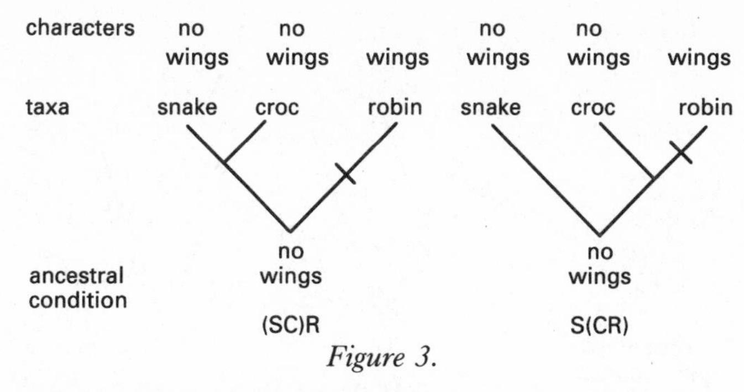

Figure 3.

do *not* count as evidence of common ancestry. When two taxa share an *ancestral* character, parsimony judges that fact to be devoid of evidential meaning.

To see why, consider the fact that snakes and crocodiles lack wings, whereas robins possess them. Does the principle of parsimony judge that to be evidence that snakes and crocodiles are more closely related to each other than either are to robins? The answer is *no*, because evolutionists assume that the common ancestor of the three taxa did not have wings. The (SC)R tree displayed in Figure 3 can account for this character distribution by assuming a single change; the same is true for the S(CR) tree.

In summary, the idea of phylogenetic parsimony boils down to two principles about evidence:

Derived similarity *is* evidence of propinquity of descent.
Ancestral similarity is *not* evidence of propinquity of descent.

It should be clear from this that those who believe in parsimony will reject *overall* similarity as an indicator of phylogenetic relationships. For the last twenty years, there has been an acrimonious debate in the biological literature concerning which of these approaches is correct.

Why should derived similarities be taken to provide evidence of common ancestry, as parsimony maintains? If multiple originations were impossible, the answer would be clear. However, no one is prepared to make this process assumption. Less stringently, we can say that if evolutionary change were very improbable, then it would be clear why a hypothesis requiring two changes in state is inferior to a hypothesis requiring only one. But this assumption also is problematic. Defenders of parsimony have been loathe to make it, and critics of parsimony have been quick to point out that there is considerable evidence against it. Felsenstein (1983), for example, has noted that it is quite common for the most parsimonious hypothesis obtained for a given data set to require multiple changes on a large percentage of the characters it explains.

I have mentioned two possible assumptions, each of which would

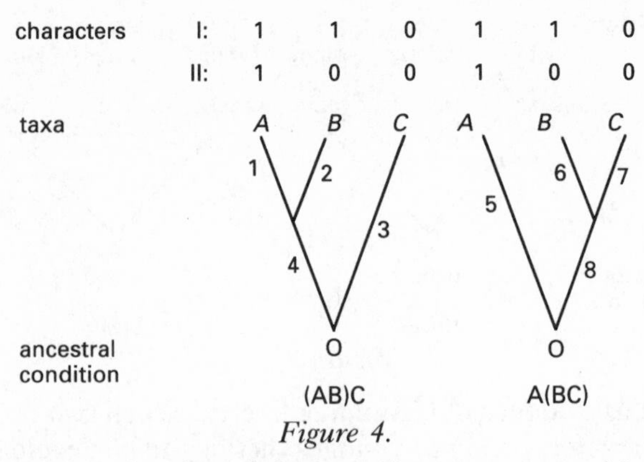

Figure 4.

suffice to explain why shared derived characters are evidence of common ancestry. No one has shown that either of these assumptions is *necessary*, though critics of parsimony frequently assert that parsimony assumes that evolutionary change is rare or improbable.

There is a logical point about the suggestions just considered that needs to be emphasized. A phylogenetic hypothesis, all by itself, does not tell you whether a given character distribution is to be expected or not. On the other hand, if we append the assumption that multiple change is impossible or improbable, then the different hypotheses do make different predictions about the data. By assuming that multiple changes are improbable, it is clear why one phylogenetic hypothesis does a better job of explaining a derived similarity than its competitors. The operative idea here is *likelihood*: if one hypothesis says that the data were hardly to be expected, whereas a second says that the data were to be expected, it is the second that is better supported. The logical point is that *phylogenetic hypotheses are able to say how probable the observations are only if we append further assumptions about character evolution*.

Figure 4 allows this point to be depicted schematically. The problem is to infer which two of the three taxa *A*, *B* and *C* share a common ancestor apart from the third. Each character comes in two states, denoted by '0' or '1'. It is assumed that 0 is the ancestral form. Character I involves a derived similarity that *A* and *B* possess but *C* lacks.

The branches in each figure are labelled. With the *i*th branch, we can associate a transition probability e_i and a transition probability r_i. The former is the probability that a branch will end in state 1, if it begins in state 0; the latter is the probability that the branch will end in state 0, if it begins in state 1. The letters are chosen as mnemonics for 'evolution' and 'reversal'.[11]

[11] The complements of e_i and r_i do not represent the probabilities of stasis. They represent the probability that a branch will end in the same state in which it began. This could be achieved by any even number of flip-flops.

Given this array of branch transition probabilities, we can write down an expression that represents the probability of obtaining a given character distribution at the tips, conditional on each of the hypotheses. That is, we can write down an expression full of *e*'s and *r*'s that represents P[110/(AB)C] and an expression that represents P[110/A(BC)]. Which of these will be larger?

If we assume that changes on all branches have a probability of 0.5, then the two hypotheses have identical likelihoods. That is, under this assumption about character evolution, the character distribution fails to discriminate between the hypotheses; it is evidentially meaningless. On the other hand, if we assume that change is very improbable on branches 1, 2 and 4, but is not so improbable on branches 5 and 6, we will obtain the result that the first character supports (AB)C *less well* than it supports A(BC). This paradoxical result flies in the face of what parsimony maintains (and contradicts the dictates of overall similarity as well). As a third example, we might assign the *e* and *r* parameters associated with each branch a value of 0.1. In this case, (AB)C turns out to be more likely than A(BC), relative to the 110 distribution.

With different assumptions about branch transition probabilities, we get different likelihoods for the two hypotheses relative to character I. A similar result obtains if we ask about the evidential significance of character II, in which there is an ancestral similarity uniting *B* and *C* apart from *A*. Parsimony judges such similarities to be evidentially meaningless, but whether the likelihoods of the two hypotheses are identical depends on the assignment of branch transition probabilities.

If parsimony is the right way to determine which hypothesis is best supported by the data, this will be because the most parsimonious hypothesis is the hypothesis of maximum likelihood. Whether this is so depends on the model of character evolution one adopts. Such a model will inevitably rest on biological assumptions about evolution. One hopes that these assumptions will be plausible and maybe even commonsensical. But that they are biology, not pure logic or mathematics, is, I think, beyond question.

I will not take the space here to explain the results of my own investigation (in Sober (1988c)) into the connection between phylogenetic parsimony and likelihood.[12] The more general philosophical

[12] I will note, however, that assigning e_i and r_i the same value for all the *i* branches is implausible, if the probability of change in a branch is a function of the branch's temporal duration. In addition, there is a simplification in the treatment of likelihood presented here that I should note. The (AB)C hypothesis represents a family of trees, each with its own set of branch durations. This means that the likelihood of the (AB)C hypothesis, given a model of character evolution, is an *average* over all the specific realizations that possess that topology. See Sober (1988c) for details.

point I want to make is this: if parsimony is the right method to use in phylogenetic inference, this will be because of specific facts about the phylogenetic process. The method has a no *a priori* subject matter neutral justification.

By now it is an utterly familiar philosophical point that a scientific hypothesis (H) has implications (whether deductive or probabilistic) about observations (O) only in the context of a set of auxiliary assumptions (A). Sometimes this is called Duhem's thesis; sometimes it is taken to be too obvious to be worth naming. It is wrong to think that H makes predictions about O; it is the conjunction $H\&A$ that issues in testable consequences.

It is a small step from this standard idea to the realization that quality of explanation must be a three-place relation, not a binary one. If one asks whether one hypothesis (H_1) provides a better explanation of the observations (O) than another hypothesis (H_2) does, it is wrong to think this as a matter of comparing how H_1 relates to O with how H_2 relates to O. Whether H_1 is better or worse as an explanation of O depends on further auxiliary assumptions A.

Why should a more parsimonious explanation of some observations be better than one that is less parsimonious? This cannot be a matter of the logical relationship between hypothesis and observation alone; it must crucially involve auxiliary assumptions.[13]

4. Concluding Remarks

Philosophical discussion of simplicity and parsimony is replete with remarks to the effect that smooth curves are better than bumpy ones and that postulating fewer entities or processes is better than postulating more. The natural philosophical goal of generality has encouraged the idea that these are methodological maxims that apply to all scientific subject matters. If this were so, then whatever justification such maxims possess would derive from logic and mathematics, not from anything specific to a single scientific research problem.

Respect for the results of science then leads one to assume that general principles of simplicity and parsimony must be justified. The question is where this global justification is to be found; philosophers have been quite inventive in generating interesting proposals. As a

[13] Of course, by packing the auxiliary assumptions into the hypotheses under test, one can obtain a new case in which the 'hypotheses' have well-defined likelihoods without the need to specify still further auxiliary assumptions. My view is that this logical trick obscures the quite separate status enjoyed by an assumed model and the hypotheses under test.

fallback position, one could announce, if such proposals fail, that simplicity and parsimony are *sui generis* constituents of the habit of mind we call 'scientific'. According to this gambit, it is part of what we mean by science that simpler and more parsimonious hypotheses are scientifically preferable. Shades of Strawson on induction.

Aristotle accused Plato of hypostatizing The Good. What do a good general and a good flute player have in common? Aristotle argued that the characteristics that make for a good military commander need have nothing in common with the traits that make for a good musician. We are misled by the common term if we think that there must be some property that both possess, in virtue of which each is good.

Williams argued that we should prefer lower-level selection hypotheses over group selection hypotheses, since the former are more parsimonious. Hennig and his followers argued that we should prefer hypotheses requiring fewer homoplasies over ones that require more, since the former are more parsimonious. Following Aristotle, we should hesitate to conclude that if Williams and Hennig are right, then there must be some single property of parsimonious hypotheses in virtue of which they are good.

Maxims like Ockham's razor have their point. But their force derives from the specific context of inquiry in which they are wielded. Even if one is not a Bayesian, the Bayesian biconditional provides a useful reminder of how parsimony may affect hypothesis evaluation. Scientists may assign more parsimonious hypotheses a higher antecedent plausibility; but just as prior probabilities cannot be generated from ignorance, so assignments of prior plausibility must be justified by concrete assumptions if they are to be justifed at all. Alternatively, scientists may assert that more parsimonious hypotheses provide better explanations of the data; but just as scientific hypotheses standardly possess likelihoods only because of an assumed model connecting hypothesis to data, so assessments of explanatory power also must be justified by concrete assumptions if they are to be justified at all.[14]

By suggesting that we razor Ockham's razor, am I wielding the very instrument I suggest we abandon? I think not. It is not abstract numerology—a formal preference for less over more—that motivates my conclusion. Rather, the implausibility of postulating a global criterion has two sources. First, there are the 'data'; close attention to the details of how scientific inference proceeds in well-defined contexts of inquiry suggests that parsimony and plausibility are connected only because some local background assumptions are in play. Second, there

[14] The principal claim of this paper is not that 'parsimony' is ambiguous. I see no reason to say that the term is like 'bank'. Rather, my locality thesis concerns the *justification* for taking parsimony to be a sign of truth.

is a more general framework according to which the evidential connection between observation and hypothesis cannot be mediated by logic and mathematics alone.[15]

Admittedly, two is not a very large sample size. Perhaps, then, this paper should be understood to provide a *prima facie* argument for concluding that the justification of parsimony must be local and subject matter specific. This sort of circumspection is further encouraged by the fact that many foundational problems still remain concerning scientific inference. I used the Bayesian biconditional while eschewing Bayesianism; I offered no alternative doctrine of comparable scope. Nonetheless, I hope that I have provided some grounds for thinking that razoring the razor may make sense.

References

Carnap, R. 1950. *Logical Foundations of Probability* (University of Chicago Press).

Dawkins, R. 1976. *The Selfish Gene* (Oxford University Press).

Edwards, A. and Cavalli-Sforza, L. 1963. 'The Reconstruction of Evolution', *Ann. Human Genetics*, **27,** 105.

Edwards, A. and Cavalli-Sforza, L. 1964. 'Reconstruction of Evolutionary Trees' in V. Heywood and J. McNeill (eds.) *Phenetic and Phylogenetic Classification* (New York Systematics Association Publications), **6,** 67–76.

Felsenstein, J. 1983. 'Parsimony in Systematics', *Annual Review of Ecology and Systematics*, **14,** 313–33.

Fisher, R. 1930. *The Genetical Theory of Natural Selection* (Reprinted New York: Dover, 1958).

Haldane, J. 1932. *The Causes of Evolution* (Reprinted New York: Cornell University Press, 1966).

Hempel, C. 1965. *Aspects of Scientific Explanation and Other Essays in the Philosophy of Science* (New York: Free Press).

Hennig, W. 1966. *Phylogenetic Systematics* (Urbana: University of Illinois Press).

Jaynes, E. 1968. 'Prior Probabilities', *IEEE Trans. Systems Sci. Cymbernetics*, **4,** 227–41.

Jeffreys, H. 1957. *Scientific Inference* (Cambridge University Press).

Miller, R. 1987. *Fact and Method* (Princeton University Press).

Popper, K. 1959. *The Logic of Scientific Discovery* (London: Hutchinson).

Quine, W. 1966. 'On Simple Theories of a Complex World', in *Ways of Paradox and Other Essays* (New York: Random House).

Rosenkrantz, R. 1977. *Inference, Method and Decision* (Dordrecht: D. Reidel).

[15] I have tried to develop this thesis about evidence in Sober (1988a, b and c).

Rosenkrantz, R. 1979. 'Bayesian Theory Appraisal: A Reply to Seidenfield', *Theory and Decision*, **11**, 441–51.
Salmon, W. 1984. *Scientific Explanation and the Causal Structure of the World* (Princeton University Press).
Seidenfeld, T. 1979. 'Why I am not a Bayesian: Reflections Prompted by Rosenkrantz', *Theory and Decision*, **11,** 413–40.
Sober, E. 1975. *Simplicity* (Oxford University Press).
Sober, E. 1981. 'The Principle of Parsimony', *British Journal for the Philosophy of Science*, **32,** 145–56.
Sober, E. 1984. *The Nature of Selection* (Cambridge, Mass.: MIT Press).
Sober, E. 1987. 'Explanation and Causation: A Review of Salmon's *Scientific Explanation and the Causal Structure of the World*', *British Journal for the Philosophy of Science*, **38,** 243–57.
Sober, E. 1988a. 'Confirmation and Law-Likeness', *Philosophical Review*, **97,** 617–26.
Sober, E. 1988b. 'The Principle of the Common Cause', in J. Fetzer (ed.) *Probability and Causation: Essays in Honor of Wesley Salmon*. (Dordrecht: Reidel), 211–28.
Sober, E. 1988c. *Reconstructing the Past: Parsimony, Evolution and Inference* (Cambridge, Mass.: MIT Press).
Sober, E. 1988d. 'What is Evolutionary Altruism?', *Canadian Journal of Philosophy Supplementary Volume*, **14,** 75–99.
Sober, E. forthcoming. 'Contrastive Empiricism', in W. Savage (ed.) *Scientific Theories*, Minnesota Studies in the Philosophy of Science, Vol. 14.
Van Fraassen, B. 1980. *The Scientific Image* (Oxford University Press).
Wade, M. 1978. 'A Critical Review of Models of Group Selection', *Quarterly Review of Biology*, **53,** 101–14.
Williams, G. C. 1966. *Adaptation and Natural Selection* (Princeton University Press).
Wilson, D. 1980. *The Natural Selection of Populations and Communities* (Menlo Park: Benjamin/Cummings).
Wright, S. 1945. '*Tempo and Mode in Evolution*: a Critical Review', *Ecology*, **26,** 415–19.

Singular Explanation and the Social Sciences*

DAVID-HILLEL RUBEN

1. Difference Theory

Are explanations in the social sciences fundamentally (logically or structurally) different from explanations in the natural sciences? Many philosophers think that they are, and I call such philosophers 'difference theorists'. Many difference theorists locate that difference in the alleged fact that only in the natural sciences does explanation essentially include laws.

For these theorists, the difference theory is held as a consequence of two more fundamental beliefs: (1) At least some (full) explanations in the social sciences do not include laws; (2) All (full) explanations in the natural sciences do include laws. For example, Peter Winch criticizes and rejects Mill's view that 'understanding a social institution consists in observing regularities in the behaviour of its participants and expressing these regularities in the form of a generalization . . . [the] position of the sociological investigator (in a broad sense) can be regarded as comparable . . . with that of the natural scientist'.[1] Daniel Taylor argues that 'Scientific explanations involve universal propositions . . . if you think of an historical event and try to imagine explaining it in a way which fulfils these criteria you will see the difficulties at once.'[2] Michael Lesnoff claims that, quite unlike the explanations of physical science, 'at whatever level of detail a social phenomenon is investigated, the correct explanation may or may not conform to general laws'.[3]

* I am grateful to Peter Lipton and Marcus Giaquinto, whose comments made this paper less bad than it would have otherwise been. I also profited from the contributions to the discussion made by many of those who heard the paper read in Glasgow, and have revised it in the light of that discussion. The paper has also been published in *Midwest Studies in Philosophy* XV (1990),

[1] Peter Winch, *The Idea of a Social Science* (London: Routledge, 1967). Quote from p. 86.

[2] Daniel Taylor, *Explanation and Meaning* (Cambridge University Press, 1973). Quote from p. 75.

[3] Michael Lesnoff, *The Structure of Social Science* (London: Allen and Unwin, 1974). Quote from p. 106.

Finally, A. R. Louch opposes 'the univocal theory of explanation that all explanation consists in bringing a case under a law. This view has an initial plausibility when developed within the domain of the science of mechanics . . . but as applied to human performance it is totally irrelevant . . .'.[4] For these writers, and many more, explanations of natural phenomena include laws; explanation of social phenomena may not, or do not. Hence, explanations in the two types of science fundamentally differ.

These difference theorists are not really interested *per se* in there being a difference between explanations in the two types of science. Their basic belief is (1), that some full explanations in the social sciences do not include laws. Since they sometimes accept rather uncritically the covering law theory of explanation for the natural sciences, a view which does require the inclusion of laws in any full explanation, they are driven to being difference theorists. But if they could be convinced that full explanations in the natural sciences can do without laws too, that is, that (2) is false, they would not mind the defeat of difference theory. They would think that their most important view, (1), had been further vindicated.

'Full explanations . . . can do without laws' might be ambiguous. Some of these difference theorists believe that there are full explanations in the social sciences which include no laws, because there are in principle no such laws to be had. On their view, no laws (at any level) governing these phenomena exist. That is, they believe that all or part of the subject matter of the social sciences is radically anomic. I am not concerned with their view here (although I think that it is of great interest). I assume, for the purposes of this paper, that everything that occurs, occurs nomically, is governed by some law or other. The question before us is not whether there *are* laws at some level governing everything, but whether every full scientific explanation must contain the statement of a law.

I want to look at some old literature, which I think contains insights not yet fully appreciated. Michael Scriven rejects the difference theory, on the grounds that (1) is true but (2) is false.[5] Moreover, Scriven never, as far as I am aware, supposes that anything occurs anomically. Rather, the question he sets himself is that of the place, if any, for those laws in full explanation. This paper argues against the difference theory, when it is held as a consequence of (1) and (2). Like Scriven, I

[4] A. R. Louch, *Explanation and Human Action* (Oxford: Blackwell, 1966), 233.

[5] Michael Scriven wrote extensively on explanation, but I will confine my remarks to his 'Truisms as the Grounds for Historical Explanations', in *Theories of History*, Patrick Gardiner (ed.) (New York: The Free Press, 1959), 443–75.

think that (1) is true and (2) is false, so no difference between the logic of explanation in the natural and social sciences has been demonstrated. Of course, it is consistent with my thesis that the difference theory may be true for some other reason.

However, there is a sting in my view, that should cool the ardour of the proponent of (1) to some extent, and comfort his covering law opponent. Explanations in both natural and social science need laws in other ways, even when not as part of the explanation itself. Surprisingly perhaps, the covering law theorist, in his zeal to demonstrate that laws are a part of every full explanation, has tended to neglect the other ways in which laws are important for explanation. These other ways make explanations in both types of science somewhat more like the explanations required by the covering law theory, than (1)'s truth and (2)'s falsity might have otherwise led one to suppose. The distance between my view and the covering law theory is not as great as it might at first appear to be.

2. The Covering Law Theory

By 'the covering law theory', I refer to any theory of explanation which requires that every full explanation include essentially at least one law or lawlike generalization, whose role in the explanation is to 'cover' the particular event being explained. Perhaps the best known statement of such a theory is that by Carl Hempel. In a series of important writings, beginning with 'The Function of General Laws in History' (1942), followed in turn by the jointly authored (with Paul Oppenheim) 'Studies in the Logic of Explanation' (1948), 'Aspects of Scientific Explanation' (1965), and *The Philosophy of the Natural Sciences* (1966), Hempel developed an account of singular explanation which has come to be widely known as the covering law theory. It is only singular explanation (e.g. the explanation of why some token event or particular phenomenon occurred or has some charateristic) which shall be the focus here; in particular, I will not discuss Hempel's account of the explanation of laws or regularities.[6]

[6] I do not think that it is always easy to see when an explanation is an explanation of a singular event, and when it is an explanation of a class of similar events, or a regularity. Grammatical form by itself may be a poor guide. Of course, there is a sense in which a token is *always* explained as a representative token of some specific kind. In a rough-and-ready way, though, I think we can see the intention behind the distinction between singular explanation and the explanation of a regularity.

Why did Vesuvius erupt in 79 C.E.? Why is there this unexpected reading in the light emission from such-and-such star? Why did Protestantism arise in

Briefly, Hempel's covering law theory is that there are two distinct 'models' of explanation: Deductive-Nomological and Inductive-Statistical.[7] What both models have in common is agreement about the logical form that all explanations have, and, as a consequence, the place of laws in explanations of both kinds. On Hempel's view, all full explanations (and hence all full singular explanations) are arguments. Let us call this view 'the argument thesis'. Some full explanations are deductively sound arguments; others are inductively good arguments. Hempel does not assume that there can be only one full explanation for an explanandum; there may be indefinitely many sound or good explanatory arguments for a single explanandum.[8]

Any plausible version of Hempel's argument thesis must have recourse to some distinction between partial and complete explanation, since Hempel must recognize that actual explanations often fail to measure up to the requirements his models set for full explanation. On his view, sometimes the explanations which are actually offered are less than full: elliptical, partial, explanation-sketches, merely enthymemes, incomplete, and so on. There can be good pragmatic or epistemic grounds for giving partial or incomplete explanations, depending on context. On any theory of explanation, we sometimes do not say all that we should say if we were explaining in full. Laws may be omitted entirely from a partial explanation. Sometimes we assume that the audience is in possession of facts which do not stand in need of repetition. At other times, our ignorance does not allow us to fill all of the gaps in the explanation. In such cases, in which we omit information for pragmatic or epistemic reasons, we give partial explanations. But, on the argument thesis, all *full* explanations are sound or good arguments.

[7] He sometimes counts Deductive-Statistical as a third, but I ignore this here.

[8] This assumption is made by David Lewis, since his conditions for something's being a full explanation are set so high: 'Every question has a maximal true answer: the whole truth about the subject matter on which information is requested, to which nothing could be added without irrelevance or error. In some cases it is feasible to provide these maximal answers'. David Lewis, *Philosophical Papers* II (Oxford University Press, 1986), 229.

---

northern Europe in the sixteenth century? These are requests for singular explanation. On the other hand, when a student explains the results of mixing two chemicals in a test tube in the laboratory, he is unlikely to be engaged in giving a singular explanation of the particular mixing at hand. He is in fact explaining why mixings of those types of chemicals have that sort of result. The claims of this paper are relevant only to genuinely singular explanations, not to ones which might have the grammatical form of a singular explanation but are in truth the explanation of a regularity.

Different theories of explanation will disagree about what counts as a full explanation. Some will hold that explanations, as typically given in the ordinary way, are full explanations in their own right; I call the holder of such a view 'an explanatory actualist'. Others (like Hempel) are explanatory idealists, who will argue that full explanations are only those which meet some ideal, rarely if ever achieved in practice.[9] Most of those who have rejected the argument thesis have also been explanatory actualists, inclined to accept our ordinary explanations as full as explanations need ever to be: 'Now I have an alternative description of what Hempel calls explanation-sketches . . . I regard them as explanations as they stand, not incomplete in any sense in which they should be complete . . .' (Scriven, p. 446).

If Hempel's argument thesis is true, then all full explanations must essentially *include* at least one law or lawlike generalization amongst its premisses. Let us call this 'the law thesis': 'an explanatory statement of the form "q because p" [is] an assertion to this effect: p is . . . the case, and there are laws . . . such that . . . q . . . follows logically from those laws taken in conjunction with . . . p and perhaps other statements.'[10] Roughly, if we restrict ourselves to full explanations in the natural and social sciences, the law thesis is equivalent to the denial of (1) and the assertion of (2).

True, there are valid arguments with singular conclusions, among whose premisses there is no lawlike generalization. But, given some additional fairly uncontroversial assumptions about what explanations are like (e.g. that one cannot explain p on the basis of the conjunction, p&q), the argument thesis will imply the law thesis.[11] One will need a law to go, in a valid or good argument, from the singular explanans information to the singular explanandum.

[9] If, as Lewis asserts, a complete explanation is a maximally true answer, it is not easy to see how it could *ever* be feasible to provide one. Lewis' theory is bound to drive one into an extreme form of explanatory idealism. The causal history of each thing to be explained stretches into the indefinite past; surely I can *never* provide all of that. On my view, the whole truth about something will typically contain many full explanations for the same thing, so that one can proffer a full explanation about a thing without having to state the whole relevant truth about it. David Lewis, *ibid*.

[10] C. G. Hempel, *Aspects of Scientific Explanation and Other Essays in the Philosophy of Science* (New York: Free Press, 1965), 362.

[11] Winch's criticisms of the covering law account of explanation were certainly not confined to the law and argument theses. He also argued against the view that having a reason is or can be the cause of the action it helps to rationalize. I consider the battle against this to be lost, principally due to the work of Donald Davidson. Once issues about causation, explanation, and law are disentangled, I can see no convincing argument against the claim that the having of reasons can be a type of cause.

It is important to see that the law thesis, as I have explained it, is not just a commitment to there *being* laws that in some sense or other lie behind or underpin the very possibility of explanation. Scriven and Hempel might have no disagreement about this. More specifically, the law thesis requires that a law or lawlike generalization be included in every full explanation.

What is a law? There are widely different responses to this question in the literature, and I do not want the argument of this paper to turn on the answer to this question. In what follows, I assume (like Hempel) that the 'orthodox' answer is correct: a necessary condition for a sentence's stating a deterministic law of nature is that it be a true, universally quantified generalization. There are unorthodox, 'stronger', conceptions of law in the literature, but I stress that nothing in my argument, as far as I can see, turns on the adoption of the orthodox view. If my argument works using the latter view, it would also work if the stronger view is adopted instead.

On the orthodox view, sentences which state deterministic laws of nature typically have or entail something with this form: $(x)\ (Fx \supset Gx)$.[12] Although the universally quantified conditional might be more

[12] No orthodox theorist would consider this condition by itself sufficient. Accidental generalizations have this form, too. Further, universally quantified material conditionals are true when their antecedent terms are true of nothing. So, if this condition were by itself sufficient for lawlikeness, and if nothing in the universe was an F, then both of the following would be laws of nature: $(x)\ (Fx \supset Gx)$ and $(x)\ (Fx \supset -Gx)$.

There are various proposals for adding further conditions to the one above. Some are proposals for strengthening the generalization by adding a necessity-operator: laws of nature are nomically necessary universally quantified generalizations. See for example: William Kneale, 'Natural Laws and Contrary-to-Fact Conditionals', *Analysis* **10** (1950), 121–25; Karl Popper, *The Logic of Scientific Discovery*, (London: Hutchinson, 1972), Appendix *10, 420–41; Milton Fisk, 'Are There Necessary Connections in Nature?', *Philosophy of Science* **37** (1970), 385–404. Others ascribe to the universally quantified generalization an additional special epistemic status, or a special place in science, or impose further syntactic requirements. See Richard Braithwaite, *Scientific Explanation* (Cambridge University Press, 1964), Ch. IX, 293–318; Ernest Nagel, *The Structure of Science* (New York: Harcourt, Brace & World, 1961), Ch. 4, 47–78; D. H. Mellor, 'Necessities and Universals in Natural Laws', *Science, Belief, and Behaviour*, D. H. Mellor (ed.) (Cambridge University Press, 1980), 105–19. Hempel himself is unclear about how to complete the list of sufficient conditions for something's being a law: 'Though the preceding discussion has not led to a fully satisfactory general characterization of lawlike sentences and thus of laws.. . .', (*Aspects*, 343).

complicated than this (e.g. the consequent might also be existentially quantified), it will make no difference to the argument if we only consider sentences with this simple conditional form. I recognize that there are stochastic natural laws; for the sake of simplicity, I restrict myself to the deterministic case.

Hempel expressly applied his models of explanation, and hence the argument and law theses, to the social as well as the natural sciences. Most notoriously, this occurred in his 1942 'The Function of General Laws in History' and the 1959 'The Logic of Functional Analysis'. But the same eagerness for application of the models to social and behavioural science is apparent in *Aspects of Scientific Explanation*: witness the section of the last mentioned, entitled 'The Concept of Rationality and the Logic of Explanation by Reasons'.

Why does Hempel subscribe to the argument and law theses? One reason is this. His theory of explanation is in essence a development of this idea: 'The explanation . . . may be regarded as an argument to the effect that the phenomenon to be explained . . . was to be expected in virtue of certain explanatory facts.'[13] An explanatory 'argument shows that . . . the occurrence of the phenomenon was to be expected; and it is in this sense that the explanation enables us to understand why the phenomenon occurred.'[14] This idea is summed up by Hempel in his symmetry thesis: every full explanation is a potential prediction, and every full prediction is a potential explanation.

Call the information used in making the prediction 'the predictans'; the event predicted, 'the predictandum'. There is no doubt that we need laws in order to make well-grounded predictions, because in making a prediction we do not have the predictandum event available prior to the prediction, as it were. We have to get to it from some singular event mentioned in the predictans, and the law is required to bridge the gap, to permit our moving from one singular event to another. If the symmetry thesis were true, the need for laws in predictions would transfer to explanation.[15]

[13] Hempel, *Aspects*, 336.
[14] Ibid., 337.
[15] In any event, Hempel himself, in *'Aspects'*, more or less abandons the view

---

Suppose, though, that the orthodox view does not provide even a necessary condition, let alone a sufficient one, for a sentence's stating a law of nature. If so, my argument would have to proceed somewhat differently. I am sympathetic to some of these non-orthodox views, but I do not deal with any of them here, nor with how their acceptance would alter my argument. See Fred Dretske, 'Laws of Nature', *Philosophy of Science* **44** (1977), 248–68; for a reply to Dretske, see Ilkka Niiniluoto, *Philosophy of Science*, **45** (1978), 431–9; David Armstrong, *What Is A Law of Nature?* (Cambridge University Press, 1987).

Although I reject the symmetry thesis, I would not like to break entirely the connection between explanation and prediction. It is not true that every full prediction is a potential explanation, nor is it true that a full explanation (since, on the view I develop below, it can omit all laws) is itself a potential prediction. But I will argue that, if one has and *knows* that one has a full explanation, then laws play a crucial role in that knowledge, and that therefore anyone with this knowledge will be in a position to be able to make predictions.

3. Opponents of the Covering Law Theory

I believe that there are strong reasons to dispute the argument thesis; I do not believe that the idea of logical dependence of a conclusion on a set of premisses can capture the idea of explanatory dependence, which is the essential contention of the argument thesis. I have argued elsewhere that explanations are typically not arguments, but sentences, or conjunctions thereof. I would not want to rest my case for this on the overtly non-argument grammatical form that explanations have, as they occur in ordinary speech, but I cannot rehearse my arguments for the rejection of the argument thesis here.[16] A non-argument view of explanation has been developed by Achinstein, Ryle and Salmon (the list is not meant to be exhaustive).

I argued above that the argument thesis, in conjunction with additional uncontroversial assumptions, leads to the law thesis. But the inverse is not true. Suppose that the argument thesis is false. It does not follow from the fact that not all full explanations are arguments, that a law is not a part of every full explanation. It only follows that, if laws are a part of full explanations which are not arguments, the idea of their parthood in such cases is not to be cashed out as that of a premiss in an argument. So the absence of laws from even some full explanations does not follow from the fact that some full explanations are not arguments.

However, the idea that full explanations do not always include laws has been argued for by many of the same philosophers who reject the

[16] I am convinced by the argument of Tim McCarthy, 'On An Aristotelian Model of Scientific Explanation', *Philosophy of Science* **44** (1977), 159–66. See my *Explaining Explanation* (London: Routledge, 1990), Ch. 6, where I develop the point at considerable length.

---

that every prediction is a potential explanation, in his discussion of Koplik spots and measles. Another case worth considering is this: suppose c and d both occur. Both are potential causes of e, but, since c occurs and actually causes e, d is merely a potential but pre-empted cause of e. One can predict but not explain e's occurrence from the occurrence of d, its pre-empted cause; e's explanation must be in terms of c, its actual cause, and not in terms of its pre-empted cause.

argument thesis. In numerous papers, Michael Scriven has said things similar to what I would wish to maintain about the role of laws or generalizations in explanation (although I do not need to agree with any of his specific examples). In 'Truisms as the Grounds for Historical Explanations', he defended the view that the following was a perfectly *complete or full* explanation as it stood (what he actually said was that it was not incomplete; he was sceptical about the idea of a complete explanation): the full explanation of why William the Conqueror never invaded Scotland is that 'he had no desire for the lands of the Scottish nobles, and he secured his northern borders by defeating Malcolm, King of Scotland, in battle and exacting homage'.[17] The explanation is a conjunctive statement formed from two singular statements and contains no laws. Explanations which lack laws are 'not incomplete in any sense in which they should be complete, but certainly not including the grounds which we should give if pressed to support them'.[18]

Scriven's example above is an explanation of a human action. It is sometimes argued in the case of human actions that they are explicable but *anomic*. As I indicated at the beginning, the thought here is different. Human actions might be, perhaps must be, nomic. The first of Scriven's claims is that although or even if human actions are always nomic, sometimes the laws or 'truisms' which 'cover' them form no part of their full explanation.[19] Similarly for explanations in the natural sciences. Natural occurrences, on his view, are covered by laws or truisms, but which may form no part of their explanation: 'abandoning the need for laws . . . such laws are not available even in the physical sciences, and, if they were, would not provide explanations of much interest . . . When scientists were asked to explain the variations in apparent brightness of the orbiting second-stage rocket that launched the first of our artificial satellites, they replied that it was due to its axial rotation and its asymmetry. This explanation . . . contains no laws.'[20]

What is a 'truism'? Scriven's distinction between a genuine law and a truism is similar to Donald Davidson's distinction, between homonomic and heteronomic generalizations.[21] Scriven's examples of truisms include these: 'If you knock a table hard enough it will probably cause an ink-bottle that is not too securely or distantly or specially

[17] Scriven, 444.

[18] Ibid., 446.

[19] See Thomas Nickles, 'Davidson on Explanation', *Philosophical Studies* **31** (1977), 141–5, where the idea that 'strict' covering laws may be 'non-explanatory' is developed.

[20] Scriven, 445.

[21] See Davidson, 'Mental Events', reprinted in his *Essays on Actions and Events* (Oxford University Press, 1980), 207–27. Also in *Experience and Theory*, Foster and Swanson (eds) (London: Duckworth, 1970), 79–101.

situated to spill over the edge (if it has enough ink in it)'; 'Sufficient confidence and a great desire for wealth may well lead a man to undertake a hazardous and previously unsuccessful journey'; 'Power corrupts'; 'Strict Orthodox Jews fast on the Day of Atonement'. Scriven says that truisms are *true*: 'The truism tells us nothing new at all; but it says something and it says something true, even if vague and dull.'[22]

How shall we represent a truism? Do they have the form: $(x)(Fx \supset Gx)$? If they did, they would be false, for, so construed, there are exceptions or counterexamples to each of them. Scriven says that the discovery of a devout Orthodox Jew who is seriously ill and has received rabbinical permission to eat on the fast day does not disconfirm the truism. Truisms are, for him, a type of 'normic statement': 'they have a selective immunity to apparent counterexample'.[23]

> The normic statement says that everything falls into a certain category except those to which certain special conditions apply. And, although the normic statement itself does not explicitly list what count as exceptional conditions, it employs a vocabulary which reminds us of our knowledge of this, our trained judgment of exceptions. (p. 466)

'Other modifiers that indicate normic statements are "ordinarily", "typically", "usually", "properly", "naturally", "under standard conditions", "probably". But . . . no modifier is necessary. . . .'[24]

4. Are Laws a Part of All Full Explanations?

So, Hempel is an upholder of the law requirement for all full singular explanations; Scriven a rejector for at least some full singular explanations in both the natural and the social sciences. Neither is a difference theorist. But if Scriven is right, what many difference theorists really care about will have been vindicated.

Who is right, and why? Notice that the rejector certainly does not have to maintain that laws form no part of *any* singular explanation (see Scriven, pp. 461–2), only that laws do not necessarily form a part of *all* full singular explanations in either type of science. It seems obvious that many full singular explanations do include laws, and this seems to be especially so in the natural sciences. Explanations in natural science frequently include relevant laws, although when this is so, their inclu-

[22] Scriven, 458.
[23] Scriven, 464.
[24] Scriven, 465.

sion in the explanation will not necessarily be as a major premiss of an argument: 'o is G because o is F and all F are G' is a (contingently true) explanatory *sentence* which includes a law, but is not an *argument*.

As we have already seen, even without the support of the argument thesis, the law thesis might still be true. Let us assume, for the remainder of the discussion, that the argument thesis (and the symmetry thesis) about explanation is false; this means that the upholder of the law thesis cannot argue in its favour on the grounds that without laws, there is no derivability of explanandum from explanans, and no potential prediction. In fact, I hope that the plausibility of my rejection of the law thesis will itself provide an independent reason to reject the argument thesis, thereby rendering that latter rejection somewhat less than a mere assumption. Is there anything convincing that can be said for or against the law thesis directly, independently of the argument and symmetry theses?

I think that there is something that can be said *against* it. It is this: in a singular explanation that is sufficiently full, there may simply be no work that remains for a law to do by its inclusion in the explanation. The inclusion of a law in a sufficiently full singular explanation may be otiose. Let me elaborate.

The non-extensionality of explanation is well-known. Events or phenomena explain or are explained only as described or conceptualized. The point derives from Aristotle, who saw so much so clearly so long ago: what explains the statue is Polyclitus *qua* sculptor, but not Polyclitus *qua* the musical man or *qua* the pale man. A more modern terminology in which to make the same point is in terms of descriptions and the different properties they utilize. Any token event has an indefinitely large number of descriptions true of it. Suppose some token event is both the F and the D. Under some descriptions ('the F'), it may be explanatory, but under others ('the D') it may not be. If so, it is not the token event tout court that explains, but the event *qua* an event of type F. Hempel himself makes a point very similar to this; he says that we never explain concrete events, but only sentential events.[25]

To borrow and expand on an example from John Mackie: the disaster that befell poor Oedipus is explained by the event of his marrying his mother. Precisely the same event may have been his marrying the prettiest woman in Thebes, or his participating in the most sumptuous wedding in all Greece, but so described, the event does not explain the disaster that befell him. The event explains the disaster only as an event of the type, a marrying of one's mother. So properties matter to explanation, and in part account for its non-extensionality. Two prop-

[25] Hempel, *'Aspects'*, 421–3.

erties, like being renate and being cordate, can be coextensive, but an animal's being renate can explain things that is being cordate cannot touch, and vice versa.[26]

Suppose the law thesis rejector (I count myself amongst their number) claims that (A) *can* be a full explanation: (A) object o is G because o is F. The law thesis upholder will deny that this is possible, and say that the full explanation is (B): object o is G because o is F and (x) ($F \supset Gx$). If so, then he has to motivate the thought that (A) could not really be a full explanation, by showing what of importance it is that (A) omits, which is not omitted by (B).

Can we pinpoint what it is that the law is meant to add to (A), as far as explanatory impact is concerned? What has (B) got that (A) lacks in so far as explanation is what is at issue? Of course, someone who held the argument thesis, would say that there is no derivability of o's being G, from o's being F by itself, without the addition of a law. And he may no doubt wish to add something about the requirements for prediction. But the strategy here was to see what could be said for the law thesis, without assuming the argument thesis, or the symmetry of explanation and prediction, and so both the (purported) explanations with and without the law have been represented by sentences, not arguments.

Return to the thought, adumbrated above, that what matters to explanation are properties. When o's being F does fully explain o's being G, it is not (to put it crudely) that *o*'s being F explains *o*'s being G; there is nothing special about o in any of this. Rather, it is o's *being F* that explains o's *being G*. Explanatory weight is carried by properties, and properties are, by their very nature, general. If o's being F fully explains o's being G (and given our assumption that everything that happens, happens nomically), then for any other relevantly similar particular which is both F and G, its being G will also be explained by its being F. There is, to be sure, generality in singular explanation, but that generality is already insured by the properties ascribed to the particulars in (A), without the addition of laws.

Of course, there is one obvious sense in which an explanation of o's being G, in terms of o's being F, could be only partial. The explanation might fail to specify or cite all of the explanatorily relevant properties or characteristics of o. *But all of the explanatorily relevant properties of o can be cited without inclusion of any law or lawlike generalization.*

[26] I do not say that nothing but properties matter; names sometimes seem to matter to explanation too. Naming in one way can be explanatory, naming in a different way can fail to be so. If I do not know that Cicero=Tully, I can explain why Cicero's speeches stopped in 43 B.C.E., by the fact that Cicero died in that year, but not by the fact that Tully died in that year. But in any case, properties matter to explanation, even if names do too.

Suppose, for the sake of argument, two things: first, that it is an exceptionless law of nature that (x) $(Fx \supset Gx)$; second, that the *only* property of some particular o, relevant to explaining why o is G, is o's F-ness. So, in the law or generalization in (B), the only information that could be relevant to the explanation of o's being G is already given by the property linkage between *being F* and *being G* which is expressed by (A). The additional information in the generalization which is about (actual or possible) F's other than o which are also G, is simply irrelevant to the explanation of o's being G. At most, what (B) does that is not done by (A) is merely to explicitly apply the connection between properties F and G, already expressed by (A), to cases other than o. And this cannot have any additional explanatorily relevance to o's case. The case of temporally and spatially distant F-objects which are G is surely not relevant to o. One might ask about explanation the question Hume asked himself (but believed he could answer) about his constant conjunction theory of causation: 'It may be thought, that what we learn not from one object, we can never learn from a hundred, which are all of the same kind, and are perfectly resembling in every circumstance.'[27]

The law thesis would make relevant to the explanation of o's being G, the F-ness and G-ness of other actual or possible particulars, a, e, i, u, etc. To borrow and amend an example from Mill's discussion of deduction and real inference, it is hard to see how, if the Duke of Wellington's manhood cannot fully explain his mortality on its own, introducing the manhood and mortality of people other than the Duke (whether by a generalization or by the enumeration of other particular instances) could explain it. What is the explanatory relevance to the good Duke's mortality of the mortality of men spatially and temporally far distant from him? It is only his being a man that explains his being mortal. If the world is nomic, then other men will be mortal too, and it may be that the very possibility of explaining the Duke's mortality on the basis of his manhood presupposes this. But even if this were a presupposition of the explanation of the Duke's mortality by his manhood, that presupposition contributes nothing additional to the explanation of the Duke's mortality, and has no place in it.

5. Laws Get Their Revenge

But laws are still important, even to those cases of explanation which do not include them, in other ways. Indeed, the law requirement view, by

[27] David Hume, *A Treatise of Human Nature*, L. A. Selby-Bigge (ed.) (Oxford University Press, 1965), 88. I think that much of the motivation for the inclusion of a generalization in every full explanation stems from the Humeian analysis of causation.

insisting that laws are a *part* of every full explanation, has tended to neglect the other ways in which laws are essential to explanation. Let me add some remarks about how laws are still crucial for the explanation of the world about us, all consistent with my above claim; the remarks will also permit me to sharpen my view somewhat on the role of laws and generalizations in explanation.

Why did the match light? I struck it, and my striking of the match was, let us suppose, the penultimate thing that ever happened to the match. Or, my striking of the match was the event that caused the match to light. Why, then, can I explain the fact that the match lit by the fact that the match was struck, and not by the different facts that the penultimate thing that ever happened to the match occurred, or that the cause of its lighting occurred, even though these three singular facts (the fact that the match was struck, the fact that the cause of the match's lighting occurred, the fact that the penultimate thing that every happened to the match occurred) are all facts about the same event, but differently described? We saw before that properties matter to explanation, and, because they do matter, it is redundant to include laws in all singular explanations. But in virtue of which of the features or properties of an event is that event fully explanatory of some other? How do we determine which properties of a thing matter for the purposes of explanation?

Aristotle's reply would have been that the explanatory features are the ones linked in a law (whether deterministic or stochastic).[28] In the case at hand, there is a law that links the features, being a striking and being a lighting, of the two token events, but not other of their features, e.g. being a penultimate occurrence and being a lighting. Nor is there a *law* which links being a cause of lighting and being a lighting, even though it is a (contingent, if the scope is read correctly) general truth that causes of lightings cause lightings. On Aristotle's view, it is *not* that the laws are always part of the explanation (he does think that this is true of scientific explanation, but not of explanation generally); rather, the laws provide the properties for determining under which descrip-

[28] Aristotle gives us a definition of the incidental or accidental in *Metaphysics* **V**, 30, 20–5: ' "Accident" means (1) that which attaches to something and can be truly asserted, but neither of necessity nor usually . . . for neither does the one come of necessity from the other or after the other, nor . . . usually . . . And a musical man might be pale; but since this does not happen of necessity nor usually, we call it an accident. Therefore, since there are attributes and they attach to subjects, and some of them attach only in a particular place and at a particular time, whatever attaches to a subject, but not because it was this subject, or the time this time, or the place this place, will be an accident.' For a defence of my interpretation of Aristotle, see my *Explaining Explanation*.

tions a particular fully explains another. Laws provide the appropriate vocabularly for full singular explanation.

The above allows me to make a closely related point about the role of theories in explanation. Scientists often cite theories in explaining a phenomenon. For example: the theory of gravity explains why the moon causes the earth's tides; the law of inertia explains why a projectile continues in motion for some time after being thrown; subatomic particle theory explains why specific paths appear in a Wilson cloud chamber. And theories consist (perhaps *inter alia*) of generalizations. *But (a) it does not follow that theories are explanatory in virtue of their generality, (b) nor does it follow that the way in which they are explanatory is in all cases by being part of the explanation*. I have already argued for (b). But I now wish to argue for (a). Theories typically help to explain singular facts, in virtue of supplying a vocabulary for identifying or redescribing the *particular* phenomena or mechanisms at work, which are what explain those singular facts.

The examples of 'syllogistic explanation' that I have been using might have struck the reader as exceedingly artificial: whoever would have thought, the reply might go, that the Duke of Wellington's mortality could be explained by his manhood and the generalization that all men are mortal? And, I have been working with this example form: 'o is G because o is F and all F are G'. These generalizations are 'flat', in the sense that they are simple generalizations that use the same vocabulary as do the singular explanans and explanandum descriptions. Flat generalizations do not contribute at all to singular explanation.

However, from the fact that flat generalizations are explanatorily useless, it hardly follows that all are. What is needed, so the reply might continue, are generalizations which employ a theoretical vocabulary with greater depth than 'man' and 'mortal'. Perhaps the vocabulary should be in *deeper* terms that refer to the fragility of hydrocarbon-based life forms. To explain why o is G, in terms of o's being F, if a law is to be included, typically a scientific explanation will cite a law employing a theoretical vocabulary which is different from and deeper than the vocabularly of which 'F' and 'G' are part. Only as such could the generalizations be explanatory.

And such a reply is correct. But it confirms rather than disconfirms my view. If generalizations or laws were always *per se* explanatory, then flat ones ought to help explain (perhaps not as well as deep ones, but they should help to explain to some extent nonetheless). The fact that *only* ones that are deep, relative to the vocabulary of the explanans and explanandum singular descriptions (in general, theories), will help explain *at all* is an indication that laws are explanatory only in virtue of offering a deeper vocabulary in which to identify or redescribe mechanisms, but not just in virtue of being generalizations. And even so, to return for a moment to (b), the generalizations that make up the wider or deeper theory may help to explain by offering that alternative vocabulary, but without being part of the explanation itself.

I have been arguing that full explanations do not always include laws, but that they sometimes may. Here perhaps is such an example (although it is not a singular explanation as it stands): 'There are tides because it is a law that all bodies exert gravitational forces on all other bodies, and because of the fact that the moon exerts this force on the earth's seas.' When laws are included within an explanation, as they sometimes are, the purpose of the inclusion is often to introduce a vocabulary different from the one used in the common or ordinary descriptions of the explanans and explanandum events and the particulars to which they occur ('the tides', 'the earth', 'its seas', 'the moon').

On the one hand, if the common and less deep vocabulary used to describe the the particular phenomena were wholly expendable, the theoretical vocabulary could be explicitly used to describe them, and any mention of the law in the explanation would be redundant. If on the other hand no deeper vocabulary were available, there would be no purpose for a law to serve in the explanation. Laws find their honest employment within singular explanations in situations between the two extremes: when the less deep vocabulary used to describe the singular explanans and explanandum is to be retained at the shallower level, but a deeper vocabulary is available, and needs introduction.

One important role that theories play in science is to unify superficially diverse phenomena.[29] In virtue of a unifying theory, what seemed like different phenomena can be brought under one set of deep structural laws:

> By assuming that gases are composed of tiny molecules subject to the laws of Newtonian mechanics we can explained the Boyle-Charles law for a perfect gas. But this is only a small fraction of our total gain. First, we can explain numerous other laws governing the behavior of gases . . . Second, and even more important, we can integrate the behavior of gases with the behavior of numerous other kinds of objects . . . In the absence of the theoretical structure supplied by our molecular model, the behavior of gases simply has no connection at all with these other phenomena. Our picture of the world is much less unified. (p. 7)

[29] See Michael Friedman, 'Theoretical Explanation', in *Reduction, Time and Reality*, Richard Healey (ed.) (Cambridge University Press, 1981), 1–16. See also his 'Explanation and Scientific Understanding', *Journal of Philosophy*, **71** (1974), 5–19, and the reply by Philip Kitcher, 'Explanation, Conjunction, and Unification', *Journal of Philosophy*, **73** (1976), 207–12.

On my view, there is a difference between unification and explanation. Unification of a phenomenon with other superficially different phenomena, however worthwhile a goal that may be, is no part of the explanation of that phenomenon. It is true that a good theory unifies apparently disparate phenomena, and unificiation is a highly desirable goal for science. But that some particular mechanism is common to both what is being explained and other apparently disparate things is not relevant to the explanation of the first thing in terms of that mechanism. If other men's mortality could not explain why the good Duke is mortal when his own manhood does not, then the fragility of other hydrocarbon-based life forms (which will 'unify' men's mortality with that of other plants and animals) could not explain the Duke's fragility or mortality when his own hydrocarbon constitution does not. It does not matter, from the point of view of explanation, whether there are any other phenomena which get explained by the deeper vocabulary; the point is that the vocabulary gives a new and more profound insight into the phenomenon at hand and the mechanisms at work in its case, whether or not the vocabulary unifies it with other phenomena.

Consider a, perhaps not perfect, analogy. I want to explain both why Cicero was bald, and why Tully was bald (thinking that they are two different men). I am informed that Cicero is Tully. There is a clear sense in which there has been unification, for there is now only one thing left to explain. Our view of the world is more unified. But has there been any explanatory gain? Only in the rather derivative sense that the identity explains why there are not two different things to be explained. If that is the substance of Friedman's claim, I accept it. But surely nothing else has been explained at all.

6. Explanatory Idealism and Acutalism

So far, it might seem that Scriven has won the toss against Hempel. On the specific question of the law thesis, I submit that he has won. But on the issue of explanatory actualism and idealism, I think he must be wrong, and some of the motivation behind Hempel's covering law account right after all. That is, I believe that few, if any, of the singular explanations commonly offered are full explanations. I, like Hempel am an explanatory idealist.

There are two questions worth distinguishing: first, what is a full explanation; second, how can we justify a claim that some explanation is a full one? Let us look at the second question of how we might justify a claim that some singular explanation is full or complete, and at the role that truisms could play in such a justification.

Scriven addresses the second question, in his remarks on laws as providing role-justifying grounds. Does some explanans really fully explain some explanandum? Perhaps it is not adequate to fully explain it; something may be missing. How can I justify my claim that the explanans fully does the job it is meant to do?

Suppose I claim that 'e because c' is a full explanation. On Scriven's view (p. 446), if I am challenged about the adequacy or completeness of my explanation, I can justify my claim to completeness, and thereby rebuff the challenge, by citing a law (or truism), e.g. that all C are E (c being a C; e being an E). This is what Scriven calls the 'role-justifying grounds' that laws and truisms provide, in support of a claim that an explanation is not incomplete.

The law or truism plays an epistemic role in explanation. It can justify an assertion that c is the full and adequate explanation of e, without being any part of that explanation. On the basis of the law, I can justify my claim that no explanatorily relevant information about the particular explanans has been omitted. Although Scriven does not say so, there can be no objection to offering the full explanation and the justification for its fullness in a single assertion, but if this is done, we should be clear that what we have is a full explanation *and* something else, and not just a full explanation.

Scriven offers the following possible reply to his own views: 'It may appear that we are quibbling over words here, that it is of little importance to decide exactly what is included in an explanation and what in its justification.'[30] Like Scriven, I do not accept the sentiment expressed by such a reply. The distinction is important. There is quite generally a distinction between the correct analysis of an expression, and the grounds or evidence that one might have for believing that it is true. In particular, there is a difference between the roles that Hempel and Scriven assign to laws. On Hempel's view, the law is part of the full explanation itself; on Scriven's, the law provides the criterion for determining, concerning an explanation, when it is a full one.

Perhaps an analogy will help.[31] On Hempel's account, to know that an explanatory argument is a full one, one must know that the explanatory argument is valid. And the criterion for deciding validity is given by the rules of deductive and inductive logic. But neither the assertion that the explanatory argument is valid, nor the criteria for assessing validity, is to be included as a further premiss in the full explanation. *Pari passu*, on Scriven's account, to know that an explanation is a full one, one must know that no explanatorily relevant singular information has been omitted, and the criterion for assessing whether any such information has been omitted is given by the relevant covering laws or truisms. But neither the assertion that the explanation omits no such

[30] Scriven, 448.

[31] The analogy was suggested to me by Peter Lipton.

singular information, nor the criterion for assessing whether this is so (the law), is to be included in the full explanation.

Suppose once again that o is G because o is F (Cortes sent out a third expedition to Baja, California because of his considerable confidence in his own leadership and his prospect of gigantic booty). One can cite (1) grounds for thinking that o is F (or that o is G), (2) grounds for thinking that o's being F is *an* explanatory factor in o's being G, and (3) grounds for thinking that o's being F is the full explanation of o's being G. In the case of (1), one might cite various sorts of observational evidence; historical documents make o's being F highly likely or probable. In the case of (2), one might cite experimental evidence that showed that, in the absence of their being F, things otherwise relevantly similar to o failed to be G (Mill's Method of Difference).

It would be an error to include this historical or experimental evidence in the analysis of 'o is F' or 'o is G at least partly because o is F'. To do so would involve a kind of verificationism which conflates analysis and evidence. It seems to me a similar kind of mistake to include one's evidence for 'o is G because o is F' in its analysis.

But notice that even if this sentiment about the unimportance of the distinction between 'what is included in an explanation and what in its justification' were accepted, it would cut more against the law thesis upholder than against its rejector. The law thesis upholder must think that it is an important insight that full explanations include laws; to hold that it is arbitrary whether one says that they do or do not include laws is contrary to the spirit of the thesis. At the very worst, the thesis of this paper would have to be that it is of little importance whether full explanations are said to include laws or not, and this is something the law thesis upholder should resist.

I agree with Scriven that a law can provide role-justifying grounds for full explanation, but I do not agree that a truism can do so. This raises the question of explanatory actualism, and on this question, I part from his company. A full explanation of o's being G is the fact that o is F, *only if* it is a law that all or a certain percentage of F's are G's, sans further exception. Suppose the law in question is a more complex law which says: $(x)\ (Fx \& Kx \& Hx \& Jx \supset Gx)$. A full explanation of why o is G would be the singular fact that o is F&K&H&J. In this way, my view of full explanation is, in at least one way, closer to Hempel's than to Scriven's. Full explanations, on my view as on Hempel's, may well be close to ideal things; if almost no one ever gives one, that tells us a lot about the practical circumstances of explanation-giving, but provides no argument whatsoever against such an account of full explanation.

There may be perfectly good pragmatic reasons why we are entitled to give a partial explanation of o's G-ness; it may be that o's being K&H&J is so obvious, that one never needs to say anything more than that o is F. But the law provides the criterion for what a complete or full explanation is. To turn to Scriven's second example, I am sure that the explanation of the variations in apparent brightness of the orbiting second-stage rocket that launched America's first artificial satellite, in terms only of its axial rotation and asymmetry, cannot be its full explanation. I agree that its full explanation, whatever it is, need not include a law, but since the explanation Scriven offers fails to contain any particular information about, for instance, the source of light that was present, it could not be a full explanation. Scriven's own remarks about the role-justifying grounds that laws provide help make this very point. The particular explanation Scriven offers as full can be seen to be merely incomplete, not because it does not include a law, but because the law provides the test for fullness which Scriven's explanation of the rocket's apparent variations in brightness fails.

On Hempel's (but not Lewis') view of full explanation, which I have espoused, o's being F can explain o's being G, when it is a law of nature that $(x)\ (Fx \supset Gx)$. But suppose (as Elliott Sober reminds me) that this is a case of derivative or non-immediate explanation, in the sense that there is some property H such that o's being F fully explains o's being H, and o's being H fully explains o's being G. The explanation with the greater information works in virtue of two laws: both $(x)\ (Fx \supset Hx)$ and $(x)\ (Hx \supset Gx)$. In this way, the explanation tells us more, since it tells us about the process or pathway *via* which being F explains being G.

One might capture this greater information in a single explanation in one of two ways: o is G because o is H and o is H because o is F; o is G because o is F and both $(x)\ (Fx \supset Hx)$ and $(x)\ (Hx \supset Gx)$ are laws of nature. I have no reason to resist this point, if it is sound. First, I have allowed that some explanations do or may include laws, even though not necessarily as premisses in an argument. Second, it seems a matter of pragmatic importance only, in which of the two ways the information is captured. As Sober suggested to me, 'Whenever adding a law would improve the explanation, the same improvement can be effected by adding something that is not a law.'

I certainly do not defend this Hempelian notion of full explanation on the grounds that this is the concept of full explanation that emerges from ordinary language, or from common usage.[32] I have no interest in disputing about the ordinary sense of the term. If someone wishes to argue that there is a sense of 'full explanation' in ordinary language, on which Scriven-style explanations based on truisms can count as full, I will concede that this may be so.

[32] This Hempelian notion of full explanation is to be contrasted with an even stronger sense of full explanation that one finds in the work of David Lewis. See notes 8 and 9 above.

The sense of 'full explanation' that I, following Hempel, urge is a technical one. I think that this technical sense is one that becomes plausible, when we reflect upon the sort of understanding that science poses as a possibility. A full explanation is an ideal (as Hempel himself says). For such an ideal, a true law which 'backs' the full explanation of some token of a type must back the full explanation of any other token of that type. The amount of singular information in the explanans required in the full explanation of any token of a type is pulled upwards, as it were, to the level of the greatest amount of singular information required to fully explain any token of that type.

I do not offer this as a definition of 'full explanation' (if it were intended as such, it would be circular). Rather, it is a *constitutive* principle of what full explanation (in this technical sense) is. Such an ideal is constitutive, rather than regulative, to borrow a Kantianesque distinction. Ordinary or actual explainers are not required to try to offer, as nearly as possible, full explanations. Actual explanation-giving is governed by pragmatic considerations. The ideal is constitutive; it constitutes for us what, upon reflection, we intend by 'fully explaining why something happened'.[33]

But what this means is that one cannot justify the claim that one has produced a full singular explanation, in the absence of knowledge of the appropriate deterministic or stochastic covering law. It may be that one has fully explained without knowing it; but knowledge that one has fully explained implies knowledge of the law. I am not supposing that one might explain without knowing that one had explained at all; whether or not that is possible, all that I am supposing is that one might know that one had explained at least in part, without knowing whether one had fully explained, and even though one had in fact given a full explanation.

7. Truisms and the Social Sciences

This brings us back to the question of explanations making use of truisms, a question of special importance for the social sciences (the examples of Cortes and William the Conqueror employ such truisms). How could a truism be the backing for a complete or full explanation? It seems to me that when we use truisms, as of course we frequently do, we can at best claim partial or incomplete explanation. The completeness of the explanation can only be shown to be so when we see that all of the explanatorily relevant properties of William the Conqueror and

[33] This is the Hempelian ideal of full explanation. I describe in notes 8 and 9 above David Lewis' even more demanding ideal of full explanation.

Cortes have been included in their descriptions. For this, one must have available the relevant 'strict' covering law.

It may be that these truistic, rough-and-ready generalizations are small fragments of fuller empirical generalizations. Ayer argues a case for this in 'Man as a Subject for Science', and I do not think that his case has ever been effectively answered.[34] If so, then we see what we must do to get full explanations: find the fuller and strictly true generalizations of which the truisms are only fragments.

But it is more fashionable to believe that Ayer's hypothesis is wrong, that the truisms will not reappear as fragments of fuller generalizations. Let us suppose that Scriven's truisms, like Davidson's heteronomic laws, are *incompleteable* within the terms of the vocabulary they use. As Davidson says,[35]

> The generalizations which embody such practical wisdom are assumed to be only roughly true, or they are explicitly stated in probabilistic terms, or they are insulated from counterexample by generous escape clauses (p. 93).
>
> . . . there are generalizations which when instantiated may give us reason to believe that there is a precise law at work, but one which can be stated only by shifting to a different vocabulary. We call such generalizations heteronomic (p. 94).

On my view, singular explanations which rest on heteronomic (incompleteable within their own vocabulary) generalizations must fail to be complete explanations. Notice that 'incompleteable' for Davidson does not presuppose that a generalization is complete only when it is universally quantified. There are, or may be, stochastic laws which are less than universally quantified generalizations, and which may be as perfectly complete as any law need be. 'Or perhaps the ultimate theory is probabilistic . . . in that case there will be no better to be had' (Davidson, p. 94). Heteronomic generalizations (and Scriven's truisms) do not just fail to be fragments of exceptionless laws; they fail, on their accounts, to be fragments of any kind of law at all, deterministic or stochastic.

I pretended earlier that it was a law that striking a match caused it to light. In truth, there is no law of nature that links strikings and lightings of matches. 'Striking a match will cause it to light' is also at best a truism, incompleteable within the terms of its own vocabulary.

I do not say, in the case of truisms such as this, that they cannot be explanatory *at all*. Clearly, it can be explanatory of a particular match's

[34] A. J. Ayer, 'Man as a Subject for Science', *Philosophy, Politics and Society*, third series (Oxford: Blackwell, 1967), 6–24.

[35] Donald Davidson, 'Mental Events', *Experience and Theory*, 79–101.

lighting to say that it was struck. Presumably, this is because 'striking' and 'lighting', although themselves no part of a law, are linked in some appropriate way with the vocabulary of the relevant underlying laws of physics or chemistry; Aristotle's criterion for determining when a description is one under which a particular is explanatory of another must therefore be amended to include this point. I only assert that such truisms, like the one about striking a match causing it to light, cannot provide a *complete* or *full* explanation of a singular occurrence, like the particular match's lighting.

If Ayer is wrong, and *if it is desirable to have complete explanations*, even of human action, then I think the drive towards homonomic laws, 'correctible within [their] own conceptual domain' (Davidson, p. 94), even if it requires a switch to the vocabulary of brain neurophysiology, is irresistible. Many of these explanations will be similar to the ones I discussed above, in which the old terminology is retained at the level of the singular explanans and explanandum, but a law is included within the explanation, as a way of introducing a new, and deeper, vocabulary. To be satisfied with our practical lore as it stands for the explanation of human action, with its vocabulary that either is correctible in its own conceptual domain but stands in need of correction, or is uncorrectible in that domain, is to be satisfied with incomplete explanation.

It is no part of my view that it is always wrong to be satisfied with incomplete explanations, that it is always desirable that science look for complete explanations.[36] Perhaps these reflections pose the possibility of another angle from which to argue for a difference theory. Natural science, not strictly always but anyway typically, with its emphasis on prediction and control, strives for completeness or fullness of its explanations. Recall what I mentioned in passing above. Explanations are not potential predictions, and predictions are not potential explanations (the symmetry thesis is false in both directions), but to know that one has fully explained requires knowing the relevant law, and the surest sort of prediction requires knowing the relevant law too. If one knows that one's explanation is full, one has all one needs in order to predict accurately. So predicting, and knowing that one's explanation is a full explanation, have an overlapping requirement.

The social sciences lack this interest in control and prediction beyond the rough-and-ready, and as a consequence may not be as interested in pursuing complete or full explanations of things. Methodologically, partial explanations serve the needs and interests of social scientists tolerably well.

[36] I have discussed this briefly in 'Marx, Necessity, and Science', in *Marx and Marxisms*, H. R. Parkinson (ed.) (Cambridge University Press, 1982), 39–56, see 54–6.

Explanation and Understanding in Social Science

JOHN SKORUPSKI

I

Hempelian orthodoxy on the nature of explanation in general, and on explanation in the social sciences in particular, holds that

(a) full explanations are arguments
(b) full explanations must include at least one law
(c) reason explanations are causal

David Ruben disputes (a) and (b) but he does not dispute (c). Nor does he dispute that 'explanations in both natural and social science need laws in other ways, even when not as part of the explanation itself' (p. 97 above). The distance between his view and the covering law theory, he points out, 'is not as great as it may first appear to be' (p. 97 above).

Ruben quotes the following example from Michael Scriven. William the Conqueror never invaded Scotland because 'he had no desire for the lands of the Scottish nobles, and he secured his northern borders by defeating Malcolm, King of Scotland, in battle and exacting homage' (quoted Ruben, p. 103 above). Scriven holds this to be a complete, or at least not incomplete, explanation. I agree that on a normal use of the word explanation it could be complete. Equally, on that same use, it might not be complete. It would not be reckoned a complete explanation, for example, if William was known to be a quarrelsome megalomaniac ever itching for further opportunities to lay countries to waste and demonstrate his superiority in battle. In that case the fact that he refrained from invading Scotland just because he did not want Scottish lands and had secured his northern borders might come as a surprise. We would want to know why he had acted in so unexpectedly sensible a way. Perhaps on this one occasion an influential adviser prevailed on him to be prudent?

So one thing we need to know, in order to know that we have a complete explanation, is William's normal disposition—or in other words some approximate but incipiently nomic general truths about him. He was a level-headed and prudent type, not a rash and reckless one. The general truth is incipiently nomic but approximate; it corresponds to Scriven's 'truisms', though it need not always be truistic—it might be unexpected.

So far then, I am with Ruben and Scriven. However Ruben is more demanding about full explanations than Scriven. He argues that the explanatory relation holds between events under descriptions; that is, between events in virtue of their properties. The explanatorily relevant properties are those which are linked by law, and to know that we have a complete singular explanation we must, he holds, know that law. It is the vocabulary of the relevant law which determines the terms in which a complete singular explanation must describe its explaining and explained events. Futhermore, what constitutes a *full* or *complete* singular explanation of o's being G will be determined by the exhaustive specification, provided in the law, of a particular sufficient set of antecedents for events of type G (p. 114). Ruben accepts that full explanations, on this account, 'may well be close to ideal things', and he also accepts the possibility that laws, and hence full explanations, may not be stateable at the interpretative or 'intentional' conceptual level at which the social sciences operate.

Certainly the approximate generalizations which we can supply about William the Conqueror's—or anyone else's—character are not laws. William may be a level-headed statesman but not when the sugar level in his blood stream is abnormally low. Then should the fact that the sugar in his blood stream was at a particular level—within the normal band—be mentioned in a complete explanation of his decision not to invade? On normal ideas of what consitutes a *sufficient* explanation, it need not be mentioned. It might be relevant to mention that his sugar level was low if his action was unusual, but it is not relevant to mention that his sugar level was normal if he acted normally, that is, in character.

The concept of a full or complete singular explanation, according to which all nomically relevant antecedents would have to be mentioned, is clearly philosophical.[1] It does not express a criterion applied in ordinary assessments of explanations—what we consider rather is

[1] Incidentally I would not accept, as I think Ruben does, that a law, properly so-called, must specify sufficient conditions for the event-type described in its consequent exhaustively or absolutely. This is unduly restrictive. Many causal generalizations qualify as laws even though they remain explicitly or implicitly relativized to an unspecified causal field, or to an assumption that unspecified exogeneous factors do not interfere. To refuse them the title of laws is too quick a way of concluding that only fundamental and fully universal principles of physics are laws, and only fundamental physical explanations are explanations. On the less restrictive conception of law, there can be social scientific laws. However the proper use of 'law' is not the main point at issue here.

whether an explanation is sufficient for the purpose in hand or whether it requires supplementation, leaves material questions unanswered, etc. (If, in such ordinary contexts, I insist on a 'full' or 'complete' explanation of someone's behaviour, say, I do not mean what either Hempel or Ruben mean). Ruben recognizes this, of course, and he notes that his notion is a 'technical' one (p. 115 above).

So we can properly ask what purpose it serves. Ruben answers that it expresses a constitutive ideal—'it constitutes for us what, upon reflection, we intend by "fully explaining why something happened"' (p. 115 above).

I find this not entirely helpful, since I am not persuaded that there *is* anything that I pretheoretically intend by 'fully explaining why something happened'. However I agree that there is still a worthwhile philosophical end in view. Ordinarily, what constitutes a sufficient explanation is context-dependent. Moreover, whether to give a singular explanation in Ruben's sense at all is context-dependent. An explanation in terms of dispositions or laws may be more appropriate—'What brought about the second world war was Hitler's rashness and megalomania', 'These radiator hoses always burst when the water temperature rises to that level.' It is when we take note of this diversity that we begin to appreciate the philosophical point of introducing a synoptic concept of 'complete' explanation—an account which will perspicuously gather together everything that could count as explanatorily relevant, in this or that context, into a comprehensive and coherent formal scheme. The object of the scheme will be to illuminate the general logic of explanation, not to set a target for explainers.

However, what this yields is a rationale for the *Hempelian* conception of explanation. The scheme includes laws because laws *can* figure in explanations of events—though there are also explanations in which they do not figure. A Hempelian explanation-schema stands to ordinary explaining in somewhat the way that a Quinean eternal sentence stands to ordinary speech.

How then does Ruben's account of explanation fit into this? It seems to me that it falls between two stools. It gravitates towards ordinary notions of explanation in recognizing singular explanations as proper explanations, and distinguishing, in their case, between what forms part of the explanation, and what we need to know to know *that* the explanation is good. On the other hand, it is the Hempelian's formal goals which influence Ruben when he gives his 'ideal' account of a 'full explanation'. But if we are pursuing those formal goals, I see no good reason to stop anywhere short of the full subsumptive-nomological schema of explanation. In relation to those goals, I cannot see that Ruben's technical concept of singular explanation, as against Hempel's technical concept of subsumptive explanation, is the one that becomes plausible 'when we reflect upon the sort of understanding that science poses as a possibility' (p. 115 above).

II

So far I have been considering the important and interesting, but purely general, questions Ruben raises about the relation between concepts of explanation, cause and law. I now turn to the social sciences. And here it seems to me that Ruben is mistaken in suggesting that these general questions are mainly, or even largely, at stake in the philosophical debate about whether explanation in social science is or is not of a kind with explanation in natural science.

According to Ruben, 'difference theorists', as he calls them, are centrally concerned with the question whether explanations in social science 'require laws' (p. 96 above). He suggests that the main point such theorists want to establish is that explanations in social science do not require laws: if they had a less uncritical view of the subsumptive conception of explanation in the natural sciences, they would feel no need to be difference theorists.

I must confess that this strikes me as a somewhat parochial vision of a great and historic debate. Ruben holds that laws do not have to form part of complete explanations. But he does hold that they must underlie explanations. Where there is a complete singular explanation there must be a law. Moreover if we know the explanation to be a complete singular explanation we can read off the law from it, since it will be couched in the appropriate vocabulary and will contain a complete specification of the properties which appear in the antecedent and the consequent of the law.

Now it is certainly true that a central element of the historic debate is the question whether human beings are subject to scientific laws. It is, for example, the question which John Stuart Mill feels himself compelled to consider as a preliminary to the discussion of the moral sciences in his *System of Logic*—a discussion which remains the classical statement of naturalism about the moral sciences. And Mill assumes that in order to still doubts about the very possibility of moral science he must show that human beings can be free even though human phenomena are subsumable under laws. It would be little consolation to those who disagree with Mill's naturalism to be told that while laws do not, strictly speaking, form part of singular explanations they are nevertheless indispensable in underpinning them. That is a correction which could be fitted quite unproblematically into Mill's essential conception.

The real root of the debate is the belief that man is a free being not subject to natural law, and—closely associated with that—the conviction that if feelings and thoughts are entirely subsumable under general laws in a unified theory of nature, human existence is drained of meaning. These beliefs are a part of the romantic reaction to enlightenment. It was Kant, the pivotal figure between enlightenment and romanticism, who set the philosophical terms of the subsequent debate, in his handling of the third antimony, the antimony of freedom, and in his combination of empirical realism with transcendental idealism.

On Kant's view human beings considered as phenomena are subject to natural science; but as noumena, free agents, they are not. From these seeds the German debate about the *Geisteswissenschaften* grew. I hope I may therefore be allowed to review quite briefly the main elements of the Kantian critique of naturalism.

There is, in the first place, an epistemological critique—establishing the need for *a priori* categories if knowledge is to be possible, together with a further step purporting to establish transcendental idealism as a precondition for the possibility of such *a priori* categories. On Kant's argument it seems that a natural scientific understanding of the *a priori* categories themselves, or of the mind which imposes them, is not possible. If the very possibility of science requires *a priori* categories; if *a priori* categories can only be the 'product' of the scientific inquirer, and not of the domain into which he inquires, how can there be a scientific account of the inquirer himself?

Of course Kant does think that there can be a scientific account of the inquirer himself—taking the inquirer as a phenomenon among phenomena. The inquirer's behaviour and internal states are a part of the natural world just as much as are the behaviour and internal states of any other object in that world. Thus they will be entirely subsumable under natural laws. Yet *as* an inquirer he is not a part of the natural world but an exerciser of autonomous reason. The same applies to any other of his actions in so far as they are autonomous—informed by reason—and not heteronomous. Viewed in this perspective—in his intelligible rather than his empirical character—he is not a relatively isolable causal system belonging to a larger causal order but an autonomous agent. The central claim, to borrow a phrase from Hilary Putnam, is that reason cannot be naturalized.

Such is the Kantian background to the hermeneutic view of the moral sciences. It readily translates into the following position. In the perspective of natural science human beings can and should be studied as causal systems subject to law—from their biochemistry to their anatomy and physiology and up to the cognitive science level. But the moral sciences seek to grasp them not in their empirical but in their intelligible character, as beings which recognize and respond to rational norms. In their understandings and explanations they inherently apply the categories of reason and autonomy. Explanations of this kind—interpretative explanations—do not subsume human actions under laws but interpret their meaning. They are not nomothetic and are irreducible to the nomothetic naturalistic level. They have a standing of their own which can only be elucidated by a Diltheyan critique of historical reason.

I need not rehearse the familiar difficulties which dog Kant's attempt to resolve the antinomy of freedom. My object in the rest of this paper is to bring out the important elements of truth contained in these classical lines of thought. Certainly there would be no point in pretending that Kant's transcendental idealism could provide a tenable setting for them; the larger philosophical framework into which they can be solidly set remains to be described. However these larger issues will not be tackled here.

III

It may be said that the questions we are discussing are rather distant from the practical concerns of most social scientists. Social science is mostly concerned with documenting the social world, its character and trends, in conceptually clear and statistically accurate ways. That is difficult, time-consuming—and worthwhile in its own right: not just as a basis for theory but as a powerful instrument for correcting conventional thinking about politics and society. It does not however raise hermeneutic problems in any direct way.

But as soon as we introduce social categories which imply explanation and theory we find ourselves directed towards the analysis of human beings' motives and beliefs and the meaning of their institutions. Obviously the unforseen consequences of human actions as well as their intelligible antecedents, and the latent functions of institutions, as well as their received meanings, are important. However the base level of historical, economic or sociological analysis must be an account of human action which is, in Weber's famous terms, meaningfully as well as causally adequate.

Geisteswissenschaften, it will be remembered, is a German translator's rendering of Mill's 'moral sciences'. 'Moral science' is a useful expression, which I use to cover not just the social sciences proper but history, archaeology and the like, in the way that *Geisteswissenschaften* does. The moral sciences are interpretative; and that means that they are continuous with philosophical reflection in a way in which physical theory is not (there are other continuities in that case). They are, namely, *reflexive*: our interpretations of human action are guided by—and reflect back on to—our understanding of normative canons of thought and feeling by which we ourselves are guided.

That corresponds to the hermeneutic claim that interpretation deals with the 'intelligible' character of actions and institutions. In this section I shall try to illustrate the point of the claim; in section IV I shall turn to the related question, whether interpretative explanations of human phenomena, which treat of them in their intelligible character, are reducible to a non-interpretative level which treats of them—still in Kant's terms—in their 'empirical' character.

Suppose you want to explain why a picture is hung at a particular place on the wall. Why put it there? Well the explanation is quite simple—it balances the wall and looks right. Since the owner of the room is an aesthetic *normalmensch*—just as William the Conqueror was a politico-military *normalmensch*—that explanation is perfectly good. It deploys a common aesthetic judgment: one that the actor and the inquirer share, or more weakly, one that they have a common internal understanding of. The inquirer has insight into the actor's aesthetic sense. Moreover deviation requires special explanation and deviations are determined in relation to a norm set by the inquirer's own norm. If the picture is in an extraordinarily odd position, down in the bottom left-hand corner of the wall, some special explanation is called for. Is it covering a wall safe?

But could it be that such an odd position appeals to the actor's own aesthetic sense? To find this a satisfactory explanation we must, in the first instance, seek insight into that judgment. We want to be able ourselves to grasp its aesthetic appeal. That may be a discovery for us, reflecting back on to and developing our own aesthetic sense—like the 'discovery' of primitive art at the turn of the century.[2] Understanding others means finding a community of response with them, and this may mean the development of new—but organic, not arbitrary—responses in ourselves. Coming to understand them may involve an evolution of our own norms towards theirs. One objective of understanding, as Gadamer and Habermas among others have stressed, is dialogical—or simply, philosophical. We enrich our own responses in dialogue with others.

[2] It would be an illuminating study of the complexities of interpretation to examine whether the aesthetic appeal primitive art has for us is in any way part of what it meant to its producers, the connection of that question, in turn, with the question whether it is straightforwardly 'art', or whether its aestheticization is an artefact of modernist sensibility, and the further question whether this itself imposes on the data a misleadingly modern-European distinction between art and non-art.

But there is also the possibility that we may be utterly unable to find any appeal in the positioning of the picture. Could it nevertheless be there because the actor just thought it looked good there? Could that be the right explanation of his act, even though we can achieve no internal understanding of it? Only within limits. There must be other judgments of his which justify ascribing to him an aesthetic sense, and these must overlap despite the failure in this particular case. Of course it is also possible that there is a defect in the inquirer, which he comes to recognize precisely through his failure to achieve spontaneous insight on judgments and responses on which others converge. The underlying relationship between convergence and objectivity is then what justifies him in concluding that he lacks a faculty which others have.[3]

I would generalize this as follows: in all cases hermeneutic understanding posits the possibility of a common access to underlying norms. The basis of the moral sciences is philosophical anthropology. I do not mean, in saying this, to endorse what Evans-Pritchard stigmatized as the 'if I were a horse' fallacy![4] Of course we must not rashly assume that others have the same beliefs, attitudes, cultural contexts that we do. On the other hand the fact is that we are not misguidedly trying to understand horses (or lions), in the particular sense of 'understanding' which is at stake here: it is other human beings we are trying to understand.

This hermeneutic thesis applies to all moral sciences, not just to anthropology—although of course anthropology is the moral science in which problems of interpretation are most evidently posed. Just for that reason, however, it is hard to provide realistic case studies from anthropology which can be handled briefly enough.[5] So I shall turn to the branch of economics known as 'decision' or 'rational choice' theory, which provides an example in which the issues I am concerned with strike me as being particularly clear.

The dominant position among decisions theorists is the normative view that one should maximize expected utility ('Bayesianism').[6] The corresponding descriptive theory is that that is what people actually do. Empirical tests however, as well as introspection, show a series of cases in which they do not do that. Consider in particular the case given by

[3] I consider the issue in 'Objectivity and convergence', *Proceedings of the Aristotelian Society* **84** (1986), 235–50.

[4] E. E. Evans-Pritchard *Theories of Primitive Religion* (Oxford: Clarendon Press, 1965), e.g. pp. 24, 43.

[5] I have discussed the problems of interpreting primitive religion and magic in *Symbol and Theory. A Philosophical Study of Theories of Religion in Social Anthropology* (Cambridge University Press, 1976).

[6] The expected utility of an action is calculated by discounting the utility of each of its possible outcomes by the probability of that outcome's occurring if the action is done, and summing the results.

Table 1.

	s_1 probability of s_1: 0.01	s_2 probability of s_2: 0.10	s_3 probability of s_3: 0.89
a_1	£500,000	$500,000	$500,000
a_2	$0	$2,500,000	$500,000
a_3	£500,000	$500,000	$0
a_4	$0	$2,500,000	$0

Maurice Allais in Table 1 (all the figures in the table should by now have a nought or two added!).[7] The table presents two possible situations of choice: between the pair of actions a_1 and a_2, and between the pair of actions a_3 and a_4. The monetary value, to the decision maker, of the various possible outcomes is represented in dollars; and the outcome depends on the action he chooses, and on whether the state of the world in which he is choosing turns out to be s_1, s_2, or s_3. The probabilities of these respective states, as assessed by him, are shown.

Most people, given the choice between a_1 and a_2, prefer a_1 because they prefer the certainty of $500,000. On the other hand, given the choice between a_3 and a_4, most people prefer a_4, because both a_3 and a_4 involve a gamble which may produce no return, but in a_4 there is some chance of winning $2,500,000, while in a_3 there is only a very slightly higher chance of winning $500,000. Yet this pair of choices (call it 'the Allais response') is inconsistent with maximization of expected utility, and more specifically, with one of the axioms in L. J. Savage's formalization of Bayesian theory, the 'sure-thing' principle'. It is inconsistent with maximizing expected utility, because if the expected utility of a_1 is greater for the agent than the expected utility of a_2, then the expected utility of a_4 cannot be greater than the expected utility of a_3. It is inconsistent with the sure-thing principle—which holds that the choice between two actions cannot be influenced by consideration of states of the world in which both would have the same payoff—because the pair

[7] Originally in Maurice Allais, 'Le comportement de l'homme rationnel devant le risque: critique des postulats et axiomes de l'école américaine', *Econometrica* **21** (1953), 503–46. See also Allais, 'The so-called Allais Paradox and Rational Decision under Uncertainty' in M. Allais and O. Hagen (eds), *Expected Utility Hypotheses and the Allais Paradox* (Dordrect: Reidel, 1979) (the collection in which this appears also contains an English translation of the full-length memoir which Allais 1953 presents in summary). There is a useful collection of papers, together with a survey of the issues, in Peter Gärdenfors and Nils-Eric Sahlin (eds.) *Decision, Probability and Utility* (Cambridge University Press, 1988.)

a_1, a_2 is differentiated from the pair a_3, a_4 only in the third column, where, as can be seen, payoffs are equal within each pair.

In an example as pared down to essentials as this, the relationship between the normative and the normal is very perspicuous. Certainly one thing is clear: the Bayesian cannot maintain his claim that maximizing expected utility is the rational thing to do while simultaneously accepting that the Allais response is the normal, fully considered one—the one on which informed and reflective judges would converge.

I am not saying that such a position would be strictly self-contradictory—the rational is not *defined* in terms of the normal, fully considered response. After all the response in question is a response as to what *is* rational. The concept 'rational choice' figures within its content. In asking ourselves what the rational thing to do would be we are asking ourselves *that* question—we are not asking what other people's answer to that question would be. Nor can 'rational' be operationally reduced to 'fully-considered normal'—for the usual reasons which bar operational 'definitions'. Both the term 'normal' and the term 'fully considered' express, as used here, infinitistic or undecidable notions. No finite set of steps enables me, as an individual inquirer, to guarantee that my response is *fully* considered, because I can never rule out the possibility that further thought or new considerations would bring me to a new point of view. *A fortiori*, one can never definitely establish that a particular empirical consensus expresses a fully-considered normal response—there is the further possibility that enlarging the group of individuals among whom consensus is reached might produce a different equilibrium view. In the search for the rational there can be no pre-determined limit to our voyage of discovery.

But though the position considered in the last two paragraphs would not be strictly self-contradictory, it would of course be quite ineffectual and void. What the Bayesian must try rather to do is to explain the Allais response in a way which shows that it contains some element of irrationality—an error of reasoning, or something not fully thought through. That is certainly possible—since people do act irrationally—but the theorist who takes this line must give it plausibility; so he must provide an *internal* insight into the illusion or misconception or trick of perspective which produces it. Savage attempts to do exactly that in his discussion of the case.[8] He tells us that his own first reaction conformed to the Allais response, and then he goes on to describe a way of presenting the example which caused him, on reflection, to reject that first reaction as wrong.

[8] L. J. Savage, *The Foundations of Statistics* (New York: John Wiley & Sons 1954), 101–3.

Table 2

		Ticket number		
		1	2–11	12–100
Situation 1	a_1	5	5	5
	a_2	0	25	5
Situation 2	a_3	5	5	0
	a_4	0	25	0

The two situations described in the table could also be realized, he points out, 'by a lottery with a hundred numbered tickets and with prizes' as in Table 2. Here each action is the purchase of a ticket in the particular lottery specified on the corresponding row, the prizes being stated in units of $100,000. Savage now argues thus:

> if one of the tickets numbered from 12 through 100 is drawn, it will not matter, in either situation, which gamble I choose. I therefore focus on the possibility that one of the tickets numbered from 1 through 11 will be drawn, in which case Situations 1 and 2 are exactly parallel. The subsidiary decision depends in both situations on whether I would sell an outright gift of $500,000 for a 10-to-1 chance to win $2,500,000—a conclusion that I think *has a claim to universality, or objectivity*. Finally, consulting my purely personal taste, I find that I would prefer the gift of $500,000 and, accordingly, that I prefer [action] 1 to [action] 2 and (contrary to my initial reaction) [action] 3 to [action] 4 (p. 103, my italics).

Savage's strategy is exactly right. He searches for a perspective which 'has a claim to universality, or objectivity', and from which the reversal of his original preference for action 4 over action 3 seems, as he says, the correction of an error. The interpretation in this case is self-interpretation, but the principle is the same: interpretation necessarily appeals to universal and objective norms. The agent, as well as the observer, must be assumed to respond to those norms, but he may be responding to them from a clouded or misleading perspective and they may fail to influence his action as they should. Savage instances an uncontroversial example of such perspectival error:

> A man buying a car for $2,134.56 is tempted to order it with a radio installed, which will bring the total price to $2,228.41, feeling that the difference is trifling. But, when he reflects that, if he already had the car, he certainly would not spend $93.85 for a radio for it, he realizes that he has made an error (p. 103).

Now although Savage's strategy is exactly right, that does not mean that it is successful. The proper conclusion from the Allais response, and from analysis of other risky decisions, may be that Bayesianism is not an adequate characterization of the underlying norms which for us constitute practical rationality. In fact that seems to me to be so; I believe that rationality in decision-making can only be captured by a more complex and vaguer rule. By this more complex rule one maximizes expected utility subject to a threshold-constraint which rules out actions with possible outcomes falling below it in value. (The levels at which thresholds are set are contextually determined by, among other things, one's wealth and certain benchmark levels of well-being or personal capacity.)[9] But this is not the place to pursue that particular disagreement; our concern here is only to observe how the search for a correct account weaves together, in a dialogue, the strands of interpretation and normative deliberation.

Again there is a further possibility to consider: systematic and final disagreement with no possibility of convergence. May there not be more than one rule for rational decision-making under risk, each with its own champions, who can find no effective Savage-style strategy for convincing others? These rules might include, for example, Maximize expected utility; Maximize subject to threshold constraints; 'Leximin'.[10]

At this point we are beginning to broach issues which ramify well beyond the scope of this paper. What kind of situation are we envisaging? One possibility is that every party in the dispute acquires every other party's normative insight. All come to see for themselves the equal tenability of all three rules. Many rules continue to seem irrational; but these three now seem, to all, to make equal claims of 'universality, or objectivity'. All parties could now agree that the choice within the shortlist of three is rationally indeterminate. They would all have enlarged their own normative stance, and they would be able to interpret someone else's action by reference to a principle which they might not themselves adopt for purposes of action, but which they could share, in the sense of grasping its equal normative claim.

At the other extreme, the parties might achieve no insight into each others' norms. They might be unable to see those other norms as rational options, or even to identify quite what they actually were. This

[9] I discuss this in a more general context in 'Value and Distribution' in M. Hollis and W. Vossenkühl (eds.), *Moralische Entscheidung und rationale Wahl* (forthcoming).

[10] In the context of decision-making under risk, 'leximin' will hold that one should choose the action whose worst possible outcome is better than the worst possible outcomes of other actions; where there is more than one such action, one prefers that action within the set whose second worst outcome is better than the second worst outcome of the other actions in the set . . . and so on.

is relatively less likely, but not impossible, in the fairly humdrum case we are considering—but it becomes much more likely when it is a matter of understanding the thought behind, say, the recipes of an ancient Egyptian magician. Here interpretations are likely to vary wildly, and not all of them will be strictly *interpretations*, in the sense of identifying some explanatory pattern of thought which we ourselves can begin to understand from within. Or if we understand them, it may be that we understand them only partially, and only through painfully recovering in ourselves dispositions of thought covered over by layers of culture which have been laid down by equally spontaneous, but conflicting, and eventually more successful, dispositions.

Consider a race of Riemannians: they take it as obvious that all straight lines must eventually intersect twice, finding it quite impossible to imagine how they could fail to do so; they can make no sense at all of an actual spatial infinity. They have severe difficulty in understanding our spatial notions from within—perhaps they try to explain them as very natural confusions, growing out of over-hasty generalization from observed data about surveyable short straight edges and the angles they form. (Somewhat along the lines of explanations of primitive magic and religion as over-hasty generalisation from this or that obervable order.) The Riemannians' geometry engages with the more successful physical theory and we find ourselves having to adapt to theirs. But for us it is the Euclidean concepts which accord with primitive geometrical intuition—even if they produce difficulties about actual infinity; and in their interpretation of us the Riemannians have not realized that. Perhaps their own primitive Euclidean intuition has been overlaid by layers of culture; or perhaps they are innate Riemannians who never had that primitive intuition.

Was it irrational on our part to take Euclidean principle as default assumptions in our inquiries? It was not. Would it have been rational for *innate* Riemannians to do so? It would not. As between innate Riemannians and ourselves the core of rationality differs. That places a certain limit on our capacity to interpret each other, though only in a quite weak sense. Our understanding of each other does not go so far as to converge on norms to which we both *spontaneously* (i.e. not through 'domestication') respond. Nevertheless we are able to characterize each other's norms, even though we cannot experience them in the specific way that is needed to give them epistemically authoritative 'self-evidence'. The limit on understanding from within corresponds to a divergence in our primitive norms. But on the spectrum from complete internal insight to total bafflement we are close enough to the pole of insight for dialogue to progress without difficulty; even though we travel from different starting points.

IV

The reflexiveness of interpretation is one thing, its irreducibility to naturalistic terms is another. The second element in the Kantian legacy is the claim that interpretative explanations are not in principle reducible: specifically, not reducible because the *a priori* category under which they fall is not that of causation but that of rational autonomy.

Against this I agree with David Ruben that interpretative explanations are causal. In Max Weber's terms, which we have already noted, they have to be causally as well as meaningfully adequate. Else how do we choose, among a plurality of meaningfully adequate explanations, the one that is right?

On the other hand, to recognize that they are causal is not yet to hold that they are nomothetic[11] or that they are after all reducible to naturalistic terms. No one has supplied a convincingly worked-out naturalistic reduction of the concept *acting for a reason*, nor, on the other hand, do we have a clear grasp of how the naturalistic stance would be compatible with its *ir*reducibility. The difficulty involved is a very general one, of which the difficulty of finding a perspicuous philosophical understanding of the nature of rule-following is but a special case.

In thinking of ourselves as agents and reasoners, we think of ourselves as autonomous followers of objectively given norms. It is here that a clash arises with our view of ourselves as causal systems in a larger causal order. Inference is a causal process but, as we have seen in section III, it also seems something more than, or incommensurable with, a causal process. It seems to involve the acausal recognition of a rule of reason. Precisely the same can be said for the relation between motive or deliberation on the one hand, and action or choice on the other.

When I act because I have a belief and an objective which give me reason to act, my action is caused by my having that belief and that objective. If it is not, then I did not act for that reason. Yet those mental states could cause that self-same behaviour, without the behaviour being a case of intentional action at all. To take an example of the kind made familiar by Donald Davidson:[12] suppose I dislike you so much

[11] Causal but not nomothetic: I am not envisaging a form of mental causation which breaks the link between causality and nomic uniformity. The point is simply that it remains unobvious that, if one accepts that the explaining relation between beliefs/purposes and actions is causal, one must also accept that there are uniformities—stateable at that, interpretative, level—which are substantive and sharp enough to warrant the name of laws.

[12] Donald Davidson, 'Freedom to Act', in Ted Honderich (ed.), *Essays on Freedom of Action* (London: Routledge & Kegan Paul 1978). The mention of Achilles and the tortoise in the next paragraph refers to Lewis Carroll, 'What the Tortoise Said to Achilles', *Mind* **4** (1985), 278–80.

that, seeing you on a pedestrian crossing, I have an urge to run you over. It occurs to me that all I have to do is leave my foot on the accelerator. I am so distracted by the attractiveness of doing that that I fail to take my foot off the accelerator, with the result that I run you over—quite unintentionally.

I do not act, for, or on the basis of, a reason, even though my behaviour is caused by mental states which do indeed provide me with a reason for just that behaviour. The analogous point holds in the case of inference, as we can learn from Achilles and the tortoise. One may be caused to believe that Q by the beliefs that P and that $P \rightarrow Q$, without that causal process constituting an inference. Nor will adding the belief that (P and $P \rightarrow Q$) entails Q into the causal antecedents guarantee that the process constitutes an inference: that the conclusion was *drawn on the basis of a reason*. Nor does adding in indefinitely many further premises as causally relevant antecedent beliefs. So too in the case of analysing intentional action: adding further events or states into the causal chain—such as 'volitions'—will not help.

A naturalistic *analysis* of inference or action would have to supply sets of logically necessary and sufficient conditions. It may be unreasonable to expect that they could be couched in a directly causal-physicalistic terms; but naturalism also has the resources of the functional or 'teleological' level of description, and it is less obvious that an analysis of inference and action in those terms is impossible.

Yet what still seems to make it impossible, even with those resources taken into account, is the difficulty we have just touched on: that acting for a reason or drawing an inference are, precisely, responses to reason. To interpret an entity as an intelligible agent is to ascribe to it a disposition to do what there is reason to do: a responsiveness, one wants to say, to the 'reason function'—*(o, p) gives reason to do z*—which takes suitable pairs of objectives and propositions into recommendations. If an intelligible agent endorses o and believes that p, it is disposed to do z. However it does not just act in line with that function, it does so *because* it recognizes and responds to it. Responsiveness to the function seems to enter into the *explanans* of the action—but not as an event or state.

I see no way round this difficulty—if our search is for an analysis which provides naturalistic truth conditions. We could nevertheless hope to state, in naturalistic terms, the circumstances in which we are ready to apply hermeneutic categories: in which we are ready to treat a natural object as a reasoner and a free agent. We are ready to do so when behaviour and dispositions to behaviour sustain a hermeneutic representation; that is, when they fit the idea of an agent whose actions and beliefs can be represented as sensitive to what we recognize as good reasons for acting and believing. If reduction is impossible, that representation will never be strictly entailed, because it is conceptually richer—just as talk about common *recognition* of a norm of reason is richer than talk about idealized convergence; but it could still be warranted *a priori*. The relationship would be like that between talk of constant conjunctions and talk of causes.

There is a great deal more to be said about these issues—I do not pretend to have done more than scratch the surface. But I hope to have said enough to sustain the conviction which I think really fuels the claims of hermeneutics. The conviction is that moral science has its own *a priori* categories, grounded on humanly intelligible norms of understandable feeling and rational response; and its own distinctive goals, of which a most fundamental one is that of representing common reason in a way which is concrete but unconditioned by particular prejudice. It pursues that goal by searching for the underlying unity of common reason across the whole diversity of material conditions and social forms in which common reason is historically refracted.

Explanation*

MICHAEL REDHEAD

1. Introduction

In what sense do the sciences explain? Or do they merely describe what is going on without answering why-questions at all. But cannot description at an appropriate 'level' provide all that we can reasonably ask of an explanation? Well, what do we mean by explanation anyway? What, if anything, gets left out when we provide a so-called scientific explanation? Are there limits of explanation in general, and scientific explanation, in particular? What are the criteria for a *good* explanation? Is it possible to satisfy all the desiderata simultaneously? If not, which should we regard as paramount? What is the connection between explanation and prediction? What exactly is it that statistical explanations explain? These are some of the questions that have generated a very extensive literature in the philosophy of science. In attempting to answer them, definite views will have to be taken on related matters, such as physical laws, causality, reduction, and questions of evidence and confirmation, of theory and observation, realism versus anti-realism, and the objectivity and rationality of science. I will state my own views on these matters, in the course of this essay. To argue for everything in detail and to do justice to all the alternative views, would fill a book, perhaps several books. I want to lead up fairly quickly to modern physics, and review the explanatory situation there in rather more detail.

2. Why-questions: the D-N model as providing a necessary condition for explanation

'Why' is ambiguous.[1] If someone asserts that the earth is round, and I ask 'Why', I may be asking for a reason why you believe that the earth is round. What is your evidence for the assertion? Typically you may refer to *consequences* of the earth being round, horizon phenomena, the shape of the shadow of the earth on the face of the moon during an

* I acknowledge the warm hospitality of the Wharfedale farmhouse where this paper was written.

[1] Cf. Hospers (1967), 240.

eclipse, and so on. The evidence you cite supports or confirms the hypothesis that the earth is round, justifies you in believing that the earth is round. But I may also be asking for an explanation: why *is* the earth round? Is it just a brute fact or can you *explain* it, in a way which confers understanding and removes my perplexity in the face of just reciting the brute fact? Typically you will cite the roundess of the earth as itself a *consequence* of other propositions in a manner which contributes to my understanding. This is the famous deductive model of explanation.[2] Is deducibility either sufficient or necessary for a scientific explanation? Well, it is obviously not sufficient. Most trivially X is deducible from X, and we can hardly allow X to explain itself. But I am going to claim that it is necessary. This is to fly in the face of most of the post-Hempelian literature on explanation.[3] There are two sorts of counter-examples to necessity commonly cited. First, volcanoes—we cannot deduce when they will erupt, but, so it is said, we can explain an eruption *post hoc* in terms of known geological process. But that is to confuse an explanation-sketch with a full-blooded scientific explanation.[4] If we knew enough about the distribution of stress in the rocks and the laws governing mechanical rupture we could make the prediction. If we do not possess all the relevant information then we are not in a position to give a scientific explanation, that is full or complete. That is not to say that partial explanations may not confer a measure of understanding, but they do not measure up to the scientific ideal involving strict deducibility, i.e., in the volcano example, predictability.

There is an important ambiguity here we must be clear about. Many physical systems are governed by deterministic laws in the sense that *exact* specification of initial and boundary conditions fixes the later physical state uniquely, but the prediction is unstable in the sense that any error, *however small*, introduces divergent behaviour in specifying the future state of the system. So, in practice, and even in principle, *we* cannot compute the prediction. But from a 'God's eye' point of view everything is fixed. In this paper we are not concerning ourselves primarily with the pragmatics of explanation, and hence when we talk of predictability, we mean this in the ontological 'God's eye' sense.

But some events cannot be predicted. Does this mean they cannot be explained? I think the appropriate answer to this question is to bite the

[2] The classic statement of the D-N model of explanation is Hempel and Oppenheim (1948)—reprinted in Hempel (1965).

[3] Influential critiques of the Hempelian approach include Scriven (1962) and Salmon (1971). A comprehensive survey of the literature on explanation is provided by Achinstein (1983).

[4] Cf. Hempel (1965), 416.

bullet and say 'yes'. So what do probabilisitic or statistical explanations achieve? Well, they enable us to deduce and hence to explain the limiting relative frequencies with which events of a given kind turn up in a long-run repetition of the set-up producing the phenomenon. But they cannot explain what happens on a particular occasion. It is as simple as that!

In particular they cannot explain in the strict deductivist sense the relative proportions in a *finite* sample. Suffice it to say that these proportions can be used to provide a rational estimation of what can be explained, i.e. the probabilities, in accordance with the usual statistical procedures.

There is an enormous literature on statistical explanation, arguing whether the probability of the explanandum has to be high in the presence of the explanans[5] or merely higher than it would be in the absence of the explanans.[6] In the strict *scientific* sense, I regard these discussions as irrelevant. Again I am not denying that there are senses of promoting understanding, or removing perplexity, other than that produced by the scientific ideal. But it is with the scientific ideal that we are concerned.

But if deducibility is admitted to be necessary for scientific explanation, how are we to fill out the conditions to achieve sufficiency. That is a much more difficult task, to which I now turn.

3. The Circularity Objection

In the deductive-nomological (D-N) model of Hempel the explanans cites one or more scientific laws. In the usual schematic fashion adopted by philosophers of science, let us represent a typical scientific law in the universally quantified form $\forall x(Px \rightarrow Qx)$—succinctly all Ps are Qs.

If *a* is a P, i.e. Pa is true, then we seek to explain why *a* is a Q by deducing Qa from the premisses

$$\forall x(Px \rightarrow Qx) \quad (1)$$
$$Pa \quad (2)$$

Thus: from (1) by Universal Instantiation

$$Pa \rightarrow Qa \quad (3)$$

Whence, from (2) and (3) by *modus ponens*

$$Qa$$

Why is this thought to be explanatory? In (1) the implication as we have written it is material implication. On a Humean (regularity) view of

[5] See Hempel (1965), 376ff.
[6] Cf. Salmon (1971).

laws that is all there is to (1).[7] It is true in virtue of all its instances being true. But if (1) depends for its truth on the truth of (3), and this, given the premiss Pa, must turn on the truth Qa. So is not the argument completely circular? The truth of Qa, given Pa, is grounded in the truth of a universal stratement, whose truth is grounded in the truth of Qa, the very fact we are trying to explain. What this amounts to is that (1) is nothing more nor less, on the Humean account, than a compendium of all the instances (3). (In the case of a finite variety of instances the universal law is indeed nothing else than the conjunction of its instances.) On the Humean account the instances are 'loose' (there is no cement!) so effectively the Hempelian model, under this interpretation of law, amounts to the assertion that facts only explain themselves.

But this whole argument hinges on the assumption that the universal law is only supported or confirmed by the totality of evidence which would make the law (deductively) true! If we think that (1) may be supported in some *inductive* sense by instances other than (3) then the circularity would be satisfactorily mitigated. But here we are backing ourselves straight into the problem of induction. That induction is not *deductively* valid as a mode of inference, committing as it does the fallacy of affirming the consequent, must of course be admitted. So in our example, if we are to avoid circularity, we must acknowledge that the explanans is never known definitely to be true, and this is an obvious, but unavoidable, defect in scientific explanations. For a Popperian, of course, the matter is much worse than that. Finite evidence never supports to any degree the conjectural truth of universal laws quantified over infinite domains. So a Popperian expects, in so far as he allows himself any expectations, that an essential part of the explanans, in any scientific explanation, is definitely false (although not currently known to be false!). Does this render scientific explanations irrational?—a conclusion much advertised by the critics of Popperism. The answer depends very much on what we count as rational. Is it rational to aim at the impossible? Can we not rationally *accept* scientific laws, without believing them to be true, provided they have been subject to the severest available *criticism*?[8]

Let us turn from these deep methodological issues to consider whether a necessitarian account of natural laws helps with the question of explanation. On the question of evidence not at all. After all, that was

[7] The Humean view of laws is most recently defended in Swartz (1985). More sophisticated Humeans of the Ramsey-Lewis stripe incorporates a 'more for less' criterion for natural laws that links closely with our later discussion of Friedman's views on explanation and unification. A good exposition of the Ramsey-Lewis approach is provided in Armstrong (1983), chapter 5.

[8] For a sustained defence of neo-Popperian rationality see Watkins (1984).

Hume's original point. We have no epistemological access to the idea of necessary connection. But let us take an ontological standpoint. If laws of nature involve nomological necessity, however analysed, over and above the merely factual material implication, would this not account for how Pa being the case is the ground for Qa being the case? The ground for Qa being grounded in Pa is the necessary connection between P'ness and Q'ness on one popular understanding of these matters.[9] Of course, we can press on to query the ground for *that* ground. We shall return to the question of ultimate or self-supporting explanations presently. But there are other matters we want to deal with.

Firstly, superfluous aspects of explanation: we deduced Qa from (1) and (2), but we could also deduce it from (1), (2) and

God's in his Heaven (4)

We must check that (4) is idle, that we could equally deduce Qa from (1), (2) and the negation of (4).

Then there is the question of explanations which could be correct but as a matter of fact are not. Example: Fred who is shot through the head, after he has died as a result of being stabbed through the heart! The shot could have been the correct explanation of his death, but in deciding whether it actually is, we need to attend to all the relevant circumstances. The law linking shooting through the head with death is more accurately rendered as linking the shot with a transition from life to death. As Fred is already dead, this more fully amplified version of the law is no longer applicable and cannot be cited in explaining Fred's death.[10] Again the scientific ideal assumes that all the relevant circumstances are being cited.

So far we have given some indication of what constitutes an explanation in science, but what constitutes a *good* explanation, when is one explanation better than another?

4. Good Explanations: Unification

There is first of all the element of suprise, of unexpectedness, the 'Aha' factor.[11] That is no doubt related to the criterion of non-circularity, that the explanandum in no way presents itself as what we take to be the

[9] Cf. Armstrong (1983).

[10] For a critical discussion of whether *ceteris paribus* clauses and provisos can in principle or in practice be spelled out in the required detail see Grünbaum and Salmon (1988).

[11] See Feigl (1949).

evidence for the explanans. In practice good explanations, by this dimension of appraisal, arise at the intersection of several universal laws, all of which are necessary to deduce the explanandum.[12] Iron sinks in water, not just because all solids with density greater than that of water sink in water, or even with density greater than that of liquid in which it is placed, but better, because the difference in density is associated via Archimedes Principle with an imbalance of weight versus buoyancy, which shows, from the laws of mechanics, the direction in which the iron will move. The sinking of the iron is not just a special case of a more inclusive law, and so would not be cited as direct evidence for any of the laws stated separately.

But there is another important aspect that this example brings out. What we described as the better explanation involves a series of laws, which can be used to deduce and hence explain many other facts in the field of hydrostatics or with appropriate extension of the theoretical apparatus, the enormous richness of hydrodynamic phenomena from waterfalls to the breaking of waves on the seashore, from whirlpools to the lift and drag of an aerofoil. This points to the very important unification aspect of explanation. The world at the surface level of immediate experience appears very complicated, very rich in diverse phenomena with no apparent connection. But at a 'deeper' theoretical level can all this diversity get reduced to a few interlocking explanatory principles? This has always provided an ideal of theoretical progress in science, the ideal of unification. There is no doubt that the history of modern physics has provided examples of increasing unification in our fundamental theories. But it is important to be clear as to what is being claimed. There are a number of distinct senses of unification that need to be distinguished.[13]

Firstly, there is the question of the inter-relatedness, or 'working together' of the explanatory nexus. Supose we have two sorts of phenomena, P_1 and P_2, which stand for the sets of lawlike regularity governing the phenomena (at the immediate 'empirical' level of everyday physicalist discourse) and suppose that P_1 and P_2 are explained by theories T_1 and T_2. Then $P_1 \cup P_2$ is certainly explained by $T_1 \wedge T_2$ since any member p_1 of P_1 can be deduced from T_1 and any member p_2 of P_2 can be deduced from T_2. In a trivial sense there are new predictions that can be deduced from $T_1 \wedge T_2$ but not from either theory separately, viz. conjunctions like $p_1 \wedge p_2$. But there are no *interesting* novel predictions that arise from $T_1 \wedge T_2$. In one sense of unification, a unified explanation of $P_1 \cup P_2$ would arise from a theory T in which there was no partition of the axioms which separately yielded all the (interesting)

[12] See Nagel (1961), 34ff.
[13] Cf. Redhead (1984).

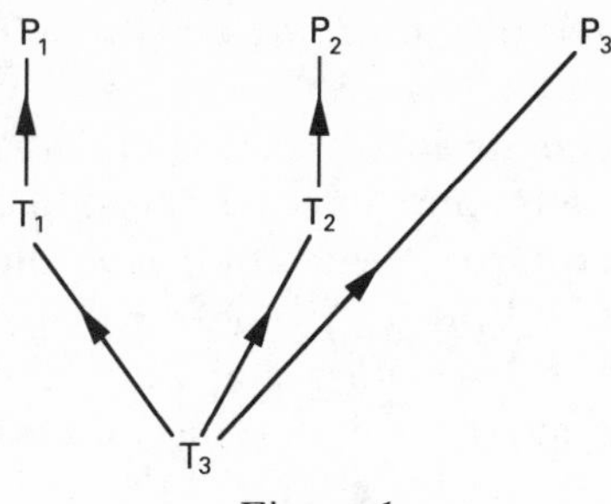

Figure 1

predictions. There would be predictions that required the interlocking working together of the axioms. Typically a unified explanation T of $P_1 \cup P_2$ would also predict new phenomena arising from the interactive effect of the axioms comprising T. Such an explanation would not only be unified but could be in a sense 'deeper' than T_1 and T_2. Suppose we denote the new phenomena by P_3, then we can illustrate the situation we have in mind very schematically as shown in Figure 1. T_3 not only explains T_1 and T_2 that originally accounted for P_1 and P_2, but makes new prediction P_3 and this is done in such a way that T_3 provides a unified account of $P_1 \cup P_2 \cup P_3$. There are many complications associated with working out this idea of unity-cum-depth. These are admirably treated in Watkins' 1984-monograph *Science and Scepticism*. Firstly T_3 may correct T_1 and T_2, not just unify them, and then the idea of increased empirical content becomes formally problematic. Watkins deals with this by his method of counterparts in which incompatible statements are 'matched up' according to appropriate rules. Then there is the problem that a theory such as T_3 which may be unified under one axiomatization may become non-unified under an alternative axiomatization. To deal with this problem Watkins resorts to a notion of 'natural axiomatization', satisfying rules that prevent unnecessary proliferation of axioms and defeat proposed unification-defeating reaxiomatizations. In addition to independence and non-redundancy requirements, Watkins has a rule demanding segregation of axioms containing only theoretical terms, but more importantly there is a non-molecularity requirement, stating that an axiom is impermissible if it contains a proper component which is a theorem of the axiom set, or becomes one when its variables are bound by the quantifiers that bind them in the axiom. Finally there is a decomposition requirement specifying that if the axiom set can be replaced by an equivalent one that is more numerous, without violating the other rules, it should be.

I want to use Watkins' idea of a natural axiomatization to solve a vexing problem that now arises. Why does T_3 remove our perplexity about $P_1 \cup P_2 \cup P_3$? Part of the answer lies in showing that these apparently unconnected phenomena are in fact related via the unified derivation from T_3 (under a natural axiomatization). But we also want to say that T_3 is in some intuitive sense simpler than $T_1 \wedge T_2$ together with P_3 itself. An obvious approach here is just to count the number of axioms $N_A(T_3)$ in a natural axiomatization of T_3 and check whether this is less than $N_A(T_1)+N_A(T_2)+N_A(P_3)$. (Note that Watkins assumes the underlying logic and mathematics is already 'given', so we are concerned with a *finite* number of non-logical and non-mathematical axioms. These may be regarded as part of the specification of a class of models in the Sneed-Stegmüller structuralist approach to theories, although Watkins himself seems to subscribe to the standard 'statement' view of theories'.) What do we mean by $N_A(P_3)$? Well it is the number of axioms in a natural axiomatization of the phenomena comprised in P_3. This in effect means counting the number of laws in P_3. But we must be careful that the laws are expressed in a manner allowed by Watkins' rules. For example, as Friedman (1974) noted, any statement Q can be expressed as the conjunction of N statements, for any N. Thus write

$$
\begin{aligned}
Q &\equiv P_1 \wedge (P_1 \rightarrow Q) \\
&\equiv P_2 \wedge (P_2 \rightarrow P_1) \wedge (P_1 \rightarrow Q) \\
&\rule{10cm}{0.4pt} \\
&\equiv P_{n-1} \wedge (P_{n-1} \rightarrow P_{n-2}) \wedge (P_{n-2} \rightarrow P_{n-3}) \ldots \wedge (P_1 \rightarrow Q)
\end{aligned}
$$

Where $P_1, P_2 \ldots P_{n-1}$ is a descending chain of increasingly weak logical consequences of the statement Q. Such rewritings for law statements Q would be eliminated immediately by Watkins' rules against inessential proliferation of axioms.

We can also compare $N_A(T_1)$ with $N_A(P_1)$ and $N_A(T_2)$ with $N_A(P_2)$ to see how a reduction in the number of laws left unexplained may already have been achieved at the level of introducing T_1 and T_2, and compare with the further reduction effected by T_3.

On this account the reduction in our perplexity in the face of P_1, P_2, P_3 corresponds just to the reduction to $N_A(T_3)$ of those laws of nature which we have to accept without explanation. This was the intention behind Friedman's (1974) approach to explanation, but the account he gave was technically quite incoherent, as shown by Kitcher (1976). Kitcher himself rejected the simple idea of counting laws in an explanatory framework in favour of (effectively) counting what he calls patterns of explanation (see Kitcher (1981) for details). I believe myself that the original Friedman approach is much more straightforward and perspicuous, if it can be rescued from the Kitcherian strictures by employing the Watkins natural axiomatization approach in the way I have described.

There is one important matter we must attend to. We have spoken of laws of nature comprised in the axioms of T_3 explaining laws of nature comprised in the phenomena P_1, P_2 and P_3. But we must be clear as to what we are going to allow as a law of nature. It expresses a regularity, that all items of a certain kind, the subject S of the law, subject to certain conditions C, behave according to some property P. The problem in counting laws is concerned with the question of how general S needs to be and how specific C. If we allow for the maximum generality in S and the minimum specificity in C we could easily rule out electromagnetic laws, for example, other than: all charged bodies obey Maxwell's equations! Achinstein in his (1971) argues that we may refer to a universal generalization as a law if at some period in the history of science it is not known (or believed) that generalization is possible, to a more inclusive law. But this approach introduces what many philosophers regard as an unacceptable historical relativism in place of an objective criterion. Perhaps we should just accept that all generalizations that can be deduced as theorems in an axiomatic-deductive framework should count as laws. The question at issue is also related to the vexed notion of natural kinds, that the extension of the subject-predicate in a law should not be the result of some arbitrary and conventionally imposed system of classification, that the subject-predicate should correspond to a 'genuine' universal. At all events, while there may be borderline cases, such as the much-discussed example of Kepler's laws, where the decision in these matters is somewhat controversial, I believe that sufficient liberality in identifying generalization as laws will permit the notion of unification to be a useful and important one in discussions of the scientific explanation of laws. The essential point is that unification delimits what we must accept without explanation.

In many discussions of explanation the point is made that the explanans must comprise some distinguished set of explanatory principles. On some accounts[14] the explanans must have an analogy with phenomena with which we are familiar in everyday experience, that perplexity in the face of the explanandum is removed by exploiting our familiarity with the analogy. Crudely we may require the explanans to be in some sense picturable. This requirement, while clearly exemplified in examples such as the nineteenth century mechanical aether models, is not characteristic of explanations in modern theoretical physics. While pictures, such as the Feynman diagrams in elementary particle physics, are an aid to keeping track of complicated computational procedures, they are actually potentially very mislead-

[14] See Campbell (1920) and Hesse (1966).

ing in affording what we may term physical understanding of what is going on.[15]

On other accounts the privileged set of explanatory principles is relativized to what in a particular theoretical-conceptual scheme are regarded as 'natural', as not requiring explanation. We shall return to this requirement in a moment in discussing the relevance of causality to questions of explanation. But for the moment, we would merely note that the superiority of the unification approach of Friedman is that an arguably objective criterion can be presented that characterizes progress in the explanatory endeavour in science.

One of the most favoured approaches to unification in modern physics has been through the process of micro-reduction. This raises a number of issues which we now turn to.

5. Micro-reduction

The micro-reduction programme sees the microscopic world as composed of unobservable 'atomic' units whose properties serve to explain in a unified way the whole range of macroscopic phenomena. Unification now incorporates the *additional* requirement that the explanatory principles only cite properties of the microscopic entities involved in the explanans.

The question of the terminus of explanation now presents itself as the alternative between ultimate atoms and an infinite regress of 'Chinese boxes', in which micro-entities are themselves resolved into micro-micro constituents *ad infinitum.* Molecules are resolved into atoms, atoms into electrons and atomic nuclei, nuclei into protons and neutrons, nucleons into quarks . . . One must not think that micro-reduction necessarily eliminates the autonomous branches of macroscopic physics, in favour of a physics of elementary particles. The identity statements characteristic of scientific reductions may best be understood as contingent, but law-like coextensionalities. One can imagine worlds where hot bodies exist, but temperature is not a measure of mean energy of the constituent molecules. That may not be our world, but it is a possible world, whose very conceivability shows that the identity statements in reduction schemes are not analytic in character.[16]

While it is true that the major trend in modern theoretical physics has been micro-reductive, there are some caveats that must be entered.

(1) On the orthodox Copenhagen interpretation of quantum mechanics, the macroscopic world of classically described apparatus and

[15] See Redhead (1988a) for further discussion of this point.

[16] The view expressed here should be contrasted with that canvassed by Causey (1977).

experimental set-ups must be presupposed in analysing the properties of those very micro-entities which make up or constitute the macroscopic objects. In a sense the reduction instead of descending linearly towards the elementary particles, moves in a circle, linking the reductive basis back to the higher levels.[17]

(2) Even on non-orthodox interpretations of quantum mechanics of the hidden-variable variety, there is a choice between non-local action-at-a-distance between elementary particles and a holistic conception of non-separability, that in so-called entangled states it does not make sense to attribute properties to separate 'particular' entities, independently of a prior understanding of the properties of the whole composite system.[18]

(3) In resolving the problem of measurement in quantum mechanics some have held that human consciousness plays an essential role in resolving the ambiguity in 'pointer readings' predicted by straightforward application of the quantum-mechanical formalism. This would again make the reductive hierarchy Psychology → Biology → Chemistry → Physics → Elementary Particles bite at its own tail. We shall return to the question of human consciousness in scientific explanations shortly in discussing so-called anthropic explanations.

(4) In the bootstrap approach to elementary particle physics that was very popular in the 1960s, a democratic approach to composite entities, constituent entities and 'force-carrying' entities was proposed. Every entity could potentially play any of these three roles, so that the analysis of composite entities into a few 'aristocratic' constituents in the typically micro-reductive fashion was rejected.[19]

Despite these reservations, micro-reduction still seems a viable option, if taken as in point (2) with unorthodox interpretations that allow some sort of action-at-a-distance between micro-entities. This poses potential problems in reconciling quantum mechanics with special relativity. But perhaps only if the action-at-a-distance is of a causal character. This is the first point at which we have mentioned the notion of cause, and we want now to say something about the view that the only 'genuine' explanations of events occurring in the physical world are causal explanations.

6. Causal Explanations

What are causes? There is no consensus among philosophers, although they generally regard them as an important matter in any metaphysical

[17] For further discussion see Redhead (1987a).
[18] Cp. Redhead (1987b).
[19] For details see Redhead (1980a).

comprehension of the world. The surprising thing is that physicists long ago gave up the notion of cause as being of any particular interest![20] In physics the explanatory laws are laws of functional dependence, how one physical magnitude is related in a regular (and law-like on the necessitarian account) fashion with another physical magnitude. The pressure and volume of an ideal gas are related at constant temperature by Boyle's law. The pressure does not cause the volume, or the volume the pressure, they just coexist in a manner regulated by Boyle's law. But of course there are dynamical laws showing how the state of physical system depends on the time. Consider an object falling from rest with a constant acceleration g. After time t the distance s it has fallen is given by Galileo's law, $s = \frac{1}{2}gt^2$, so at t_1 the distance is $s_1 = \frac{1}{2}gt_1^2$ and at a later time t_2 the distance is $s_2 = \frac{1}{2}gt_2^2$. In what sense does the event consisting in the object having fallen a distance s_1 at t_1 cause the object to reach s_2 at t_2?

All we have is the relation

Position at t_1 + Galileo's law $\Rightarrow$ Position at t_2

The position at t_1 can hardly be cited as the cause of the position at t_2. Surely the cause must be something that is introduced to account for why the position of the object has changed. Well, is not the answer to cite the acceleration? But the acceleration is just *defined* by the kinematic relationship expressed in Galileo's law. So can the cause be Galileo's law? But Galileo's law is not another event which happens at t_1, 'causing' the particle to move with constant acceleration and so to reach s_2 at t_2. It is just an expression of that acceleration. So should we not retreat to citing the force, such as gravity, which 'causes' the acceleration. But the idea that forces 'cause' bodies to move is a very anthropomorphic notion. What we actually have in physics is a force law, such as the inverse square law of gravitational attraction, which relates, via Newton's second law, the acceleration of the body to the relative location of other bodies such as the earth.

Instead of $s = \frac{1}{2}gt^2$, we have, in idealized approximation, $s = \frac{1}{2}(GM/R^2) \cdot t^2$, where M is the mass of the earth, R its radius and G is a new gravitational constant. So we are back with a regularity connecting s with t, but also now with M and R. But the force of gravity has been eliminated between the force law and Newton's second law. Many metaphysicians want to resist this elimination. There really is a force which then causes the motion. But the sense in which they mean this is that the force causes the body to change its *natural state of motion*, as

[20] For the historical background to the changing attitude of physicists to the notion of cause see Kuhn (1971), reprinted in Kuhn (1977). Compare also the classic paper of Russell (1917).

specified by Newton's first law, that if a body is not moving, its natural state is one of perpetual rest. There is a definite sense in which the accelerated motion is not 'natural' and hence has to be caused. But modern physicists would reject this distinction. Acceleration under gravity is just as 'natural' a change, as perpetual rest or uniform motion in the circumstances appropriate to those descriptions. All of this is vividly illustrated in general relativity where all motion is geodesic motion (in four-dimensional space-time). If space-time is not flat this geodesic motion is a close approximation to the old Newtonian motion with acceleration induced by a force. But there is no such thing as a non-natural motion. To most physicists the old-fashioned idea of cause arises from the idea of our interfering in the natural course of events, pushing and pulling objects to make them move and so on. In modern physics there are just regularities of one sort or another. They are all 'natural' and hence leave no room for causation. It is true that some natural regularities can be analysed in the manner of one event impinging on a contiguous one, so transmitting action in a continuous train of contiguous action. This is true of many field theories, which play such a prominent role in modern physics. But this causal method of articulating what is going on in a field theory should not be taken seriously by philosophers. It is, at best, a crutch to the understanding of non-physicists!

But what, you may say, of causality principles employed by physicists? That effects cannot precede their causes and so on. What physicists mean here is that law-like regularities in nature must not be such that if we, *as free agents*, interfered in the natural course of events, we could alter past events, in such a way as to induce inconsistencies in the specification of present states of affairs. But backwards causation and forward causation into the past, via closed time-like world lines in general relativistic space-times, are not *a priori* prohibited, only in so far as they could lead to causal loop paradoxes. Theories which exhibit these 'anomalous' features express constraints on present states of affairs, that must make them consistent with 'causal loop' restrictions. There is less contingency in the world than we had suspected, there are events like shooting our own grandmothers that we are not free to bring about, on pain of rendering our physical theories logically inconsistent.

7. Quantum-mechanical Correlations and Causation

In the preceding section I have put forward the view that I believe most physicists would have concerning causation. This is reinforced by the fact that causal processes in the old classical sense are not contemplated in the orthodox interpretations of our most fundamental theory, quantum mechanics. But what about the unorthodox hidden-variable interpretations? Since these are inspired by classical intuitions, many philosophers feel that questions of causality need to be discussed and evaluated. Perhaps the world really is ultimately to be understood, to be explained, in the language of causal networks.[21] I want to discuss this possibility briefly in the face of EPR correlations and the Bell arguments.

How should one explain correlations, understood in the sense of the probabilistic dependence of one event on another? On a causal view of explanation, there are two options. Firstly, a direct causal link, and secondly, a common cause which screens off one event from the other. Of course these options are not mutually exclusive—a common cause may be overlaid by a direct cause. Now, in the case of the correlations contemplated in the Bell version of the Einstein–Podolsky–Rosen experiment in its Bohm spin-correlation version, a common cause explanation of the correlations leads to the constraint known as the Bell inequality which is violated by the predictions of quantum mechanics (and also by experiment). So the common cause must be overlaid by a direct causal interaction, but since the events comprised in spin-components in specified directions taking values on the two 'wings' of the experiment may be at space-like separation, this poses problems for special relativity as mentioned at the end of section 5.

A way out of this impasse has been proposed by the present author (1988b) in terms of an explicit explication of Shimony's idea of passion-at-a-distance.[22] What has been shown is that the Bell-type correlations can be understood in terms of a probabilistic connection between the two events, which lacks the 'robustness' necessary for a causal connection. By robustness one means here that either event screens off the other from the way it was produced. The Bell correlations are not robust in this sense. The situation is rather like a stick which maintains a correlation, namely the relative distance of the two hands of someone holding on to the two ends of the stick, but which is such that if one end of the stick is 'jerked' by someone else, it mysteriously adjusts to a new length! Of course robustness may be restored by citing who jerked the end of the stick. If Jack jerks the stick it goes to one length, if John jerks it to another. But suppose the length depends also on who knocked Jack, with the result that he jerked the stick, and who knocked Bill who knocked Jack *ad infinitum* . . . This would be a marshmallow world where no adequate notion of causation as an explanation of the correlation between the ends of the stick could be rescued. The Bell correla-

[21] Cf. Salmon (1984).
[22] Shimony (1984).

tions are not like that; robustness is lost 'at a stroke' so to speak. But is talk of passion-at-a-distance in this situation genuinely explanatory? Is it not just giving a name to the brute fact of the correlations? I think not, because we are pointing here to a precise characterization that is exemplified in the Bell situation, and hence enables us to get a grip on what sort of situation we are dealing with. In the last section we saw how physicists have recourse to laws of functional dependence, but the philosopher may well be interested in classifying these laws into the coexistence of properties of natural kinds, dynamical laws, causal laws, 'passion' laws, and so on. Additional understanding of the physical situation may result from philosophical analysis of the category of law statements involved in the explanans.

8. Anthropic Explanation

In the recent literature, there has been considerable discussion of the explanatory virtues of the so-called anthropic principle in understanding the values of fundamental physical constants, such as the mass-ratios of elementary particles, or the dimensionless couplings characterizing the strengths of interactions.[23]

The basic idea is very simple. It is a remarkable fact that if the values of such fundamental constants differed significantly from the values they actually have, then the necessary conditions for the emergence of carbon-based life would not obtain, and we could not now exist, to speculate on the explanation of why these constants have the values they do! Or, arguing contrapositively, the fact of our own conscious existence implies within a very narrow 'window' the values of the fundamental constants. It is now argued that the fact of our existence is the explanation of why these constants have the values they do, modulo, of course, the fundamental laws governing the behaviour of micro-entities, other than those specifying the values of the constants in question.

Let us immediately distinguish a weak and strong version of this anthropic principle. The weak version says just that: our existence allows us to infer the values of certain fundamental constants. Our existence is, if you like, an *indicator* of what values these constants have. But the strong version of the anthropic principle claims that our existence not only allows us to infer the values of the constants, but is the explanatory ground for the values they have. We have already discussed briefly the relationship between explanation and prediction

[23] See Barrow and Tipler (1986) for a very extensive discussion of the physics involved.

(or more generally indication if we do not want to stress the temporal aspect). We have claimed that all complete scientific explanations *were* indicators. But what about the converse? Does mere indication qualify as explanation? The answer to this question is generally agreed to be no. Counter-examples commonly cited include the length of the shadow of a flagpole that indicates for us, but does not explain, the length of the flagpole; the period of a pendulum which indicates the length of the string, but does not explain why the string has that length; the barometer that indicates, but does not explain, the weather, and so on. There has also been much discussion as to the exact reasons that underlie the immediate intuition that the examples cited are not explanatory. The unification approach to explanation can help us here. Science looks for unified patterns of explanation.[24] In the case of the flagpole and the string of the pendulum, we explain their length by appeal to the way they were produced, i.e., on a micro-reductive account, citing the molecular structure of the component materials and how this allowed division by a saw or pair of scissors or whatever, to the dimension in question. And this pattern of explanation is available for all questions of the form 'Why is such-and-such the size it is?' Similarly for the barometer, explanation of the weather and all other hydrodynamic phenomena follow the unified pattern of deriving consequences from the macroscopic laws of hydrodynamics together with appropriate initial conditions, or at a deeper level, from the kinetic theory of fluids. Now, in discussing the implication afforded by microreductions we have already stressed that the explanatory apparatus comrises the micro-entities and their properties, thus coralling off a few deep fundamental principles that, at any rate for the time being, are not susceptible of explanation.

But in the anthropic mode of explanation we substitute the orthodox and unified explanatory scheme with another non-integrated scheme, the fundamental laws at the deep theoretical level plus something up at the macroscopic level, viz. our own existence. And notice furthermore that the existence of dragonflies or crocodiles would do just as well as our own existence to provide this sort of explanation. Indeed generalizing the pattern of anthropic explanation could lead us to cite any fact as the ground for any other fact that can be deduced from it, thus trivializing the notion of explanation in the same way that the simple-minded D-N model allows. In addition, on the basis of counting the number of constituents in the explanatory scheme as a measure of unification, we can argue that the existence of a living organism unpacks into a very large number of facts, the instantiation of all those law-governed biochemical processes necessary to life, as compared with

[24] Cf. Kitcher (1981).

the relatively small number of constants whose values are being explained.

The fact of the matter is that the strong anthropic principle arises from an initial prejudice in favour of reconstituting Man at the centre of the Universe, rather than examining the whole matter dispassionately in the light of those desiderata we have detailed for a satisfactory scientific explanation.

Notice carefully I am not saying that the orthodox scientific account, that the fundamental laws *plus* the constants of nature are the ontological ground for the *possibility* of our existence, is true in a way which has been established beyond any shadow of doubt, and that the anthropic account, that our existence modulo the laws of nature is the ontological ground for the constants having the values they do, is false. I am arguing only what it is rational, scientifically speaking, to accept. Scientific rationality in no way depends on establishing indubitable truths about the world. In the context of the truth-status of scientific laws, rather than questions of what is the ontological ground for what, as we penetrate below the surface level of appearance the whole account becomes essentially conjectural, but these are not conjectures made in some intellectual vacuum, they are controlled and made criticizable by the possible disagreement of their empirical consequences with experiment. This is what makes the acceptance of these conjectures scientifically rational, not their demonstrated and certifiable truth. It is rather like the creationist versus evolutionist controversy. The objection to creationism is not that science has shown it to be false, but that it is not consonant with scientific rationality to accept it. In the explanation case the situation is admittedly a little different, since even the negative control of experiment is not available to bear on the issue. The decisive point is that the orthodox unified account does confer a greater measure of understanding in the sense we have explained, whereas the anthropic account could only be regarded as superior on an assumption of teleological necessity, to which we have no empirical access. Notice that we are not attempting ourselves to argue in favour of causal necessity as opposed to teleological necessity. As we intimated in section 6, it is philosophers who are led to such considerations, perhaps even on the basis of believing, incorrectly, that physics has something to contribute in this area! We maintain that any such arguments must be essentially metaphysical in character.

9. Conclusion

It is time to sum up and answer a few remaining questions. The ideal of scientific explanation is a matter of logical deduction, given a unified set of deep explanatory principles that are themselves accepted, for the time being, without explanation. But of course the ideal of scientific explanation is one for ongoing improvement. Perhaps from the fundamental laws of microphysics, by some consistency criterion, it will turn out that the constants of nature are tightly constrained or even uniquely determined. But even then we would still have the task of explaining the laws themselves at a still more fundamental level. At some stage scientific explanations always turns into description—'That's how it is folks'—there is no *ultimate* terminus in science for the awkward child who persists in asking why! I do not believe the aim of some self-vindicating *á priori* foundation for science is a credible one.

Then there is the question of delimiting contingency in the world. On a necessitarian view of laws, contingency is usually understood in terms of the contrasted non-necessity of initial and boundary conditions. But in matters of cosmology that sort of distinction may not be the right one to draw. The no-boundary universe of Hawking and Hartle is an example, where essentially there are only laws![25]

Another matter deserving of mention is the role of symmetry principles, which are really metalaws constraining the form of laws themselves.[26] Many explanations in modern physics bypass the detailed laws by invoking directly the metalaws. But their status again is not self-vindicating. The discovery of the violation of mirror symmetry in the physics of weak interactions was one of the crowning triumphs of modern science.

Some people have argued that the orthodox ideal we have been expounding and defending should be given up, and replaced by an account which lies closer to scientific practice.[27] Now, it is quite true that the fundamental laws of microphysics are quite useless as a practical basis for deducing, and hence explaining many phenomena. The fundamental theories are mathematically much too intractable. So, in practice, physicists use all kinds of approximate and indeed inconsistent 'models', to discuss the properties of complicated systems such as atomic nuclei or lasers.[28] But I have been concerned with the scientific ideal. The fact that, for practical reasons, it cannot be implemented, does not detract from its status as an ideal, something we should try for, rather than substituting the messy 'real' physics as something which, instead of falling short of the ideal standard, should itself be elevated into the standard for scientific explanation.

[25] See Hawking (1988) for a popular exposition.
[26] See Redhead (1975).
[27] Cf. Cartwright (1983).
[28] See Redhead (1980b).

In his book *The View from Nowhere* (1986), Tom Nagel has argued that science in eschewing subjectivity, does not tell the whole story. To give a proper account of human actions and intentions we need the subjective view, the view from somewhere. But in its proper place and sphere the traditional methods of objective science have proved extraordinarily successful—there is no reason in my view to think that modern micro-reductive physics throws inevitable doubt on those methods and their motivating ideal of the progressive unification of science.

References

Achinstein, P. 1971. *Law and Explanation*, (Oxford: Clarendon Press).
Achinstein, P. 1983. *The Nature of Explanation*, (Oxford University Press).
Armstrong, D. M. 1983. *What is a Law of Nature*, (Cambridge University Press).
Barrow, J. D. and Tipler, F. J. 1986. *The Anthropic Cosmological Principle* (Oxford: Clarendon Press).
Campbell, N. R. 1920. *Physics: The Elements*, (Cambridge University Press).
Cartwright, N. 1983. *How the Laws of Physics Lie*, (Oxford: Clarendon Press).
Causey, R. 1977. *Unity of Science*, (Dordrecht: Reidel).
Feigl, H. 1949. 'Some Remarks on the Meaning of Scientific Explanation', in H. Feigl and W. Sellars (eds.) *Readings in Philosophical Analysis*, 510–14.
Friedman, M. 1974. 'Explanation and Scientific Understanding', *Journal of Philosophy*, **71**: 5–19.
Grünbaum, A. and Salmon, W. C. 1988. *The Limitations of Deductivism* (Berkeley: University of California Press).
Hawking, S. W. 1988. *A Brief History of Time*, (London: Bantam Press).
Hempel, C. G. 1965. *Aspects of Scientific Explanation*, (New York: The Free Press).
Hempel, C. G. and Oppenheim, P. 1948. 'Studies in the Logic of Explanation', *Philosophy of Science*, **15**: 135–75.
Hesse, M. B. 1966. *Models and Analogies in Science*, (Notre Dame University Press).
Hospers, J. 1967. *An Introduction to Philosophical Analysis*, 2nd ed. (London: Routledge & Kegan Paul).
Kitcher, P. 1976. 'Explanation, Conjunction and Unification', *Journal of Philosophy*, **73**: 207–12.
Kitcher, P. 1981. 'Explanatory Unification', *Philosophy of Science*, **48**: 507–31.
Kuhn, T. 1971. 'Les Notions de causalité dans le developpement de la physique', *Etudes d'Épistémologie Génétique*, **25**: 7–18.
Kuhn, T. 1977. *The Essential Tension*, (University of Chicago Press).
Nagel, E. 1961. *The Structure of Science*, (London: Routledge & Kegan Paul).

Nagel, T. 1986. *The View from Nowhere*, (Oxford University Press).
Redhead, M. L. G. 1975. 'Symmetry in Intertheory Relations', *Synthese*, **32**: 77–112.
Redhead, M. L. G. 1980a. 'Some Philosophical Aspects of Particle Physics', *Studies in History and Philosophy of Science*, **11**: 279–304.
Redhead, M. L. G. 1980b. 'Models in Physics', *British Journal for the Philosophy of Science*, **31**: 145–63.
Redhead, M. L. G. 1984. 'Unification in Science', *British Journal for the Philosophy of Science*, **35**: 274–9.
Redhead, M. L . G. 1987a. 'Whither Complementarity?', in N. Rescher (ed.) *Scientific Inquiry in Philosophical Perspective*, (Lanham: University Press of America), 169–82.
Redhead, M. L. G. 1987b. *Incompleteness, Nonlocality, and Realism: A Prolegomenon to the Philosophy of Quantum Mechanics*, (Oxford: Clarendon Press).
Redhead, M. L. G. 1988a. 'A Philosopher Looks at Quantum Field Theory', in H. R. Brown and R. Harré (eds.), *Philosophical Foundations of Quantum Field Theory* (Oxford: Clarendon Press), 9–23.
Redhead, M. L. G. 1988b. 'Nonfactorizability, Stochastic Causality, and Passion-at-a-Distance', to appear in J. Cushing and E. McMullin (eds.), *Philosophical Consequences of Quantum Theory.*
Russell, B. 1917. 'On the Notion of Cause', reprinted in *Mysticism and Logic*, (London: Allen and Unwin), 180–208.
Salmon, W. 1971. *Statistical Explanation and Statistical Relevance* (Pittsburg University Press).
Salmon, W. 1984. *Scientific Explanation and the Causal Structure of the World*, (Princeton University Press).
Scriven, M. 1962. 'Explanations, Predictions and Laws', in H. Feigl and G. Maxwell (eds.), *Minnesota Studies in the Philosophy of Science III*, 170–230.
Shimony, A. 1984. 'Controllable and Uncontrollable Non-Locality', in S. Kamefuchi *et al*. (eds.) *Proceedings of the International Symposium: Foundations of Quantum Mechanics in the Light of New Technology*, (Tokyo: Physical Society of Japan), 225–30.
Swartz, N. 1985. *The Concept of Physical Law*, (Cambridge University Press).
Watkins, J. W. N. 1984. *Science and Scepticism* (Princeton University Press).

Explanation in Physical Theory

PETER CLARK

1. Introduction: Epistemology and The Philosophy of Physics

The corpus of physical theory is a paradigm of knowledge. The evolution of modern physical theory constitutes the clearest exemplar of the growth of knowledge. If the development of physical theory does not constitute an example of progress and growth in what we know about the Universe nothing does. So anyone interested in the theory of knowledge must be interested consequently in the evolution and content of physical theory. Crucial to the conception of physics as a paradigm of knowledge is the way in which physical theory provides explanations of a vast diversity of natural phenomena on the basis of a very few fundamental principles. A central problem for the epistemologist is therefore what is theoretical explanation in physics? Here we can get good insight from what Redhead has said (this volume pp. 145–54).[1] Indeed one could agree with almost everything Redhead says and simply endorse much of his careful and extensive defence of the covering law account of explanation in the physical sciences at least as an *ideal*. However I shall, I fear, try the reader's patience by extending some of the considerations he introduced and raising those issues where we disagree, especially in the important area of statistical explanation.

Now while I am happy to accede (at least partially) in Redhead's endorsement of the much criticized covering law account of explanation in the physical sciences originating with Hempel and Oppenheim's (philosophically classic) paper of 1948[2] and its subsequent elaborations—it needs to be emphasized that no simple schema could possibly capture the enormously rich panoply of methods and techniques employed in an explanatory context in the complex human enterprise

[1] This paper owes much to the discussion following the presentation, at the Royal Institute of Philosophy Conference at Glasgow, of Redhead's and my own symposium papers. I should like to acknowledge the useful critical comments of the audience and especially those of Michael Redhead. I should also like to thank the organizers for their invitation to speak at the Conference.

[2] Hempel and Oppenheim (1948), reprinted in Hempel (1965) and Pitt (1988), 9–50. This latter provides an excellent introductory collection on the topic of scientific explanation.

we call modern physics as it is practised in laboratories, universities, journals etc. that is, in the actual day to day exercise of the discipline. Nor do I think that it is possible (and I think this for philosophical reasons) to lay down in advance constraints, *a priori*, which the development of Science must somehow *always* satisfy. I do not think philosophy could ever be in such a position. As an example imagine that this symposium on explanation in physics had been held in say 1859 (not *so* long ago) and a year before Maxwell's pioneering 'Illustrations of the Dynamical Theory of Gases' and seven years before his revolutionary 'The Dynamical Theory of Gases' of 1866.[3] There would hardly have been a mention of probability and its interpretation, of statistical inference and statistical significance and testing, either ontologically (i.e. that the laws of nature cannot be formulated without appeal to probability), methodologically (what is the appropriate way of testing statistical claims) or epistemologically (how is probability to be understood in physical contexts). Yet it is surely no exaggeration to say that after 1866, such issues posed many of the central philosophical problems concerning physical theory. Since we do not know how Science will evolve, since we do not know what methods, techniques or concepts will be introduced in an explanatory context in the evolution and development of physical theory, an account of explanation in physics cannot really lay down constraints *a priori*, not even *necessary* conditions, let alone sufficient ones. The point here is simply this, what is to be regarded as a 'good explanation' will evolve just as scientific theory itself evolves and it may very well be that conditions formally held to be necessary in explanation may come, for the very best of reasons, to be given up. As a straightforward and uncontentious example consider the idea that the relation between the explanatory premises and the explanandum must be that of logical consequence. But if one is to allow probabilistic explanation of singular events (which Hempel called in general inductive statistical explanation) then the connection between explanans and explanandum cannot be of a deductive kind. Of course one could always insist that probabilistic premises cannot provide any explanation at all of singular events, but such a constraint is far too much like pure legislation to be tolerable (see below pp. 164–66). If however the laying down of necessary and sufficient conditions is not the business of accounts of explanation in physics, what can be achieved by such accounts? What I think they can do, and what in my view is philosophically immensely revealing, is provide, by careful analysis of existing theory and method, the global characteristics and structural

[3] Maxwell (1860) and Maxwell (1866). There were of course earlier efforts to develop a kinetic theory of gases; for details see Brush (1976), 107–81.

features (the necessary conditions—in a very weak sense) common to the various branches of our current understanding of Nature.

These remarks may appear trite—true but trivial, if so all well and good—but I think they are sometimes easily missed. Let me now quickly turn to the main issue. What view is it that Redhead and I seek to defend? Hempel put it in 1966 (with his usual pellucid clarity) like this:

> What scientific explanation, especially theoretical explanation, aims at is not [an] intuitive and highly subjective kind of understanding, but an objective kind of insight that is achieved by a systematic unification, by exhibiting the phenomena as manifestations of common, underlying structures and processes that conform to specific, testable, basic principles.[4]

This is the core intuitive informal deductive nomological account of explanation in physics and it can be turned into an explicit formal account by answers to the question as to exactly how we 'exhibit the phenomenon as manifestations of . . . basic principles'. According to the classical account we do this as everyone knows by bringing the aspect of the phenomena to be explained under some description in terms of a statement 'e' and showing (1) 'e' follows deductively from a statement expressing a general law of Nature (say L) *and* sentences describing antecedently available facts (I) *viz.* initial or boundary conditions. But this alone is not the covering-law model for Hempel was careful to add certain constraints on the deduction *viz.*: (2) that the explanation (I, L) must contain general laws and these must appear essentially in the derivation of the explanandum, (3) the explanans (L) must have empirical content over and above 'e' and (4) a material condition—that the sentences constituting the explanans must be true. Now it is well known as Redhead remarks that no such set of conditions can possibly be sufficient for L to constitute an explanation of 'e'. There is a long and a short route to this conclusion, the short route is to use the familiar Salmon-type[5] counter-examples, where the conditions (1) to (4) are satisfied but utterly irrelevant laws and initial conditions are employed. The longer route is to follow the method of Eberle, Kaplan and Montague[6] and show how virtually any sufficiently general theory

[4] Hempel (1966), 83; cf. also Hempel (1965), 336. Here he remarks that an explanation may be thought of as 'an argument to the effect that the phenomenon to be explained, was to be expected in virtue of certain explanatory facts'. These latter being, 'particular facts' and 'uniformities expressible by means of general laws'.

[5] Salmon (1970).

[6] Eberle, Kaplan and Montague (1961).

yields an 'explanation' of virtually any particular fact satisfying conditions (1) to (4) if these are taken as sufficient. At the very best therefore (1) to (4) can only be regarded as necessary conditions.

2. The Laws of Physics Do Not Lie

Now there is one general objection to this schematic account which has received considerable attention but which I think is empty and that centres on what are referred to as *ceteris paribus* clauses. Cartwright has for example argued that as she puts it 'true covering laws are scarce' and that explanations in physics most often employ *ceteris paribus* laws. However *ceteris paribus* laws are simply not true, so the laws of physics being in the majority *ceteris paribus* laws, lie.[7] Thus to take a simple (the simplest possible) example, consider the explanation of the period of a simple pendulum swinging from a pivot on a beam, with small displacement 'Θ' from the vertical. We will need in order to effect a deduction of the period, to employ the gravitational law to obtain a value for the acceleration on the pendulum bob, but then the gravitational force will be the total force acting on the bob only if there are no draughts of wind, the bob is itself not responding to magnetic or electrostatic forces, there are no damping or accelerating pulses at the pivot, etc., etc. So we have to employ *ceteris paribus* clauses to that effect, in the statement L of the explanatory schema. This claim seems to me just incorrect, for there are no *ceteris paribus* laws used in physics at all. What does happen is that in the explanation of the period say of a given particular pendulum in a given experimental set-up, we will in describing the initial conditions add sentences like (if of course we have good reason to believe them to be true) the *total force* on the bob *is* the gravitational force. We will then use (or better instantiate) the well-known bridge principle which says the term on the left hand side of Newton's second law is to be identified with the total force (i.e. the acceleration experienced by the bob is to be identified with the total force divided by the mass of the bob) and using the assertion above concerning the total force and the transitivity of identity, arrive at the claim that the acceleration experienced by the pendulum bob is $-g/l \sin \Theta$. Of course what will be crucial is whether these statements expressing identity between the total force experienced by the pendulum and the gravitational force, which are absolutely essential in the deduction (if they are absent nothing follows from Newton's second law and the gravitational law when taken together) are true of the particular pendulum whose period we are trying to explain. All I wish to emphas-

[7] Cartwright (1983), especially 44–58.

ize, and I think complicating the physical examples adds nothing to the formal point at issue, is that no use of *ceteris paribus* laws is made whatsoever. What is employed is the idea that the set of sentences used in the description of the experimental situation are true, and that Newton's second law and the gravitational law are *true* descriptions of the physical facts of the case. Hempel has recently called such sentences as in our example 'The total force acting on the bob is the gravitational force' which have infinitely many deductive consequences, among which importantly are 'there are no magnetic or electrostatic forces operating', 'there is no viscous drag due to drafts of wind', etc. *proviso sentences*, which he sees as really being 'assumptions of completeness'.[8] What is interesting is that an assumption of descriptive completeness can be expressed entirely in the vocabulary of Newtonian mechanics by a single formula (in fact a sentence). I know of no case where this cannot be done. Let me add one final point about Cartwright's analysis of explanation in physics, concerning the Law of the vector addition of forces—this addition of forces she describes as a metaphor. She writes:

> The vector addition story is, I admit, a nice one. But it is just a metaphor. *We* add forces (or the numbers that represent forces) when we do calculations. Nature does not 'add' forces. For the 'component' forces are not there, in any but a metaphorical sense, to be added; and the laws that say they are there must also be given a metaphorical reading.[9]

But as a claim about the vector addition law of the addition and decomposition of forces, this seems just to be false. Alan Musgrave has given a nice counter-example;[10] imagine two teams equally matched in a tug-of-war. The net force experienced by a point mass at the centre of the rope is zero, but there is absolutely nothing metaphorical about the two component forces along the opposite ends of the rope, and the desperate heaving of the teams is hardly 'metaphorical'. Nature most certainly 'adds' the forces there (equal and opposite). The point is not trivial for in virtually all interesting cases of the employment of proviso sentences in physical explanation some use of composition of forces will be employed (or its equivalent). So if the addition of forces were merely

[8] Hempel (1988). Hempel's essay is part of the collection edited by A. Grünbaum and W. C. Salmon (Grünbaum and Salmon, 1989). It contains a number of important essays on theoretical explanation. It should be noted that Clark Glymour retained a deductivist account of explanation in his classic (1980); what he proposed which was so novel, was a new theory of 'relativized' confirmation.

[9] Cartwright (1983), 59.

[10] In a paper read at a conference on 'The Rationality of Science' in Krakow, Poland, June 1989.

metaphorical (i.e. not literally true or false), then the consequences of the proviso sentences when taken together with the addition law would presumably have no truth value at all, so they could no longer function in the way I have argued they do function in the deductive schema of explanation. I see however no reason to accept either Cartwright's criticism of covering law explanations in physics or her analysis of the law of the vector addition of forces.

3. Statistical Explanation: The Problem

In one of the most challenging parts of Redhead's paper, he rejects the idea that probabilistic or statistical explanations can explain single events. He remarks:[11]

> So what do probabilistic or statistical explanations achieve? Well, they enable us to deduce and hence to explain the limiting relative frequencies with which events of a given kind turn up in a long run repetition of the set-up producing the phenomenon. But they cannot explain what happens on a particular occasion. It is as simple as that.

Thus he rejects the class of inference schema which Hempel called Inductive Statistical Explanations as being genuinely explanatory at all. The I-S schema has the well-known form:

Initial conditions	: Pa
Statistical law	: $(\forall x)(Pr(Qx/Px)=r)$
	═════════════ [r]
Singular event	: Qa

where $r(0 \leq r \leq 1)$ is the degree of probabilistic support that the explanans confers upon the explanandum. Hempel laid down two further criteria as necessary conditions: (i) r should be 'sufficiently high' and (ii) the statistical premise must satisfy the conditions of maximum specificity to the effect that the statistic refers the frequency of Q to the broadest homogeneous reference class.[12] Now Redhead asserts that only deductive statistical explanations are genuinely explanatory—in these cases that which is to be explained is a statistical regularity which follows from a statistical law together with certain boundary conditions *deductively*. There is no doubt whatever that deductive statistical explanations, the explanation of statistical regularity as a logical consequence of very general statistical laws at a much higher level abound in physical theory. We need only recall the explanations of Avogadro's

[11] See above, 156.
[12] Hempel (1965), 381–412.

law, the viscosity of dilute gases, their thermal conductivity and specific heats, the speed of sound in dilute gases, etc. provided by the kinetic theory to see this. The challenging claim is that essentially the deductive-statistical model exhausts statistical explanation. This view it should be noted is quite compatible with Railton's claim that the correct form of statistical explanation in physics is deductive-probabilistic in the sense that what is shown to be a consequence (deductive) of statistical laws is a statement as to the *probability* of a given event.[13] What Redhead denies, is the additional component of Railton's view, that a statement concerning the probability of an event, can provide the basis of an explanation of that particular event, should it occur.

Now there are in the subject of statistical explanation a collection of issues which are often mixed together and require careful analysis and separation. They are the issues of determinism and indeterminism, randomness and law-likeness, chance and probability and I would like to comment on these separately before returning directly to Redhead's claim since it seems to me that recently quite a lot of new and surprising light has been thrown on the logical connection (and independence) of these issues.

4. Determinism and Deterministic Laws

What does the doctrine of determinism amount to in physical theory? The basic intuition seems to be this: if we consider an *isolated* (this restriction is essential; I do not think a notion of determinism can be properly defined for *open* systems—ones actively interacting with their environment) physical system and think of the history of that system as the 'graph' of the values of its state variables in phase-space (i.e. the n-dimensional space, for a system of n degrees of freedom in which the instantaneous state of the system is a 'point') through time, then the system is deterministic *iff* there is one and only one possible path consistent with the values of its state variables at any arbitrary time.[14] Essentially then, the problem of determinism in classical physics reduces to the problem of the *existence* and *uniqueness* of solutions in the theory of ordinary differential equations and it is to be recalled that it is in the form of differential equations that the laws of Nature receive their expression at least in classical physics.[15] Although quantum mech-

[13] Railton (1978).

[14] For more details see Clark (1987) and especially Earman (1986). This latter classic study of determinism contains an excellent discussion of physical probability and chaos, 137–67.

[15] Cf. Russell (1917).

anics has no phase-space representation the basic intuition above can easily be recovered for that theory in terms of the unique temporal evolution of density operators representing the instantaneous state of an isolated quantum mechanical system in Hilbert space. The general characterization of what it is for a *theory* to be deterministic in the state variables say $x_1,-,x_n$, is due to Montague and Earman and says essentially that if any two models (physical structures which make the theory true) of the theory agree in the values of the state variables $x_1,-,x_n$ at some given time then they will agree in the values of the state variables $x_1,-,x_n$ at all other times. Looking then at classical mechanics one can ask: Does every physical system which satisfies the postulates of that theory automatically satisfy the necessary and sufficient conditions for the existence and, most importantly, the uniqueness of solution of the equations of motion condition, necessary for the system to be deterministic? The answer to this question is definitely no; for there is nothing in classical mechanics itself which lays down the form of the laws of force occurring in Nature. Hence there is nothing which rules out the possibility that there be situations in which the necessary conditions for the existence and uniqueness of solution to the differential equations of motion do not apply. Indeed wherever we have as an expression of the covering law for the system a *non-linear* differential equation, the existence and uniqueness of solution is not at all a straightforward matter. The study of non-linear systems is currently a very important field of research.[16] Let me say now why I think the Montague–Earman[17] account of determinism is particularly important. It is because: (1) it captures the determinism in physics where it occurs; (2) it makes no appeal to strange and empty shibboleths like 'Every event has a cause' or 'Like causes produce like effects' and (3) it clearly separates the notions of determinism and *predictability*.[18] This latter separation is most desirable, for predictability is a thesis about effective computability or recursiveness (i.e. predictable according to the methods of science means, from computable initial conditions together with computable solutions to the equations of motion it is possible to effectively compute in advance the value of some observable at some future time). Determinism is a much weaker thesis saying nothing about computability. This is not just a logical point, but one of physical significance, for the consequences of the definite appearance of non-recursiveness in physics are just now beginning to be investi-

[16] A clear popular introduction to this area is Nicolis (1989), Ford (1989) and Ford (1983). The best technical introduction is that of Berry (1978).

[17] Montague (1962) and Earman (1971).

[18] Famously Popper (1950) identified these two notions.

gated.[19] For these structural reasons, *not* just pragmatic ones I strongly doubt the symmetry thesis concerning explanation and prediction. We may very well have explanation without even the theoretical possibility of prediction. Of course the investigation of systems which are deterministic but depend very, very finely upon initial conditions (e.g. the so-called mixing systems which possess the property of exponential divergence of trajectory in phase space with time) falsifies the symmetry claim, for there are very good explanations why these systems behave in the way they do, but because *exact* values of initial states are inaccessible to us, there is a failure of predictability of their long-term behaviour. This is an epistemic failure, but the failure of recursiveness would induce an 'in principle' failure in the thesis. So much for explanation entailing prediction. In the other direction it is well-known that prediction in general does not yield explanation; barometers predicting storms, or indeed any purely analogue computational device is sufficient to show that.

5. Randomness in Physics

My readers (if I still have any) may have wondered by now, what has happened to the topic of statistical explanation. Well, I am now about to return to it. One very well-known interpretation of physical probability (due to von Mises) says that probabilities are well-defined only over special sorts of collections of outcomes of experiments or trials called *collectives*.[20] A *collective* is an infinite sequence of trials or experiments fulfilling two conditions: (i) that the relative frequencies of the pertinent attribute within the sequence of observables has a limit and (ii) that the sequence satisfy a randomness principle, i.e. that the relative frequency of the pertinent attribute be insensitive to place selection. This means that no matter according to what computable rule we select subsequences from the main sequence, the relative frequency of the attribute in question is the relative frequency of that attribute in the main sequence. This, as is very well-known, is the Church-Von Mises characterization of Randomness.[21] But now let us for a moment recall the definition of a deterministic theory, and ask what would happen if we had a theory in which say the values of single state variable were given by an arithmetically definable but not effectively computable function of some discrete valued parameter? Would it not then be

[19] See particularly Pour-El and Richards (1979, 1981, 1983) and Kreisel's (1982) review of this work.

[20] Von Mises (1957), especially 18–29.

[21] Church (1940).

possible to satisfy both constraints in the sense that any countable sequence of values of the state variable which satisfied the theory would be *deterministic*, this being guaranteed by the fact that the state-variable value is a *function* of the parameter, while at the same time *random* in the occurrence of some particular character, in virtue of the non-computability of the functions determining the values of the state-variables. Montague conjectured that this was so in 1962 and Humphreys in 1978 exhibited such a *collective*.[22] It follows then that randomness does not require *indeterminism*, and neither does the notion of objective probability.[23] Further this has nothing to do with epistemic matters of ignorance, or putting randomness in at the initial conditions and allowing a suitably chosen deterministic theory to replicate it in later subsequent states at the observational level. Nor is it in my view merely due to choosing a weak characterization of randomness, for measures of randomness which depend upon the algorithmic complexity of a sequence (i.e. those in which the algorithms for computing or generating the sequence are of the same length as the sequence itself) preserve the essential result.[24] Deterministic systems can generate sequences of outcomes which are very complex indeed as can be clearly seen from the study of simple iterative processes (like the famous 'logistic' mapping $S(x)=4x(1-x)$ and its iterations on the unit interval). Again this is not of purely logical interest for a great deal of current research in physical theory is devoted precisely to the study of the very strong stochastic properties exhibited by deterministic systems, the study of which began with the pioneering paper of Ulam and Von Neumann.[25]

6. Statistical Mechanics and Redhead's Thesis

Now there is one area of physical theory where probabilistic explanation is paramount and that is statistical mechanics both classical and quantum. (I will concentrate on the classical theory.) Now statistical mechanics refers not to just *one* theory but a number, first of all there is

[22] Humphreys (1978).

[23] In this respect I completely reject one horn of Watkins' dilemma in his (1984), 238–9 to the effect that determinism precludes objective statistical explanation. My reasons are given in detail in Clark (1987), 202–7. As we shall see I also reject the second horn of his dilemma. A similar position to my own has been developed by Colin Howson.

[24] The various (equivalent) characterizations of algorithmic complexity can be found in G. J. Chaitin's classic study (1987), especially 107–61.

[25] Ulam and Von Neumann (1947).

a general hierarchy of kinetic theories (which pay special attention to the internal properties of gases and to explaining irreversible approach to equilibrium), there is a quite different theory generally referred to by the phrase 'equilibrium statistical mechanics' of which the most famous example is Gibbs' statistical mechanics of representative ensembles and a third class is non-equilibrium statistical mechanics. These theories all employ objective, physical probability measures, despite the fact that their underlying dynamics is deterministic. The justification for the introduction of such probability measures varies from one class of theory to another in the above groups, but in each case it is bound to pose a consistency problem: *viz*. how can it be consistent to add statistical assumptions to treat an aggregate motion, the component submotions of which, being governed by the laws of mechanics, are entirely deterministic? In at least one class of cases, that of the Gibbs' theory of representative ensembles we have in the so-called ergodic theory, I think, the best solution to the consistency problem and the best justification for the interpretation of the probability measures occurring in that theory as real objective probability (as objective as the underlying mechanics) as we could have. Now part of this justification is directly relevant to the assessment of Redhead's idea, that statistical physics only explains observed frequencies in a large number of trials (i.e. an experimental ensemble). I do not think that in general this is so. The point is this: for certain idealized model gases it can be shown that the functions that represent physical quantities of interest (the macroscopic observables) have a dispersion (calculated with respect to a dynamically grounded probability measure) which tends to zero very rapidly as the number of particles in the system increases. Thus the probability of obtaining for a given system in equilibrium a measured value for such a function different from its average across the ensemble tends to zero. For such systems then we have *the very best of reasons* for asserting that we do have an explanation of the behaviour at equilibrium of a *single system* and this despite the fact that it does not *deductively* follow that the (phase) average of a macroscopic quantity will be equal to the measured value for that observable for a single system. Why do we have this explanation? In my view because there is an intimate association between the underlying dynamical (causal) structure of system and the probability measure being used. Essentially then my claim is this, when we have, on the basis of the underlying physical description of the system, to assign a very high value (very close to 1) to the parameter [r] in Hempel's schema, we do have genuine (probabilistic) explanations of singular events.[26] If one were to general-

[26] Cf. Howson (1988), 121. Howson points out that Hempel's condition of maximal specificity should be abandoned. He remarks rightly 'The

ize this account, one would be giving a dispositional theory of physical probability. I think Professor Redhead is committed to such an account, for if he were to rely merely on the pure relative frequency theory of Von Mises I do not see how any observational evidence could possibly bear on the statistical claims of physical theory, for the obvious reason that relative frequencies in the limit are compatible with any finite initial segment of evidence of observed trials. So what empirical significance would such claims have? We need to be able to *test* claims about objective chance, and in order to do this we need, it seems to me, some sort of dispositional theory, and such a theory would if the conditions I have described were satisfied, generate an explanation of singular events.[27] Since it is not 'scientifically irrelevant' that we can explain the thermodynamics of single many particle systems at least in some special cases I cannot agree with Redhead that the explanation of singular events is scientifically irrelevant.

So much (or so little) for statistical explanation.

7. Explanation, Reduction and 'Emergence'

I am very much in sympathy with the general tenor of Redhead's remarks in respect of the enormous explanatory advances achieved by microphysical reduction and unification. Of course one has to be careful here because at least in the history of physics, some important cases of reduction have strictly speaking been cases of elimination! In particular in the classic case of the reduction of equilibrium thermodynamics to statistical mechanics, the reducing theory showed that the central *law-like* claims of classical thermodynamics, were in fact false. Not just the latter's neglect of remote theoretical possibilities, e.g. the existence of initial conditions of a very special kind, whose probability (measure) is zero but which yield periodic, non-equilibrium behaviour, nor the possibility of very improbable fluctuations nor the inevitability of dynamic recurrence but rather in the neglect of physically realizable possibilities which had clear empirical consequences, like the existence of the Brownian motion. Nevertheless it is quite proper in my view to

[27] There is a beautifully clear exposition of this issue in Howson and Urbach (1989), chapter 9.

---

maximal specificity condition is an undesirable strength, and the lack of any condition that the explanans should invoke only causally relevant conditions is, as Salmon and others have insisted, an undesirable weakness.' In the indeterministic case if the probability [r] is very close to 1, we have a definite rational expectation that the effect will occur and that is surely explanatory. So I reject the second horn of Watkin's dilemma, see note 23 above.

talk of microphysical reduction here. Thus many of the special theoretical *terms* of thermodynamics (e.g. from elementary examples like temperature and pressure, to more complex cases like entropy and the various potential functions of state like the Gibbs' and Helmholtz's free energies) were shown to be identical with the mean or average values of certain purely molecular dynamical quantities. (I should point out that since I think of all true identity statements as being necessarily true, statements of the above type are examples if true of necessary but *a posteriori* truths and not as Redhead would have them as true contingent co-extensionalities.) There was no elimination of the theoretical concepts or terms involved in thermodynamics here. Secondly, though of course, in full generality classical thermodynamics had to be abandoned (it was anyway incompatible with mechanics) it could easily be seen as a limiting case of statistical mechanics (hence the well-known trick of taking the thermodynamic limit, N the number of particles in the system tends to infinity, while N/V remains *finite* for V the volume of the system of moving molecules). This then seems to me to be a clear case of law but not term or concept elimination. Now it is often claimed that the general reductive hierarchy

quantum mechanics of elementary particles ⇨ quantum mechanics of atoms and molecules ⇨ quantum chemistry ⇨ chemistry ⇨ biology

could not possibly be successful because of the existence of so-called 'emergent' properties, properties which appear only at certain levels of complexity and organization and which cannot be reduced in the way alluded to above to functions of more fundamental entities or properties lower down the scale. Well, I for one am extremely sceptical about such arguments, especially if they are arguments to 'irreducibility in principle'. The reason I am is that we have very little knowledge of the quantum mechanics of large scale systems. In fact at present there is no real theory (there are a few *ad-hoc* rules) for the quantum-mechanics of many-particle systems, still less of self-sustaining, self-organizing systems far from equilibrium (though there are some hints). The theoretical difficulties standing in the way of such a treatment are very considerable, but so were the theoretical difficulties standing in the way of Boltzmann—none of these difficulties however suggests to me an argument in principle, to irreducibility. That great man of nineteenth-century physics and graduate of Glasgow University, William Thomson, First Baron Kelvin, in his *Popular Scientific Lectures* alluded to what he took to be an absolute partition in Nature. This was the unbridgeable partition between living and dead matter, and he claimed that laws of living matter were quite independent of the laws of dead matter. Now that arbitrary partition is gone, but only at the expense of great theoretical labour. I suspect that many so-called 'emergent laws' similarly effecting partitions in our knowledge of the world are doomed to go the same way.

8. Anti-Realism and Mathematical Physics

Finally let me allude to an issue which I think is particularly central in epistemology, that is the realism, anti-realism debate and briefly examine what I regard as an important but rather neglected aspect of it. This is the role of classical mathematics in physical explanation. At issue is the acceptability of anti-realist construals of the content of the statements of applied mathematics as they occur in physical theory. (A related controversy concerns the existence of abstract entities, like numbers but this is a matter which I shall not discuss directly since there is an enormous literature on Field's famous challenge to the Putnam–Quine argument that the explanatory success of mathematical physics gives good inductive support for—indeed constitutes the best argument for—the reality of numbers. However as we shall see what significance one can assign anyway, quite independently of Field's strictures on the Putnam–Quine argument, does rest on a contentious matter, viz.: how much mathematics does one really need to do physics?[28]) The essential issue in its simplest form is this: on the one hand the mathematical structure known as the classical continuum (or the Real Line) plays a key role in the formulation of modern physical theory, on the other hand Dummett and Wright[29] (and from a different point of view, Putnam[30]) have developed a positive theory of meaning for mathematical statements much more in line with the intuitionistic (or more broadly constructivist) position which originated at least in its modern form with the work of Brouwer and Borel in the early part of the century. The great merit of Dummett's theory and modern anti-realism in general is that it captures and at least makes intelligible the basic tenets of intuitionism without requiring any commitment to a psychological or strongly subjective account of mathematics nor any appeal to the very dubious metaphor of 'construction' which was so salient a feature of the formulation of Brouwer's ideas. At the same time however, or so its proponents argue, it resolves the epistemological difficulties which so bedevil realism (platonism) in the philosophy of mathematics. The point being that the realist's claim that we have a

[28] Field (1980). For a very extensive discussion of Field's theory see Wright (1988).

[29] See particularly, Dummett (1959, 1973) and Wright (1985, 1986).

[30] Putnam (1980). (Burgess (1984), 192–3 alludes to the problem I raise.)

truth-conditional account of the content of mathematical statements will inevitably at some point conflict with how we could come to have mathematical knowledge at all, that is, come to recognize that some particular claim is true.

Now this anti-realist programme which avoids the epistemological problem posed by realism from the very beginning by explaining the meaning of mathematical statements, in terms not of truth conditions, but by reference to what is to constitute a proof of such statements can be carried through it seems only at a certain cost. For it at least *prima facie* has the consequence that the concepts involved and the procedures adopted in the classical basis of the mathematical structure of physical theory in so far as it is based upon the Cantorian conception of the Real Line (which it virtually totally is) is actually incoherent. As a consequence, in order for physical theory to be properly founded, it would have to be rewritten based upon an anti-realistically acceptable mathematics, one which for example did without *uncountable* collections altogether. But is this possible? Before attempting even a partial answer to this question the issue itself requires a more careful examination.

Classical mathematics, in particular real and complex analysis, plays a pivotal role in the articulation and testing of physical theory at least as standardly understood. In short it is a vital component of physical explanation. (A reader who doubts this should pick up any standard mechanics or quantum-mechanics text-book or consult any reputable research journal.) Even a cursory glance at current practice in mathematical physics reveals the continuum (or real numbers and their extension in the complex field) as a core mathematical structure, from Newtonian mechanics to quantum field theory. It is difficult to exaggerate this point. Looking first at classical physics it is surely correct to describe it as a 'boolean' universe. The phase space representation of dynamical systems, which is so paradigmatic of classical physics (including electromagnetism) requires that we think of observables as associated with subsets of a phase space, the algebra of the subsets being boolean, the phase space itself forming a classical continuum. The basic theorem for such a class of physical systems which essentially guarantees that all the right properties shall be available, the Stone-representation theorem is a purely classical non-constructive result. Even in quantum mechanics where undoubtedly there is no phase-space representation the corresponding Hilbert-space theory again requires as a basis classical complex analysis, which is replete with non-constructive existence proofs. It might be objected that the natural logic for quantum mechanics is very far from a classical one because therein physical states are not identified with points but with vectors, and the natural logic of vector-spaces is not boolean (but that of a non-distributive lattice), but while this might well be true it has no bearing on the issue in question, for the Hilbert-space theory which provides the setting for such a representation of states, is a classical non-constructive theory.

Now three points emerge as being important here. They are (a) is the anti-realist committed to revision of the practice of mathematical physics in so far as it relies on classical logic and mathematics? (b) if he is, will this just reflect upon the representation (i.e. the general formulation) of physics or will it directly affect the *physical content* (the experimental predictions) and consequently the testability of physical theory and (c) how much do we really need to suppose in the way of set existence to obtain the core theorems of the ordinary practice of mathematical physics? I shall look briefly at these three points in turn.

It might be thought that the answer to point (a) is no. The argument might go by analogy with other cases, e.g. in the philosophy of mind or the philosophy of tense and time. There we have a linguistic practice, say the common sense ascription of psychological properties like moods, emotions, desires and beliefs. Such practice when construed realistically seems to induce insuperable epistemological problems like those involving knowledge of the private inner mental states of others. *Suppose* however it was possible to reinterpret that linguistic practice, without in any way rendering as a consequence part of it unintelligible or incoherent, so that the commitment to the epistemologically problematic entities like private inner mental states were no longer present, then this would certainly constitute a philosophical advance. The intimate connection between what was known, that is the objects of knowledge and how we came to know them could be re-established. Whatever the prospects for the success of such a programme in the philosophy of mind might be, in the philosophy of mathematics, at least on standard accounts of constructivist mathematics, the situation looks very different.[31] That is, the programme here is decidedly *revisionary*. If as seems most natural, accepting broadly anti-realist constraints in the theory of meaning of mathematical statements is to issue in not a reinterpretation of classical mathematics, but a radically different mathematics (i.e. intuitionism or perhaps 'recursive' mathematics like that championed by Bishop and Bridges[32]) then structures like the classical continuum are no longer available, they are simply unintelligible.[33] Of course this structure would be replaced by a 'constructive'

[31] Crispin Wright has argued that the acceptance of such constraints on the meaning of *logical* constants need not necessarily be revisionary of classical logic (Wright, 1981, 1986).

[32] Bishop and Bridges (1980).

[33] A very clear introduction to the varieties of purely constructive continua can be found in Trolestra and van Dalen (1988), 251–325.

continuum but this latter sustains results which have absolutely no classical analogue, e.g. the intuitionist continuum is not linearly ordered and every function defined on it is continuous. In my view then the benign view that we can simply reinterpret our linguistic (classical) practice, without revising it drastically and thereby showing that there is no commitment to realism therein, is not a possibility here.

Let us now examine (b) above, which is really the crucial one. The idea behind it is that classical mathematics (i.e. traditional real and complex analysis) is inessential, though no doubt a very convenient and smooth idealization, to the *physical content*, of physical theory. But the question itself seems to have a presupposition that is false, viz. that there exists a partition (sharp enough for the purposes at hand) between the purely *mathematical content* and the *physical content* of a theory in mathematical physics. But I have no more confidence in the existence of such a partition than I have in another famous divide viz. that between the purely theoretical vocabulary and the observational vocabulary of a theory, a partition beloved of early logical empiricists. The reason I do not is that it is extremely difficult to see how some of the core notions and concepts of physics can be formulated without presupposing a very definite mathematical structure in the formulation of these concepts. (To take but one well-known example, consider the key notion of a field in electromaganetism and the relation between charge and field, *pace* Faraday.)

Now Wright has recently raised the issue as to how much of the classical theory of the real and complex numbers is needed to do physics. He wrote:

> I don't know whether the applications in physics, e.g., of the classical theory of the real numbers make full and essential use of the concepts involved in its intended interpretation—indeed, whether physical theory needs to recognise *real* (rather than merely rational numbers) at all.[34]

This is, it turns out, a rather complex question and it is the core of question (c) above. But some things can be said. First the appeal to the (countable) collection of the rational numbers is no doubt motivated by the limits as to what measurement can reveal as the value of some physical observable. And indeed the best that measurement will reveal is a rational number for the value of an observable. But of course physical *theory* which does the job of connecting the value of an observable at one time (initial conditions) with the value at a later time in terms of the dynamical evolution of the states of a physical system cannot be expected to employ only rational values of the state variables.

[34] Wright (1988), 430.

Indeed even though the state of a system may have rational values at one time, in general its dynamical evolution will carry it out of the domain of the rationals (think only of simple harmonic motion). Were one to stick entirely with the rationals, it is to be noted, one could not even do elementary Euclidean geometry, for as far as the existence of number is concerned, elementary Euclidean geometry requires the so-called quadratic extension field. (Roughly speaking the surd field consists of those points that can be obtained from the rationals together with finite iterations of the operations of addition, subtraction, multiplication, division and extraction of *square roots*.[35]) The real point is that even so basic a physical theory as Newtonian mechanics requires in its formulation the use of limit concepts, it requires there to be, for example, in the motion of a particle, the existence (and uniqueness) of an instantaneous velocity at a point, but that is a limit concept and notoriously the rationals is just the sort of structure that does *not* contain as members the limit points of all convergent sequences of rationals. There is no doubt that physical theory as currently formulated requires structures which transcend the rationals.

To answer Wright's question in full it would be necessary to find out just exactly what are the minimum postulates of set existence which are needed to obtain what might be called the core theorems of the practice of mathematical physics, e.g. those theorems which govern the existence and uniqueness of solution of the theory of ordinary differential equations. And to see whether they are obtainable on say constructivist principles. This turns out to be a highly non-trivial mathematical task. It is essentially the programme of 'reverse' mathematics which is a partial realization of Hilbert's programme and has been systematically investigated by Stephen Simpson and his colleagues.[36] The basic idea of this programme is to take a very weak number-theoretic system as a basis, essentially recursive arithmetic (RCA) and to find for a given mathematical theorem say ϕ a set existence axiom φ such that

$$\mathrm{RCA} \vdash \varphi \longleftrightarrow \phi$$

Thus for example, if we take the classical Bolzano–Weierstrass theorem (every bounded [above/below] sequence of real numbers has a least [upper/greatest] lower bound) for ϕ then it turns out that φ is the comprehension scheme $(\exists x)(\forall n)\ (n \in x \longleftrightarrow P(n))$ where $P(n)$ is restricted to formulas involving only bounded existential quantification over N. To take an example directly from physics, if we take ϕ to be the famous Cauchy–Peano existence theorem for the solutions of ordinary differential equations then φ is essentially a comprehension principle

[35] Moise (1973), 220–47.
[36] Simpson (1984, 1988).

asserting the existence of all the *recursive* subsets of N. Thus it seems that for doing ordinary mathematical physics (basic classical physics) the *necessary* axioms of set existence are extremely weak viewed from the canonical classical standpoint according to which the continuum or Real Line is thought of as the collection of all the subsets of N, i.e. the full classical power set axiom is employed. It does seem however an urgent matter, though a mathematically highly non-trivial one, to ascertain exactly what set-theoretic principles are needed to obtain the mathematics of physics. It is known to be impossible to produce constructive existence proofs for some classical results in the existence of solutions to differential equations. The question of real interest now is what is the *physical significance* of negative results of this kind. Until this is completely clear any view on the ratification or otherwise of at least one principle constraint on the acceptability of a thesis in the theory of meaning (that it should be possible to give an account of our understanding of physical theory) must remain in abeyance.

Conclusion

As is plainly evident there are a number of important issues raised by Redhead's paper, anthropic explanation, the merits and account of unity of explanation, the special merits of passion-at-a-distance over causal explanation in the quantum theory, which I have not touched on in my response. I suspect that my readers have suffered enough. Broadly speaking I adopt the line Redhead pursues and I have been pleased to participate in at least a partial defence against some of its critics, of an old view.

References

Beeson, M. J. 1985. *Foundations of Constructive Mathematics* (Berlin: Springer Verlag).

Berry, M. 1978. 'Topics in Nonlinear Dynamics', *American Institute of Physics*, Conference Proceedings, Vol. 46.

Bishop E. and Bridges, D. 1980. *Constructive Analysis* (Berlin: Springer Verlag).

Brush, S. J. 1976. *The Kind of Motion We Call Heat*, Vol. 1 (Amsterdam: North-Holland).

Burgess, J. 1984. 'Dummett's Case for Intuitionism', *History and Philosophy of Logic* **5**: 177–94.

Cartwright, N. 1983. *How the Laws of Physics Lie* (Oxford University Press).

Chaitin, G. J. 1987. *Algorithmic Information Theory*, Cambridge Tracts in Theoretical Computer Science (Cambridge University Press).

Church, A. 1940. 'On the Concept of a Random Sequence', *Bulletin of the American Mathematical Society* **46**: 130–5.

Clark, P. 1987. 'Determinism and Probability in Physics', *Proceedings of the Aristotelian Society,* Supplementary Vol. **LXI** 185–210.

Davies, P. C. W. 1989. *The New Physics* (Cambridge University Press).

Dummett, M. A. E. 1959. 'Truth', in *Proceedings of the Aristotelian Society*, **LIX,** reprinted in M. A. E. Dummet, *Truth and Other Enigmas* (London: Duckworth, 1978).

Dummett, M. A. E. 1973. 'The Philosophical Basis of Intuitionistic Logic', reprinted in *Truth and Other Enigmas* (London: Duckworth, 1978).

Earman, J. 1971. 'Laplacian Determinism, or Is This Any Way to Run a Universe', *Journal of Philosophy* **68**: 729–44.

Earman, J. 1986. *A Primer on Determinism* (Dordrecht: D. Reidel).

Eberle, R., Kaplan, D. and Montague, R. 1961. 'Hempel and Oppenheim on Explanation', *Philosophy of Science* **28**: 418–28.

Field, H. 1980. *Science Without Numbers* (Oxford: Blackwell).

Ford, J. 1983. *Physics Today* **36** (4): 40.

Ford, J. 1989. 'What is Chaos That We Should be Mindful of It?' in Davies, 348–71.

Glymour, C. 1980. *Theory and Evidence* (Princeton University Press).

Grünbaum, A. and Salmon, W. C. (eds), 1988. *The Limitations of Deductivism* (Berkeley: University of Carlifornia Press).

Hempel, C. G. 1965. *Aspects of Scientific Explanation* (London: Free Press, MacMillan).

Hempel, C. G. 1966. *Philosophy of Natural Science* (Englewood Cliffs, N.J.: Prentice Hall).

Hempel, C. G. 1988. 'Provisos: A Problem Concerning the Inferential Function of Scientific Theories' in Grünbaum and Salmon (1988), 19–36.

Hempel, C. G. and Oppenheim, P. 1948. 'Studies in the Logic of Explanation', *Philosophy of Science* **15**: 135–75, reprinted in Hempel (1965).

Howson, C. 1988. 'On a Recent Argument for the Impossibility of a Statistical Explanation of Single Events, and a Defence of a Modified Form of Hempel's Theory of Statistical Explanation', *Erkenntnis* **29**: 113–24.

Howson, C. and Urbach, P. 1989. *Scientific Reasoning* (La Salle, Illinois: Open Court).

Humphreys, P. W. 1978. 'Is "Physical Randomness" Just Indeterminism in Disguise?', *PSA 1978*, Vol. 2, 98–113.

Kreisel, G. 1982. Review *Journal of Symbolic Logic*, **47**: 900–2.

Maxwell, J. C. 1860. 'Illustrations of the Dynamical Theory of Gases', reprinted in *The Scientific Papers of James Clerk Maxwell*, W. D. Niven (ed.), Vol. 1 (New York: Dover), 377–409.

Maxwell, J. C. 1866. 'On the Dynamical Theory of Gases', reprinted in *The Scientific Papers of James Clerk Maxwell*, W. D. Niven (ed.), Vol. 2 (New York: Dover), 26–78.

Moise, E. E. 1973. *Elementary Geometry from an Advanced Standpoint*, 2nd edn (Palo Alto, C.A.: Addision-Wesley).

Montague, R. 1962. 'Deterministic Theories', reprinted in *Formal Philosophy, Selected Papers of R. Montague*, R. H. Thomason (ed.) (New Haven: Yale University Press), 303–59.
Nicolis, G. 1989. 'Physics of Far from Equilibrium Systems and Self Organisation', in Davies, 316–47.
Pitt, J. C. (ed.), 1988. *Theories of Explanation* (Oxford University Press).
Popper, K. R. 1950. 'Indeterminism in Quantum Physics and in Classical Physics', *British Journal for the Philosophy of Science* **1**: 117–33 and 173–95.
Pour-El, M. and Richards, I. 1979. 'A Computable Ordinary Differential Equation Which Possesses No Computable Solution', *Annals of Mathematical Logic* **17**: 61–90.
Pour-El, M. and Richards, I. 1981. 'The Wave Equation with Computable Initial Data Such that its Unique Solution is Not Computable', *Advances in Mathematics* **39**: 215–39.
Pour-El, M. and Richards, I. 1983. 'Non-computability in Analysis and Physics', *Advances in Mathematics* **48**: 44–74.
Putnam, H. 1980. 'Models and Reality', *Journal of Symbolic Logic* **45**: 646–82.
Railton, P. 1978. 'A Deductive-Nomological Model of Probabilistic Explanation', *Philosophy of Science* **45**: 206–66.
Russell, B. 1917. 'On the Notion of Cause', reprinted in *Mysticism and Logic* (London: Allen and Unwin), 180–208.
Salmon, W. C. 1970. 'Statistical Explanation', in W. C. Salmon (ed.) *Statistical Explanation and Statistical Relevance* (University of Pittsburg Press), 29–87.
Simpson, S. G. 1984. 'Which Set Existence Axioms are Needed to Prove the Cauchy/Peano Theorem for Ordinary Differential Equations?', *Journal of Symbolic Logic* **49**: 783–802.
Simpson, S. G. 1988. 'Partial Realisations of Hilbert's Program', *Journal of Symbolic Logic* **53**: 349–63.
Trolestra, A. S. and van Dalen, D. 1988. *Constructivism in Mathematics*, Volume 1 (Amsterdam: North-Holland).
Ulam, S. H. and Von Neumann, J. 1947. 'On Combination of Stochastic and Deterministic Processes', *Bulletin of the American Mathematical Society* **53**: 1120.
Von Mises, R. 1957. *Probability, Statistics and Truth* (New York: Dover).
Watkins, J. W. N. 1984. *Science and Scepticism* (Princeton University Press).
Wright, C. J. G. 1981. 'Dummett and Revisionism', *The Philosophical Quarterly*, 47–67.
Wright, C. J. G. 1985. 'Skolem and the Skeptic', *Proceedings of the Aristotelian Society* Supplementary Vol. **LXI**, 117–37.
Wright, C. J. G. 1986. *Realism, Meaning and Truth* (Oxford: Blackwell).
Wright, C. J. G. 1988. 'Why Numbers Can Believably Be: A Reply to Hartry Field', *Revue Internationale De Philosophie* **45**: 425–73.

The Limits of Explanation[1]

RICHARD SWINBURNE

I

In purporting to explain the occurrence of some event or process we cite the causal factors which, we assert, brought it about or keeps it in being. The explanation is a true one if those factors did indeed bring it about or keep it in being. In discussing explanation I shall henceforward (unless I state otherwise) concern myself only with true explanations. I believe that there are two distinct kinds of way in which causal factors operate in the world, two distinct kinds of causality, and so two distinct kinds of explanation. For historical reasons, I shall call these kinds of causality and explanations 'scientific' and 'personal'; but I do not imply that there is anything unscientific in a wide sense in invoking personal explanation.

The deductive–nomological (D-N) model of scientific explanation[2] teaches that the causal factors involved in scientific explanation are of two kinds—states of affairs (or events, as I shall sometimes call them) occurring at instants of time and describable independently of their consequences, and laws which state that all states of affairs of one sort are followed by states of affairs of another sort. An event (E) is explained if a description of its occurrence is entailed conjointly by a description of the occurrence of a prior state (the initial conditions, or cause) (B) and a statement of a law of nature (L). The operation of a law (L_1) is explained if it is deducible from a statement of the operation of a higher level law, (L_2) perhaps together with some description of boundary conditions (i.e. a state of affairs) (B). Thus an event such as a ball hitting the base of a tower at 2.00.02 p.m., is explained by (1) the occurrence of a previous event in certain circumstances, say the ball being liberated from the top of the tower at 2.00.00 p.m., the tower being 64 ft high, and (2) Galileo's law of fall that all free bodies near the

[1] Some of the main ideas of this paper are contained in my books *The Existence of God* (Oxford: Clarendon Press, 1979) especially ch. 4, and *The Evolution of the Soul* (Oxford: Clarendon Press, 1986), especially ch. 5; though they are expressed here in new ways.

[2] Originally set out in C. G. Hempel and Paul Oppenheim, 'Studies in the Logic of Explanation', *Philosophy of Science* 15 (1948), 135–75.

surface of the earth are subject to an acceleration of approx. 32 ft/sec^2 towards the centre of the Earth. Galileo's law of fall is explained by its deducibility from Newton's laws, given the mass and radius of the Earth. D-N explanations purport to explain fully why events happened. But in the absence of deterministic laws, we have to make do with probablistic laws, which can provide only partial explanation. Hempel's I-S model[3] or Salmon's S-R model[4] purport to show in detail how such partial explanation works. They have in common with the D-N model the crucial characteristic of involving the two factors of states of affairs describable independently of their consequences, and laws. The explanation is partial because the occurrence of the explaining factors does not guarantee the occurrence of the explanandum, but give to it only some degree of probability.

I suggest, following subsequent writers, that the laws of nature of D-N explanation are to be understood as physical necessities (and so their descriptions can usefully include the phrase 'of physical necessity'); and that the probabilistic laws of I-S and S-R explanation concern physical probabilities, i.e. inbuilt propensities in nature.

All these models have the consequence that you cannot explain states of affairs except in terms which involve other states of affairs. Laws cannot explain states of affairs by themselves; if they are to explain states of affairs, they can only do so in virtue of their operation on prior states of affairs. From time to time, including very recently,[5] cosmologists have suggested that a law by itself could explain the beginning of the universe e nihilo. The suggestion (to caricature it a bit) is that the law might have the form 'nothing necessarily gives rise to something' or (a bit more precisely) 'there is a high probability that a zero energy vacuum lead to a space-time of negative or positive energy'. The first thing wrong with a suggestion of this sort is that under close examination 'nothing' never really turns out to be nothing; it is some sort of empty space in a quiescent state. If 'nothing' was really nothing, and the law really was 'nothing necessarily gives rise to something', then

[3] See C. G. Hempel, 'Deductive-nomological vs Statistical Explanation' in H. Feigl and G. Maxwell (eds.) *Minnesota Studies in the Philosophy of Science*, Vol. III (University of Minnesota Press, 1962); and Hempel's final account of explanation in 'Aspects of Scientific Explanation', in his *Aspects of Scientific Explanation* (New York: Free Press, 1965). For an elementary statement of Hempel's account of explanation, see his *Philosophy of Natural Science* (Englewood Cliffs, N.J.: Prentice Hall, 1966), ch. 5.

[4] W. C. Salmon, *Statistical Explanation and Statistical Relevance*, (University of Pittsburgh Press, 1971).

[5] See, e.g. Stephen W. Hawking, *A Brief History of Time* (London: Bantam Press, 1988), ch. 8.

since there are an infinite number of possible universes (each non-existent at some past moment of time), not spatially related to each other, then by the law they must all have come into existence. In fact an infinite number of universes must be coming into existence at each moment of time. But that is to multiply entitites beyond plausibility. A plausible law to explain the beginning of the universe would at least have to have the form 'empty space necessarily gives rise to matter-energy', where 'empty space' is not just 'nothing', but an identifiable particular. In an important sense there must have been something there already. The other thing wrong with the laws-can-do-it-alone suggestion is granted that laws could explain why empty space at time t_0 produced matter-energy at time t_1, that does not explain why there is a universe of matter-energy unless we have an explanation of why the laws took over at t_0 rather than earlier. Some cosmologists would deny that there is any sense in requesting an explanation for the latter, and hold with Augustine (*nullum tempus sine creatura*[6]) that if the Universe began, time began at that time as well. But the only content I can give to 'the universe began 15,000 million years ago', entails that at a time before 15,000 million years ago there was nothing and then there was a universe. How can something begin to exist if it has existed at all moments of past time (i.e. always)? So if there was empty time before the universe, we need to explain why it began at the moment it did rather than earlier, and laws alone will not explain that. To put it another way—give a name which rigidly designates the actual present moment, call it T, and suppose that name to name it even when it is past. Then the question arises—why did the Universe begin 15,000 million years before T, and rather than 16,000 million years? We need states of affairs as well as laws to explain things; and if we do not have them for the beginning of the universe, because there are no earlier states, then we cannot explain the beginning of the universe.

Although I believe that there is a major deficiency in the D-N and associated accounts of scientific explanation, arising from their assumption that we can describe states of affairs without being committed to any claims about the causal powers of the objects involved in them, to which I shall come in due course, what I have to say about regress of explanation in the next three sections, is, I believe unaffected by this point; my points there about how regress works could be made for any preferred model of scientific explanation.

[6] *Confessions* 11.30.

II

Whatever account we adopt of scientific explanation, personal explanation must turn out to be rather different from it—for personal explanation explains in terms of the goals intentionally sought by agents, and there is nothing like that operative in the scientific case. I give my account of personal explanation, which applies to just those cases in which agents do things intentionally.

Let us divide intentional actions in the normal way into the basic and the mediated. In a mediated action the agent (P) does something (e.g. kills a man) by doing something else (e.g. firing a gun). In a basic action the agent does what he does just like that, not by doing anything else—he squeezes his finger, say. The result of an intentional action is that state of affairs, the bringing about of which the action consisted in, but which is describable independently of the action and could have been brought about non-intentionally. Thus the result of me moving my hand, is my hand moving; the result of me killing a man is the man's death. The basic actions normally available to human beings are limb movements and doing things close to their bodies (e.g. tying shoelaces), and certain mental actions (e.g. conjuring up some mental image). An agent may perform a basic action either for its own sake, or as part of a mediated action, to achieve some further goal. The simplest pattern of personal explanation is where we explain some event, say a limb moving (E) in terms of an agent (P) acting intentionally (J) so as to bring about E, in virtue of his basic powers (C). In order to act intentionally he will need a belief that the occurrence of E is in some way a good thing; J could not occur without such a belief; but given that J does occur, we need only C as well in order to explain E. By C, the agent's 'basic powers' I understand his powers to bring about those changes of state which are such that if he tried to bring them about by a basic action, he would succeed. So the components of the explanation of E are the agent (P), his powers (C), and his state of 'acting intentionally' (J) to bring about E.

In the case of mediated action, say P shooting and killing a man, the pattern is more complicated. Here the effect (F), the man's death is caused in accord with normal scientific causality by the result of some basic action, e.g. the squeezing of a finger (E), in certain circumstances (viz when pressing on the trigger of a loaded gun pointing at the man's heart) in virtue of some law of nature (L) a consequence of which is that an event such as E is followed by an event such as F. E is caused by P acting intentionally (J) to bring about F, having the belief (B) that the occurrence of E would cause F and having the power (C) to bring about E. (Again, J could not occur without P having a belief that in some way F was a good thing). Having drawn attention to the structure of explanation in the case of the mediated action, let us now concentrate on the simpler case of the basic action in order to draw the contrast with scientific explanation. The difference turns crucially on the nature of J, which I have called so far, fairly neutrally, 'acting intentionally'. What is it? What is the mental factor which triggers off a basic action?

It cannot, current philosophical jargon notwithstanding, be any passive state in which the agent finds himself, such as having a 'desire' to bring about E, either in the sense of an inbuilt spontaneous inclination to bring about E or in the sense of a belief that the occurrence of E would be a good thing. For we often keep our spontaneous inclinations in check and sometimes act contrary to them. (In a natural sense of the term, we do what we do not desire to do.) And often too we do not act on our beliefs about what is good to do (although, as I have indicated, the belief that E is in some way a good thing is a *necessary* condition of our acting intentionally so as to bring it about). J cannot be any passive state of the agent, for whether that state leads to E depends on whether he chooses to let it lead to E. If J is to be a state of the agent, it must be an active state.

J is trying to bring about E, or rather that mental event which constitutes 'trying' to bring about E, if the bringing about is difficult or unsuccessful, but which it seems a little unnatural to call 'trying' when no effort is required to bring about E.

But what is 'trying'? It is not saying to yourself 'let E occur' or 'let me bring about E' or having a picture of E or anything like that. All these may on occasions occur when someone tries to bring about E, but they are not what the trying consists in. So what is the common element to trying to move your arm, say 'cheese', hear what the speaker is saying, stand on your head etc, etc. The word 'try' has a meaning. What are we saying about someone when we say that he is trying to do these things? The answer is this: he is doing whatever, he believes, will contribute causally to the event happening. By 'contribute causally' to E happening I mean cause E to happen or cause something to happen which in turn will increase the probability of E happening. Trying is causing or helping to cause (or so we believe, and cannot help believing if we are ever to try—there would not be any point in trying if it was not). So P having J is unlike the kinds of state of affairs which feature as the initial conditions of a scientific explanation of the patterns which I have discussed so far, such as a particle being in motion, which is causally inert unless some law of nature gets to work on it, as it were. P having J is P exercising some power; of course whether P having J will produce the intended effect depends on other things, viz C in the case of a basic action. But P having J is nevertheless pulling the causal lever which lets other factors take over. This 'pulling the causal lever' is something which in our own case we can be very conscious of doing, especially when we need to try very hard to bring about some effect. On the account of scientific explanation to which I shall come later and shall prefer to the D-N, I-S and S-R models, the 'initial conditions' are not so causally insert as in the earlier models, but the crucial difference remains that there is (by definition) no equivalent in scientific explanation for the intended, meant, deliberately brought about, exercise of power. Inanimate things exercise their powers either by chance or because they are caused to do so; they do not choose to do so.

Since the kind of causality involved in intentional action must be the same whether the action is successful or not, but talk of 'trying' does not seem in place if the action is easily successful I choose a different word both from 'trying' and from the rather less committed phrase 'acting intentionally' for the J component; my final name for it is 'purposing', which amounts to trying in cases of difficulty or failure. My moving my hand intentionally is me purposing to move my hand which in virtue of my basic powers causes my hand to move.

We come to our investigations of the causes of phenomena equipped with an *a priori* apparatus of general understanding of what is evidence for what, and within that of what is evidence for this being the explanation of that. That in turn can only be because we have some *a priori* understanding of what kind of thing (e.g. initial conditions and laws of nature) provides *explanation* of the occurrence of some event, as opposed to a mere licence to infer its occurrence. We can of course make *particular* discoveries *a posteriori* of what is evidence for what, e.g. that footprints in the sand are evidence that men have walked there, but only in virtue of possessing some general *a priori* pinciple which allows us to infer that past co-occurrences of men walking and footprints are evidence of a general connection (and, as we know from the 'grue' case[7] and many others, that principle is not simple as 'all of many A's having been B is evidence that A's are always B'). Without general *a priori* principles of evidence to get to grips with what we observe, we could never get started. Similarly, we have an understanding given *a priori* of when we observe A and our principles of evidence allow us to infer B, which among such B's are causes which explain A. And if we did not have an *a priori* understanding of what kinds of things explain, *a posteriori* investigation could never discover an explanation. We need to know what an explanation looks like before detailed *a posteriori* investigation will reveal when we have got one. Our *a priori* understanding of the possible patterns of explanation reveals, I have suggested, two such possible patterns—scientific and personal, to which philosophers have tried to

[7] N. Goodman, *Fact, Fiction and Forecast*, 2nd edn (Indianapolis: Bobbs-Merrill, 1965) chs. 3 and 4.

give more precise shape. These are potential patterns, waiting for us to discover which phenomena in the world exemplify them.

III

The causality involved either in scientific or personal explanation may be such that given its operation, the effect could not but occur. The description of all the factors entails the occurrence of the effect. P, J, and C may be such that given the instantiation of J (a purposing to bring about E) in P, E cannot but occur. C is such that if P tries to bring about E, he will inevitably succeed. In such cases we may call the explanation a *full* explanation.

Full explanations do explain, in that they involve factors making inevitable the occurrence of the explanandum. Some events may not have full explanations. To the extent to which the laws of nature are probabilistic (and nothing else controls how they operate), the events which they explain, they do not explain fully, and nothing else does. There may be no full explanation of why some photon passes through this slot rather than that slot, or human basic powers may be indeterministic, in the sense that it is only a probabilistic matter whether some purposing of mine to move a limb is successful. In these cases there is a partial explanation of E to the extent to which the factors involved make the occurrence of E probable. Even if there is a full explanation of E, you or I or even the best scientists of the future may be unable to provide it. We may be able only to provide a partial explanation, yet a full explanation may exist.

Full explanations do not always fully satisfy. We may want to go back further; to know why the factors involved in the full explanation were there in the first place. Why does P exist at all and have the powers (C) he does, and why did he purpose (J) to bring about E? Why did the initial conditions C occur at all, and why is L a law of nature? In the process of explanatory regress the existence and operation of the factors involved in a personal explanation might be explained either by a scientific or a personal explanation; and conversely, the existence and operation of the factors involved in a scientific explanation might be explained either by a personal or a scientific explanation.

Scientific explanation of the operation of some law involves, as we have already seen, its derivation from a higher-level law, possibly given certain boundary conditions. Scientific explanation of the occurrence of initial conditions will consist in deriving their occurrence from some law of nature operating on yet earlier initial conditions. Among the events for which a scientific explanation might be provided are the existence of some person P, his having powers C, and his purposing to bring about E. A full such explanation would, I believe, need to invoke, some laws of nature quite different in character from the physical laws which we have at present—for physical laws deal with the succession of public events, and purposings are mental events, events to which an agent has privileged access. Although I am highly sceptical about whether laws dealing with purposings will ever be constructed,[8] I see no grounds for deeming such laws to be impossible.

Clearly however a partial scientific explanation can be given of how people have the powers for moving their limbs that they do, in the sense that an explanation can be given of a necessary condition for their having those powers. Clearly I have the power to move my arm only if there are certain nerve connections between brain and arm, and if the laws of neurophysiology are as they are, and a scientific explanation can be provided of these phenomena. When I move my arm, I make certain neurones in my brain fire, and they set up a disturbance propagated to my arm. Of course I do not (normally) intentionally make those neurones fire, but by purposing (J) I do make those neurones fire; and a scientific explanation can be given of how under those circumstances my arm moves. But a full scientific explanation of how I have the basic powers I do would need to explain why my purposings make a difference to my brain state, and I am rather sceptical about whether that explanatory gap can be filled by a scientific explanation.

Personal explanation of the occurrence of the initial conditions which occur in some scientific explanation will explain their occurrence as due to the agency of some person, in virtue of his basic powers and a purpose to bring about such initial conditions. Personal explanation of the operation of some law of nature will explain it as due to the action of a person making matter (or whatever is involved) move as it does. His operation on the matter of the world would be like our operation on the matter of our brains; and as we can produce orderly and beautiful movements of our bodies, as in a dance, or produce through our bodies, orderly and beautiful patterns of succession in the world, as the notes of a song which we sing, so a person in control of the natural order could make the fundamental particles move so regularily as to produce our beautiful and orderly universe. (We however, as I have in effect noted, produce the orderly movements of our bodies or successions outside our bodies by bringing about brain states conducive thereto. A person in charge of the universe would produce its orderliness without there being such an intermediate link, i.e. without there being any partial explanation in physical terms of his having the capacity he does.)

Personal explanation of the existence, powers and purposings of some person would explain them as due to the operation of some

[8] See especially my *The Evolution of the Soul*, ch. 10.

(perhaps) different person bringing about the former person with his basic powers and making him make the purposing he does, in virtue of a basic power and purposing of the controlling person.

All these possibilities for regress of explanation exist. But does everything have an explanation in terms of something due? Is there any end to regress of explanation?

I now define a complete explanation of the occurrence of E as follows. A complete explanation of the occurrence of E is a full explanation of its occurrence in which all the factors cited are such that there is no explanation (either full or partial) of their existence or operation in terms of factors operative at the time[9] of their existence or operation. Thus suppose that a high tide is brought about by sun, moon, earth, water, etc. being in certain positions and by the operation of Newton's laws. Here is, let us suppose, a full explanation. Suppose too that Newton's laws operate here because this region of the universe is relatively empty of matter and Einstein's laws of General Relativity operate. These factors act contemporaneously to make Newton's laws operate. Suppose too that nothing at this instant makes sun, moon, etc. be where they are (even though some past cause was responsible for their being where they are). Nor does anything at this instant make Einstein's laws operate or this region of the universe be relatively empty. Then there is a complete explanation of the high tide in terms of the operation of Einstein's laws, the universe in this region being relatively empty of matter, and sun, moon, earth, water, etc. being where they are.

Complete explanation is a special kind of full explanation. I now delineate as a special kind of complete explanation what I shall call ultimate explanation. To speak loosely to start with, we have an ultimate explanation of some phenomenon E if we can state not merely which factors O operated at the time to bring E about, and which contemporaneous factors made O exist and operate at that time, and so on until we reach factors for the contemporaneous existence and operation of which there is no explanation; but also state the factors which originally brought O about, and which factors originally brought these factors about, and so on until we reach factors for the existence and operation of which there is no explanation. Less loosely, I define an

[9] I write, for the sake of simplicity of exposition, in my definition and subsequent use of the notion of 'complete explanation', of factors acting 'at the time of' or 'simultaneously with' or 'contemporaneously with' their effect. It may be that there are logical problems with the idea of some cause C bringing about an effect E literally simultaneously with it. In that case talk about C bringing about E 'simultaneously' is to be read as C bringing about E in such a way that there is no temporal interval between the end of C and the beginning of E.

ultimate explanation of E as a complete explanation of E, in which the factors O cited are such that their existence and operation have no explanation either full or partial in terms of any other factors. Those factors are ultimate brute facts. Suppose that there is no God, that the universe began with a bang in a state X at a time t, that it is governed by deterministic laws L (whose operation is not further explicable); and that in accord with L, X brought about a state Y, and Y brought about a state Z, and Z brought about E. The (X and L); and (Y and L), and (Z and L) are each complete explanations of E; but only (X and L) is an ultimate explanation of E.

Finally let us delineate as a special kind of ultimate explanation, what I shall call absolute explanation. An absolute explanation of E is an ultimate explanation of E in which the existence and operation of each of the factors cited is either self-explanatory or logically necessary. Other explanations cite brute facts which form the starting-points of explanations; there are no brute facts in absolute explanations—here everything really is explained.

I do not believe that there can be any absolute explanations of logically contingent phenomena. For surely never does anything explain itself. P's existence at t_2 may be explained in part by P's existence at t_1. But P's existence at t_1 could not explain P's existence at t_1. P's existence at t_1 might be the ultimate brute fact about the universe, but it would not explain itself. Nor can anything logically necessary provide any explanation of anything logically contingent. For a full explanation is, we have seen, such that the explanandum (i.e. the phenomenon requiring explanation) is deducible from it. But you cannot deduce anything logically contingent from anything logically necessary. And a partial explanation is in terms of something which in the context added to the probability of the occurrence of the explanandum, without which things would probably have gone some other (logically possible) way. Yet a world in which some logically necessary truth did not hold is an incoherent supposition, not one in which things would probably have gone some other way. These are among many reasons why it must be held that God is a logically contingent being, although maybe one necessary in other ways.

Clearly, as a matter of fact there may be complete and indeed ultimate explanations of all mundane phenomena with which we are acquainted—physical phenomena and human mental phenomena. But such explanations by their very nature invoke factors which themselves lack the kind of explanation which they themselves provide for other things. Explanation can end, but must it? Could there not be an infinite regress of explanation? I know of no good argument which rules out, *a priori*, any infinite regress of explanation. I cannot see that there cannot be possible universes, in which states of affairs have no ultimate and even no complete explanation. There might, to take a well known example, be an infinitely long railway train, each truck of which is caused to move by the simultaneous action of the next truck, but there be no first truck, no engine.

IV

However, I do not think that our universe is like that. For the physical and mental state of affairs with which we are acquainted I believe that there are not merely complete but indeed ultimate explanations.[10]

Confronted by the diversity of phenomena, we seek explanations of them; but we only have good reason to suppose that we have got a true explanation, if our purported explanation, our theory, is (1) simple and (2) leads us to expect the phenomena, (3) when these are not to be expected *a priori*.

Simplicity is a matter of few entities, few kinds of entities, behaving in mathematically simple kinds of ways. Other things being equal scientists have always preferred simple theories to complex theories—and that not as a matter of mere convenience but on grounds of probable truth. For if all purported explanations which satisfied the other two criteria were equally likely to be true, no prediction for the future would ever be more likely to be true than any other—for you can always devise an infinite number of theories satisfying the two latter criteria very well and making any given prediction for the future. *Simplex sigillum veri*. (I should add that when I write that 'simplicity is a matter of a few entities . . .' etc., I am writing about the simplicity of a theory where there is no background knowledge of how the world works in neighbouring fields. When there is such knowledge, the simplest theory of a field is that which best 'fits' with theories of neighbouring fields, and that means the theory which makes for the simplest overall theory of all the fields, in terms of 'few entities . . .', etc. But where we are concerned with very large-scale theories, there will be no such background knowledge.)

If our theory is such that from it we can deduce the occurrence of our phenomena, then criterion (2) above, which I call a theory's predictive power, will be maximally satisfied. In so far as a theory only makes

[10] Except in so far as there is a certain amount of indeterminancy within the processes of the natural world (e.g. the fundamental laws of quantum theory being merely probabilistic). There is in that case no full, complete or ultimate explanation of how they operate on a particular occasion, but my claim is that there is such explanation of why there is such indeterminancy in the natural world.

probable what it predicts, then the phenomena are less strong evidence in its favour. In so far as the theory predicts phenomena which may be expected anyway on *a priori* grounds, then again those phenomena provide less evidence of its truth. It follows that the fewer and less detailed the phenomena which a theory predicts, the less likely it is to be true, for the former might well have been brought about by chance. These criteria for the probable truth of a theory operate independently of whether the phenomena in question were discovered before or after the theory was formulated, and so whether they were discovered in the process of putting the theory to test. When phenomena are discovered cannot affect their evidential force. Newton's theory of motion was recognized to be very well supported by the data discovered before the theory was formulated and so to be very likely to be true, for sixty years before evidence of a new kind was discovered.

Among theories able equally well to yield the data (in the sense of satisfying criteria (2) and (3) equally well), the simplest is that most likely to be true. But if we have many complex phenomena to explain, the simplest theory able to yield the data may not be all that simple. If we have a purported explanation of the aberrant motion of a planet in terms of a hitherto unobserved heavenly body pulling it out of orbit in accord with the same simple laws as explain the behaviour of other phenomena, that is grounds for supposing that our explanation is true. But if we need to postulate six new heavenly bodies behaving in accord with hitherto unknown natural laws, in order to explain the aberrant motion, then there is considerable doubt whether we have got the true explanation. If we can derive many physical phenomena by postulating entities of only a few kinds (protons, electrons, photons etc.), that is grounds for supposing that we have got a true explanation. But if we need to postulate a new kind of entity to derive each new phenomenon, then there is little reason for supposing such entities exist. If postulated entities have similar ways of behaving (e.g. entities of $kind_1$ and entities of $kind_2$ both exert electrostatic attractions, but of different strength according to their kind), that is further reason for supposing that the theory is correct, for it postulates less radical differences between kinds of entity. And above all if the postulated entities have constant ways of behaving capturable by relatively simple mathematical laws, that is strong grounds for supposing that they exist.

What goes for scientific explanation goes for personal explanation too. The more phenomena of behaviour we can account for in terms of few persons with continuing powers, beliefs, and purposes having some unity to them, the more likely such purported explanations are to be true. We must not, for example, postulate separate persons at a time or over time in one body unless phenomena (such as those of cerebral commissurotomy) force this explanation upon us. Nor must we suppose that a person's beliefs change frequently in unpredictable ways if we can account for his behaviour in terms of unchanging beliefs or beliefs changing in regular ways.

The many, complex and coincidental cry out for explanation. Science looks for a general system of laws of nature which will explain not merely planetary motion, or all mechnical and gravitational phenomena, but all physical (and, hopefully mental) phenomena. Hence Kepler's laws are explained by Newton's laws, Newton's by Einstein's, and (maybe) Einstein's are explained by the laws of Grand Unified Field Theory. As we move upwards, the laws do perhaps become a little less simple, but they remain extremely simple relative to the enormous range of phenomena which they are able to explain. Science too seeks to postulate as few fundamental kinds of entity as possible. It has analysed many thousands of substances as made up of a hundred or so chemical elements, it has analysed these as composed of fundamental particles, and these as composed of a few basic kinds of quark. When we reach as simple a starting point as possible which will lead us to expect the phenomena which we find, with a reasonable degree of probability when we would not otherwise expect them (and increase of predictive power cannot be gained except at a cost of great loss of simplicity), then we must stop and believe that we have found the terminus of explanation—the final brute fact which provides the ultimate explanation of all other phenomena.

The trouble with physical explanations of the phenomena is that they must have as components not merely laws but initial conditions; and the very success of physics has shown us the operation of some very simple laws of nature, and vast numbers of constituents of the universe of a few very simple kinds (quarks) behaving in exactly the same ways over vast numbers of years of time and light-years of distance. Any ultimate scientific explanation of the Universe as a whole would seem bound to contain as well as simple laws, an initial condition involving an arrangement of vast numbers of particles of a few kinds, each of a kind being qualitatively identical to each other of that kind; or a vast chunk of qualitatively similar stuff out of which the particles evolved. For physics deals with the spatially extended, and unless there was an initial (potential or actual) simplicity above this (consisting in the similarity to each other of its many parts), it could not have given rise to the vast orderliness of the fundamental constituents of the Universe at the present time.

I have written that any ultimate scientific explanation of the universe as a whole would 'seem' bound to have the above character, for I do not underestimate the ingenuity of theoretical physicists. They might propose a system of laws which had the consequence that however random and diverse the constituents of the universe, the laws would, as it were, 'sort them out'. And this is the point at which to comment that there is something highly misleading about the D-N and similar explanatory schemes. Their picture of the universe is of a universe with two components—stuff in various states, and laws of nature which form a kind of invisible grid covering the universe and dictating how the stuff will behave; the stuff is inert and its states are passive, only the laws stir it into motion. Although this picture has some plausibility when we are considering fairly readily observable states, such as balls being liberated from towers, it loses its plausibility when we come to the level of electrons, and photons, and even further to the level of quarks. For such particles are the particles they are in virtue of their powers to interact with other particles and produce various effects. An electron would not be an electron unless it had certain powers of attracting and repelling other charged bodies, and so generally, the stuff of the universe is imbued with powers and liabilities to exercise those powers, it is not just passive. We could not say what the 'stuff' of the universe is, unless we built powers and liabilities into it. Once that point is recognized, we see that although scientific explanation cannot dispense with initial conditions, it can dispense with laws. All talk of a law that all bodies of certain kinds have to behave in certain ways under certain circumstances can be translated into talk about each body of that kind having the power to behave in certain ways, and the liability to exercise that power under certain circumstances. Once we see that we have to attribute some powers to objects, we come to appreciate that we get a simpler picture of the universe if we abandon the 'invisible grid' model and suppose that all the necessity in nature is built into the objects which make it up. Scientific explanation then consists in explaining the occurrence of events as brought about by objects in virtue of their powers and liabilities to exercise them. The 'laws of nature' are simply descriptions of the powers which belong to all objects of a certain kind and of the circumstances in which they are liable to exercise those powers. This in embryo is the P-L (powers-and-liabilities) accounting of scientific explanation.[11]

But then if we build the 'laws' into the stuff of the universe, what follows is that however 'random and diverse' the theoretical physicist claims the stuff of the universe to have been, when the simple laws of nature proceeded to sort it out, what his hypothesis more properly amounts to is hypothesis that while the stuff of universe was random and diverse in many ways, all of its vast number of parts had all the same simple sorts of powers and liabilities. Physics cannot avoid acknowledging as its starting point what is in effect a vast coincidence in the

[11] For the main idea of this account, I am of course indebted to R. Harré and E. H. Madden, *Causal Powers* (Oxford: Blackwell, 1975).

behavioural properties of objects at all times and in all places. Yet no rational inquirer, moved by the scientific spirit, can remain content with vast unexplained coincidences as his stopping point. He must go further, if he can. And, fortunately, he can.

V

If there cannot be a well-justified ultimate explanation of the scientific kind of the physical and mental phenomena which we find around us, perhaps there can be a personal one. Theism provides just such an explanation. The hypothesis of theism is that the universe exists because there is a God who keeps it in being and that laws of nature operate because there is a God who brings it about that they do. He brings it about that the laws of nature operate by sustaining in every object in the universe its liability to behave in accord with those laws. He makes the universe exist now by keeping those laws operative on previous matter (i.e. by conserving the previous matter of the universe through laws of conservation) and originally (if the universe had a beginning) by bringing into existence matter subject to those laws.[12] The hypothesis is a hypothesis that a person brings about these things for some purpose. He acts directly on the Universe, as we act directly on our brains, guiding them to move our limbs (but the Universe is not his body—for he could at any moment destroy it, and act on another universe, or live without a universe). Unlike any original 'stuff' of the Universe, a person need not be an extended thing, and the hypothesis of theism is the hypothesis of a non-extended person.

The hypothesis that there is a God is the hypothesis of the existence of the simplest kind of person which there could be. A person is a being with powers to bring about effects, beliefs about how to do so and what is good about doing so, and purposes which he seeks thereby to execute. We humans are complex persons in that we have limited powers; a limited number of beliefs, some of them false; purposes, some of them

[12] Any explanation of the present existence of the universe by the action of God is not merely complete, but ultimate, because in bringing about the laws of nature (i.e. the powers and liabilities of objects) what God brings about is not just laws that, as a matter of fact are such that, given a past state of the universe, there is a present state (i.e. matter is conserved); but that the laws of nature are such as to yield the present state (whether by making something out of a previous nothing, or by conserving in existence previous matter). Thus his action by itself suffices to explain the present existence of the universe, independently of any earlier factors (God, his purpose and capacities) which have no explanation in terms of anything else.

formed under the influence of irrational forces, while some are formed rationally, i.e., because we see something as a good to be pursued. God, however, is by definition omnipotent (that is, infinitely powerful; there are no limits to his powers), omniscient (that is, all-knowing; he has true beliefs about all things), and perfectly free (that is, free from all causes influencing him; and so forming his purposes solely in virtue of his perception of what is good to do). He is a person of infinite power, knowledge and freedom; and so a person to whose power, true belief, and rationality there are no limits except those of logic. The hypothesis of the existence of a being with zero limits to the qualities essential to a being of that kind is the postulation of a very simple being. The hypothesis that there is one such God is a much simpler hypothesis than the hypothesis that there is a God who has such and such limited power, or the hypothesis that there are several gods with limited powers. It is simpler in just the same way that the hypothesis that some particle has zero mass or infinite velocity is simpler than the hypothesis that it has 0·32147 of some unit of mass or a velocity of 221,000 km per second. A finite limitation cries out for an explanation of why there is just that particular limit, in a way that limitlessness does not.

Infinity is the notion of *zero* limits to the length of a sequence or whatever. Infinite power is power with zero limits upon it; infinite freedom is freedom subject to zero limits upon it, and so on. Infinity is the opposite of zero. Scientists prefer hypotheses which invoke zero or infinite degrees of properties to hypotheses which invoke finite degrees of those properties if they are able to yield the data equally well. Hence the preference for theories which postulate that a particle has zero rest mass or infinite velocity, in the absence of contrary considerations. It needed particular observations in the seventeenth century incompatible with it before the hypothesis that light has infinite velocity was abandoned;[13] and it needed to fit with overall theory (giving it ability to explain other kinds of data as well), before Newton's assumption that the gravitational force is propagated with infinite velocity was abandoned in the twentieth century in the General Theory of Relativity.

God has, *ex hypothesi*, the power to create an orderly universe, but why should he choose to do so? An orderly universe is a thing of beauty, and that is reason enough for God to make it. But, even more import-

[13] The hypothesis that light has infinite velocity was held by the most influential writers before Romer's discovery of its finite velocity, such as Aristotle and Descartes who held that the transmission of light was the transmission of the state of a medium, rather than of a particle. Those who adopted a particle theory were moved by considerations of 'fit' with the behaviour of other particles to postulate a finite velocity, i.e. considerations of the simplicity of an overall theory able to explain other data as well as those concerned with light.

antly, an orderly universe is a universe which men learn to control and change. For only if there are simple laws of nature can men predict what will follow from what—and unless they can do that, they can never change anything. Only if men know that by sowing certain seeds, weeding and watering them, they will get corn, can they develop an agriculture. And men can only acquire that knowledge if there are easily graspable regularities of behaviour in nature. It is good that there be men, embodied mini-creators who share in God's activity of forming and developing the universe through their free choice. But if there are to be such, there must be laws of nature. There is, therefore, some reasonable expectation that God will bring them about. For God, being perfectly free, is uninfluenced by influences (such as irrational desires) which deter him from pursuing the best as he sees it, and, being omniscient, he will always recognize the good for what it is. Although an orderly universe is good, there are no doubt many alternative good actions which God might do, which have no connection with producing an orderly universe like ours, and it is possible that it be no better that he produce an orderly universe than that he do something incompatible therewith. For that reason I do not wish to claim that it is certain that he will produce an orderly universe, only that there is a reasonable probability that he will.

So theism postulates a person who brings about an orderly universe in virtue of his infinite powers for the sake of its beauty and its utility to men in virtue of his true belief that intrinsically it will have these properties. It postulates the existence of a very simple kind of person from whose very nature it follows that there is quite a probability that he will so act. It yields the data with reasonable probability; and it is by far the simplest hypothesis that does. The data cry out for explanation in view of their complex and coincidental character. Theism is able to provide the simplest possible ultimate explanation of the existence of the universe, and therefore the one most probably true. Here is the ultimate brute fact which provides the limit to explanation.

Limited Explanations

STEPHEN R. L. CLARK

Introduction

When I was first approached to read a paper at the conference from which this volume takes its beginning I expected that Flint Schier, with whom I had taught a course on the Philosophy of Biology in my years at Glasgow, would be with us to comment and to criticize. I cannot let this occasion pass without expressing once again my own sense of loss. I am sure that we would all have gained by his presence, and hope that he would find things both to approve, and disapprove, in the following venture.

I also hope that my relationship with Flint, of friendly and outspoken disagreement in the service of a shared ideal, may be allowed to colour my opening response to Richard Swinburne's paper. Flint and I both thought of ourselves as philosophers, and as philosophers convinced of the relevance of biological inquiry to any truly informed anthropology. Richard Swinburne and I are both theists, and convinced of the relevance of reason to the proper grounding and explication of our theism. But even if I were entirely at one with him, agreement and mutual approval make for poor theatre, as well as poor philosophy. Pure and dismissive disagreement, of course, is not merely poor philosophy: it is not philosophy at all. To those, if there are any present, who see literally nothing to be said for theism, I can only offer the following aphoristic warning: 'when one does not see what one does not see, one does not even see that one is blind'.[1]

I have several problems with Swinburne's argument, which will also serve to open the way to my own development of orthodox classical theism. First of all, I am much more sceptical even than he is about the genuinely *explanatory* nature of either 'scientifically discoverable law' or 'essential causal powers'. Secondly, I am not convinced that anyone else needs to dismiss so quickly the possibility that there was or is indeed a single, simple beginning which was nonetheless wholly non-personal. Thirdly, it seems that more explanation is needed for Swinburne's conception of the 'simplest kind of person': all the persons that naturalists will admit they know are, as a matter of fact, very complex

[1] Paul Veyne, *Did the Greeks Believe in their Myths?*, trans. P. Wissing (University of Chicago Press, 1988), 118.

animal organisms. I do not have to agree with Richard Dawkins—and as a matter of fact I do not agree—that theism is an attempt to explain complexity by postulating a being of a still higher order of complexity[2]—to agree that that is how a great many people understand the hypothesis that a 'person' makes the world. Fourthly, Swinburne's anthropocentrism is certainly not mine. Finally, I am not convinced that it is right to work 'up' to theism from a simply non-religious evidential base. Swinburne has had a lot to say elsewhere about the moral and spiritual dimensions of theism, but I suspect that the impression is still left that his 'theism' is still too much like a belief, however well grounded, in alien intelligences to be quite true to what Abrahamic theologians have intended. I add at once that 'classical theism' itself has often enough been contrasted with that same Abrahamic tradition, as I suppose, unjustly.

Philo's Answer to Cleanthes

The Cleanthes of my section-heading is, of course, one of the interlocutors in Hume's *Dialogues*,[3] an advocate of rational theism whose arguments have for many years been considered plainly inadequate. Like Swinburne, he supposes that God's existence, like that of anything else, is not strictly necessary, but constitutes the best available explanation for the kinds of order we discover around us. 'Philo' is the name of Cleanthes' opponent, usually 'identified', with whatever cautions about the equation of authorial intent with the words of any one literary character, with Hume himself. Philo is, I should emphasize, a genuine agnostic who fully appreciates the compelling force of our confrontation with a world of teleonomic order and exhilarating beauty.

First of all, how can we treat scientifically discoverable laws as offering any kind of 'explanation'? How is the particular conjunction of A and B 'explained' by saying that there is a discoverable set of A-things conjoined with B-things? 'Why does this happen? Because it always does'. This is not explanation, but a refusal to take puzzles seriously at all. Even if 'its always happening' is a datum strictly deducible from some yet more extensive generalization, the thing is no nearer to intelligibility. 'The whole modern world-view is based on the illusion-

[2] Richard Dawkins, *The Blind Watchmaker* (London: Longman, 1986). See my review article, *Times Literary Supplement* **4356** (26 September 1986), 1047–9. The same point was made by J. J. C. Smart after the opening paper of the conference.

[3] D. Hume 'Dialogues Concerning Natural Religion', in *Hume on Religion*, R. Wollheim (ed.) (London: Fontana, 1963), 99ff.

ary deception that the so-called laws of nature offer explanations of natural phenomena'.[4] Or as Chesterton, a much neglected philosopher, remarked: '[Scientific men] feel that because one incomprehensible thing constantly follows another incomprehensible thing the two together somehow make up a comprehensible thing . . . It is no argument for unalterable law that we count on the ordinary course of things. We do not count on it; we bet on it'.[5] I doubt if the situation is improved by transferring our attention to causal powers. That an entity has the intrinsic power to do such and such presumably means more than that things of the same kind frequently do that, or that there seems no reason, at any rate, why they should not. 'X has the *power* to do A' cannot mean simply 'It is *possible* for X to do A' nor yet that 'Things like X do acts like A more often than not.' Those are not explanations, but descriptions and associations. If laws of nature or statements of causal power are to function as genuine explanations, they must record some real constraint on the sheer indifference of possible events. To explain the phenomena is to show how they express or represent a deeper reality, to show how what had seemed to be many distinct particulars are better seen as some one thing. Simplicity is not simply a desideratum in choosing between theories: simplification is what such theorizing is all about, finding 'the one in the many'. There are those who profess to have abandoned the *explanatory* role of science, in favour of a merely documentary or annalistic intention, but I doubt if this is true to their motives. 'No reader of the journal *Science* is likely to be content with the statement that the searching out of the ideas that govern the universe has no other value than that it helps human animals to swarm and feed. He will rather insist that the only thing that makes the human race worth perpetuating is that thereby rational ideas may be developed and the rationalization of things furthered'.[6] Philosophy, as Aristotle said, begins in wonder,[7] and that wonder is merely neutralized, not fulfilled, by the recitation of banal generalities.

So why do I find that a problem for Swinburne's analysis? After all, his main point is that a theory expressed in terms of scientifically discoverable laws or powers will always leave us with a complex bare fact. Even if there were only one possible way in which the totality of things could exist (so that everything that had seemed arbitrary or odd

[4] L. Wittgenstein, *Tractatus Logico-Philosophicus*, trans. D. F. Pears and B. F. McGuinness (London: Routledge & Kegan Paul, 1961), 6.371.

[5] G. K. Chesterton, *Orthodoxy* (London: Fontana, 1961), 50f (first published 1908).

[6] C. S. Peirce, writing in *Science* (1900), 620f. I should add that I myself think there might be other ways of vindicating humanity.

[7] Aristotle, *Metaphysics*, 1.983a11ff.

comes to seem as expectable, as comfortable, as the incommensurability of the diagonal), the fact that they do exist, like that, would be sheerly unintelligible. 'That something exists' is not a thesis deducible even from the most perfect theoretical synthesis of what that thing must be like if it does. Swinburne further suggests that the scientist will always be faced by an unexplained coincidence of powers: that everything somehow adheres to just the same blueprint. What is unimaginable about a universe such that literally no system of natural laws governs more than an arbitrary segment of reality, that there are literally no uniformly discoverable classes of things? But why does it help to bring in God's intention here? For all I can see He might as well intend just such a shifting and finitely unintelligible world. The more powerful and less limited the Creator the *less* reason we have to expect a simple, parsimonious creation.

The point is this: if Swinburne's theism is to be an explanation of order in diversity it must display the phenomenal world as an expression or communication to our senses of an intelligible whole, such that the conceived similarity of one region of space-time to another is not a contrived one, but a recognized identity. What we need is the kind of image Plato proposed in the *Timaeus*,[8] of a world that is constructed to one single blueprint: but it has always been a matter of debate whether Plato intended us to take this as a literal construction by a 'personal' being, or simply as a way of saying that a reasoned explanation of what is directs us to the interconnected, mutually intelligible structure of mathematically describable reality, one thing in multiple manifestations. What exactly is added by saying that things are so because someone wants them to be? Either there are real constraints on what that one could want or there are not: if there are not, we have no explanation that is more than a brute fact (and so no explanation); if there are, might not those self-same constraints themselves have an effect on what, physically, there is?

If things did 'begin', Swinburne suggests, we shall have to ask not only why they began, or are, at all but why they began just then after aeons of dusty eternity. His question is a strange one, for it is—as he knows—exactly the question that the pagans posed the Hebrews: why did God make the world 'just then', what was there to choose between one empty instant and another. The Augustinian answer, that it is only in the context of a changeful universe that talk of instants makes any sense at all,[9] is open, with suitable modifications, to the naturalist. Things began so and so many years ago, but there can only now be an unreal, abstract 'time' 'before' that beginning. If it began then, then it

[8] Plato, *Timaeus*, 29.

[9] Augustine, *Confessions*, 11.30.

did not exist at any period preceding that moment, but those 'preceding periods' are artefacts about which we do not need to be realistic. If we are, then we have to admit that the world has 'always' existed (as Aristotle thought).

So what have we got so far? If God is not to be questioned on 'why He made the universe just *then*', then the naturalist need not explain why the cosmos exists 'just then'. In neither case is there any real, vacant container waiting infinitely long for an occupant. Whether the world is finite but unbounded temporally as well as spatially is, so far, moot. Further, if God's intention to maintain the world is to be an adequate explanation of the world it must be mediated through a single, unifying scheme. But in that case what does 'God's intention' add to the bare judgment that things are as they are because, if they are at all, they must be so? Would it be more likely that the cosmos should exist if there were a person already, with the intention to make just such and such a cosmos, than if there were not? How can we possibly answer such a question? What 'probability' does the cosmos have of being? Either it exists within a larger sea, or not. If it does, then we know too little of that whole to guess 'how often' cosmoi lurch into being from the happy nothing—but in any case (as Swinburne says) such a larger sea would itself be the cosmos, hypercosmos, that we need to explain. If the cosmos no more exists 'somewhere' than it exists at a particular period of 'meta-cosmic' time (for there is neither real space nor time outside the cosmos), then it makes no sense at all to ask how 'probable' it is. Swinburne would probably reply that probabilities rest not only on background data, but on intrinsic characteristics such as 'simplicity'. The 'simpler' theory is more likely, and it is undoubtedly 'simpler' to suppose the ungrounded existence of a single person than the ungrounded existence of an irreducibly complex material universe. It is more 'likely' that one simple thing 'emerge' from Nothing than that many should. But there are problems about that answer (including ones that I shall consider later): in particular, it seems impossible to give any content to a notion of 'simplicitly' without acknowledging the existence of a 'natural language', a way of thinking that 'cuts reality at the joints', in Plato's phrase.[10]

Conversely: how likely would God make the cosmos? In the absence of any grip on what God might be like, what His intentions are, it does not seem easy to calculate how often He would make a world, nor whether such counting operations make much sense outside the turning circles of the world.

And what might God be like? The being that Swinburne postulates is 'the simplest kind of person which there could be, . . . a being with

[10] Plato, *Phaedrus*, 265.

powers to bring about effects, beliefs about how to do so or what is good about doing so, and purposes which he seeks thereby to execute . . . [who is] by definition, omnipotent . . . omniscient . . . and perfectly free'. It seems, by implication, that he is confronted—like Plato's imagined Demiurge—by facts about what is good and beautiful (which Swinburne also recognizes). But once again, if He does whatever it is He does as being the right thing to do, or as expressing a beauty which ought indeed to be expressed, why might we not say instead that the cosmos is what it is as expressing just that beauty? Why might things not continually arise from the happy nothing to embody, or approximately to embody, what would otherwise be a merely 'ideal' schema? If a cosmos exists that is rationally intelligible, it exists as one embodying a theory, a schema which is characterized by such qualities as order, simplicity, mathematical elegance, 'beauty'. If the cosmos that existed were not such as to do so it would not be rationally intelligible, but rather a mere aggregate or sequence of one damn particular after another. If we believe (as presumably we must if science is to be more than a passing fancy) that the cosmos that exists is intelligible we might not unreasonably suspect that it is elicited, as it were, by the mere attractive force of that formed order. So what does 'God's intention' add but an extra wheel that turns without effect?

And can we understand something as a 'person' which is so unlike the person we know? Our thought is fleeting, evanescent, submerged in an incessant chatter about this or that. 'A mind whose acts and sentiments and ideas are not distinct and successive, one that is wholly simple and totally immutable, is a mind which has no thought, no reason, no will, no sentiment, no love, no hatred; or in a word no mind at all.'[11] Our capacities are grounded in the material, animal nature of our bodily selves, and of the world. Can there be *thought* where there can never be ignorance or uncertainty? Can there be *action* when just wishing makes it so? God, by hypothesis, 'imagines' the cosmos into being, and the being it has is never opaque to Him: it is nothing but what He imagines it as being (which is why Berkeley was correct to see that theism entails the non-existence of a strictly material world). So the cosmos is nothing but what God thinks: how then can He be a person distinct from the cosmos unified in His thought? He *could* think something else? Then how does His presence either explain what is or guarantee that it will not instantly 'change' into something else?

In sum: what do we add to our understanding of why things are as they are by suggesting that the single coherent and rationally describable universe is a 'made' thing? Why not instead agree that if the cosmos is to be intelligible it must embody or refract a single coherent schema,

[11] Cleanthes speaks: Hume, *Dialogues*, 133.

and that its existence is either a bare fact (neither probable nor improbable) or else something elicited by just that schema's attractive, not effective, causality? The God we are asked to hypothesize turns out, at best, to be the theory or schema animating the cosmos: why do we need to suppose that anyone has the scheme 'in mind' until it is uncovered by creatures bred, with the appropriate powers, in the cosmos required by it?

Demea's Reply to Cleanthes:

You will all have recognized by now that I am not allowing Philo (as it were) the full range of Humean objection. That the universe 'just is' whatever it is, and that we 'just do' have the power to comprehend it as the one intelligible possibility, are not theses that I regard as rational (which is obvious) or remotely attractive (which is less so). The really serious questions must always be these. What prevents the world being a heap of disconnected systems, united only in the descriptions of language-users? And what can possibly bring it about that creatures bred to the kind of intelligence G. H. Mead spoke of, of getting a meal when hungry and avoiding being one,[12] have any ability to discern the fundamental way of things? Why should we expect the world to have a shape? Why should we expect to be able to discern that shape? Merely nominalistic, associative accounts of the kind Hume sponsored leave us with no rational confidence in any kind of science. It is time that we all admitted that, for all Hume's greatness, he would have brought science to a dead-stop if anyone had been ready to listen to him on that point.

But though I have disregarded Philo's more arbitrary flights of fancy, I have followed him in suggesting that we have no clear way of weighing probabilities as between the existence of a cosmos bodying out a formal schema and one united in the real thought of a being vastly unlike any person that we know. Swinburne needs there to be a schema, I remind you, both to ensure that God's making of the cosmos is not a radically diverse or discontinuous one, and to give Him a definite reason for acting as He does. But if the schema exists—and I may as well come out into the open and say 'if the Platonic Forms exist'[13]—it is not immediately obvious that we need a demiurge as well.

[12] See Van Meter Ames, 'No Separate Self', in *Philosophy of G. H. Mead*, W. R. Corti (ed.) (Amrisweiler Bucherei: Amriswil, 1973), 43.

[13] I must at once remind my readers that 'the Theory of Forms', so called, is not exhausted by a few remarks in Plato's *Meno*, *Phaedo* and *Republic*. A good introduction to what I intend by the theory is J. N. Findlay, *Ascent to the Absolute* (London: Allen & Unwin, 1970).

Part of the trouble, I think, is that Swinburne's God is, expressly, a brute fact. Swinburne accepts, like most philosophers over the last two centuries, that being such-and-such, *Sosein* is never the same as being, *Sein*. 'Tigers exist' means only 'there are tigers', and any kind of thing can be described without thereby being defined into existence. This is the principle which Swinburne quite fairly brings into play against those writers who suggest that a theory about what things must be like if anything exists at all can somehow itself explain the fact that any thing exists. But if this is so then bare existence is indeed the limit of explanation: we cannot explain the fact that the totality of things exist, even if we have some idea about what may exist *together* (i.e. what the totality is like). Either that totality includes, and depends internally, upon Swinburne's God, or it includes no single present 'point of view' in which the totality is unified, but the totality does nonetheless embody or is 'motivated by' the Platonic forms, or (finally) the totality has no real unity or internal reason. Whichever of these totalities in fact exist does so for no reason.

That, of course, is the problem. Arthur Prior,[14] another philosopher who died untimely, drew on an argument of Jonathan Edwards to rebut the then-fashionable suggestion that the world might be being constantly restocked by the spontaneous generation of matter. Prior pointed out that if such generation were possible at all there could be no limitation on what might be generated. Any attempt to limit the class of things that could just come into existence must identify causal factors in the preceding conditions, since it cannot be anything about the *Sosein* of non-existent things which would require them to exist. The point must also apply to the cosmos as a whole. If it is possible for a cosmos just to come into existence, without any precedent base, then any kind of cosmos could, including incoherent and ridiculous ones—perhaps like Terry Pratchett's Disc-world.[15] Why should *simple* worlds (like Swinburne's God) have any greater likelihood than infinitely complex ones? In which case we might be living in exactly such a world, and science is hopeless.

So perhaps the only cosmos we could sensibly conceive as existing 'for no reason' would be one that did not 'come into being' at all, one that always did exist? But this too allows for an infinity of universes, each infinite in temporal extent, and most wholly unreasonable. If it is sensible to suppose of anything at all that it 'just is' then anything at all could just be. What reason could we have to exclude any possibility?

[14] A. Prior, *Papers on Time and Tense* (Oxford: Clarendon Press, 1968), 59ff.

[15] See *The Color of Magic, et al.*: the Disc-world rests on the back of four elephants who are standing on a giant turtle swimming through interstellar space.

In brief, it is a lot more difficult than analytical philosophers have supposed to suppose that nothing in the *Sosein* of any being can guarantee, or probabilify, its actual existence. Swinburne himself, I suggest, accepts the *necessary* being of those standards of order, beauty, justice that provide his God with a reason for acting and a pattern to follow. In fact he has to accept the precosmic being of such standards of simplicity as secure his God's existence rather than Pratchett's Turtle. It is not possible to mean anything coherent by the denial of those fundamental standards, to say that there is no such thing as truth, no such thing as rationality, no such thing as good. Nor, so I believe, is there any coherent nominalistic account of what such terms intend. As Armstrong has pointed out,[16] nominalists seek to replace real presences that are the *same* in their various manifestations (as it might be, the same number, nature or relationship) by insisting that this is a merely verbal identity, in that we use the *same* word to refer to what are really many things (and might as easily have used another). But the *sameness* of the word is as much in doubt as the *sameness* of the things we (who?) speak about. All attempts to evade realism end in gibberish.[17] But these things, these Forms, do not strictly *exist*? If by 'exist' we mean 'have particular instances', that may perhaps be so, but there are at least two things to say in reply. First, that we beg the question against Plato by assuming that particulars are real, while Forms are not. Plato's 'great inversion', as Findlay points out, was to see and say that Forms are real, and manifest, the one in the many, in their multiple reflections. Forms have being just in virtue of their *Sosein*, and particulars 'are' only by derivation. Second, that it may after all be a requirement of the Forms' being that they be instantiated: could they be at all and not scatter their reflections through the happy nothing?

So there is after all something that must be, something which it makes no sense at all to declare unreal, namely the systemic whole of Forms centred and enlivened by the good itself, by what (exactly) ought to be. That rationally unsurpassable whole simply is, and may perhaps require its own reflection, refraction, 'scattering' through the physical and phenomenal world. Correspondingly, we 'explain' what we now, consensually, observe by seeing how it *has* to be like that. The description under which its necessity is seen, of course, may be very unlike our first, or 'natural' descriptions.

You will recognize that what I have reinserted into the argument is a version of the ontological argument, which is grossly misunderstood if

[16] D. M. Armstrong, *Universals and Scientific Realism* Vol. I: *Nominalism and Realism* (Cambridge University Press, 1978), 18ff.

[17] See R. Trigg, *Reality at Risk* (Brighton: Harvester, 1980).

its post-Platonic context is forgotten. What cannot rationally be denied must be, and must be as it is. That is the standard by which all of us are guided and attracted in our pursuit of wisdom. That is the standard to which the whole shuddering, variegated, incurably vague world is always approximating. Post-Platonists like Philo of Alexandria called it the *Logos*.

Hume's Demea endorsed the necessary existence of that than which nothing more perfect can be conceived—which is to say, of the very standards of truth and righteousness themselves. Those who purport to deny their being cannot be understanding what they say. But he also endorsed the so-called Negative Way, the rejection of all attempts to give a positive characterization of God, or to include God in the same class as any finite beings. Hume, via Cleanthes, insists that an absolutely incomprehensible Deity differs in no important respect from any unknown and unintelligible 'first cause' of the kind that atheists and agnostics might indifferently admit. The God of classical theism *has* no properties (for He cannot lose any), nor is He to be likened to anything in the earth or out of the earth. The One itself, which we may also call the Good itself, lies even 'beyond being', even beyond the *Logos* of which it is, incomprehensibly, the source.

Hume's Demea is presented as (and almost always taken to be) a doctrinaire and evasive bigot. Atheists usually prefer to deal with Cleanthes: even if he is mistaken, he appears to have a fairly clear hypothesis to offer, for which he offers empirical evidence. Swinburne himself is not quite Cleanthes—for he offers a reason, in simplicity, to postulate a god of infinitely great power and knowledge rather than a god of just so much power and knowledge as would be needed to produce the phenomena. In doing so he enters a domain where it is not easy—to put it mildly—to say what is or is not intelligible. In so far as the being God is could not conceivably lose his powers, His being cannot be well distinguished from His powers. 'It must not be thought that there are many different kinds of perfection in the First, which we designate by names [such as "the just", "the generous", "the living"], into which it could be divided and from the totality of which its substance is composed, but one should understand that one substance and one absolutely indivisible existence is denoted by those names.'[18] God is not omnipotent: 'Omnipotence' is one name for Him, and what real powers and possibilities it includes, God knows. If Swinburne is prepared to edge as close to Demea as that, might it not be worth reconsidering Hume's prejudicial judgment?

[18] Al-Farabi, *On the Perfect State*, trans. R. Walzer (Oxford: Clarendon Press, 1985), 99.

How does an agnostic, who says that the first cause is unintelligible and unknown, differ from an apophatic theist, who says that God is incomprehensible? 'God's intention' may be no true explanation, but at least it looks like one: 'whatever it is that terms like "the One", "the Void" are gesturing at' seems like no explanation at all.

Which is quite true. But we need to look again at what sort of activity 'explaining' is, and to be more cautious before we dismiss so long a tradition that the One is not to be pinned down in words.

Consider Koestler's experience in jail as he awaited probable execution. Passing the time by reconstructing Euclid's proof that there is no highest prime number, he became so enraptured by the way that 'a meaningful and comprehensive statement about the infinite (can be) arrived at by precise and finite means' that the thought of his impending death became a trifling, faintly amusing irrelevance. From that rapture he then passed into a state which cannot be represented in our language.[19] Whether that was 'the same' state as Plotinus sometimes reached does not seem to me to be a question having any clear meaning. It did at least identify for Koestler, as he pondered his experience, a way to represent our common experience.

'The narrow world of sensory perception constituted the first order; this perceptual world was enveloped by the conceptual world which contained phenomena not directly perceivable [*sic*!] . . . If illusions of the first type were taken at face value then the sun was drowning every night in the sea and a mote in the eye was larger than the moon; and if the conceptual world was mistaken for ultimate reality the world became an equally absurd tale, told by an idiot or by idiot-electrons which caused little children to be run over by motor cars, and little Andalusian peasants to be shot through heart, mouth and eyes, without rhyme or reason.' And again: 'time, space and causality . . . isolation, separateness and spatio-temporal limitations of the self were merely optical illusions on the next higher level'.[20] Nieli, interestingly, points out that Wittgenstein may well have had similar insights.[21]

Koestler knew his Plato—though whether he knew Plato at the time is something I have not had the patience to find out, nor indeed the courage (since Koestler's autobiography makes very painful reading, being a saga of tortures, humiliations and betrayals: *verstehen* is often agony). But a more exact Platonism could perhaps have helped. Let the world of sensory perception be the first order; the second is the world, grasped through sensory imagination, of physical reality; the third is

[19] A. Koestler, *The Invisible Writing* (Boston: Beacon Press, 1954), 353.
[20] Ibid., 354.
[21] R. Nieli, *Wittgenstein: from Mysticism to Ordinary Language* (State University of New York Press, 1987), 125.

the world on which Koestler entered through Euclid's proofs, of mathematical, formal reality, gradually purged of its sensual and physical dross (the sort of images that, as Michael Redhead has observed earlier in this volume, physicists tend to use heuristically, and when explaining themselves to laymen, but which they do not believe finally veridical). Climbing higher still, he discovers in himself the undying intellect which can treat our bodily, individual lives as the trivialities, the 'mortal flummery', that any Platonist must think they are. But the pinnacle of the Platonic vision is the Good, the unformalizable One, of which the *Logos*, the coherent system of forms, is the only expression.

How is that which cannot be spoken of except by indirection and innuendo different from mere nothing? Obviously (and traditionally) in the spirit with which it is approached, and which it inspires. Apophatic theology does not claim to worship something of which nothing at all is known, nor postulate an utterly unknown cause to boggle at: the claim instead is that the God revealed through worship and obedience is one whose being cannot be contained by us, whose attributes (so to call them) are such that we cannot know what it is like actually to *be* God. The 'truth' of our religious utterances rests in their evoking and maintaining that 'cordial consent of beings to Being in general' of which Jonathan Edwards spoke,[22] their being such propositions as 'make proper impressions on [our] mind producing therein, love, hope, gratitude, influencing [our] life and actions, agreeably to that notion of saving faith which is required in a Christian'.[23]

So what is 'explanation', and why do we seek it? In *not* understanding something, our life is checked, as by an obstacle, and we find ourselves confined by things opaque to us. To understand is to find things well grounded in our understanding, not to be faced by obstacles, opacities, sudden sensations as of stepping on a stair that is not there. One, spurious, way of 'understanding' is simply to make sure that we can ignore what does not fit our egoistic or our social requirements. But the world, it seems, does not long permit that subterfuge. Another is, perhaps, to find the simple, finite formula which will thereafter give us a sense of ruling the world, because things seemingly obey that maxim in our minds. But we have no such formula, and if we did we could hardly not hear the secret laughter of the gods. The final way is Plato's, or any believer's, to approach the *Logos* in the Spirit that proceeds from the One, via the *Logos*.

[22] See C. A. Holbrook, *The Ethics of Jonathan Edwards* (University of Michigan Press, 1973), 102f.

[23] G. Berkeley, 'Alciphron', in *Collected Works* Vol. 3, A. A. Luce and T. E. Jessop (eds) (Edinburgh: Thomas Nelson, 1948–56), 297.

And what is it that we seek to explain? What is it that we seek to live easily and smoothly with? The whole enterprise of arguing to God on the basis of evidential data that an atheist would accept seems flawed. If what we are faced by, and seek to live through, is mere matter in motion, any god invoked to explain the motion is no more than an additional source of momentum. Consider the following analogy: if I am insistent first of all in seeing in human or animal bodies nothing at all but mechanical motion the hypothesis of a moving spirit dedicated to keeping the mechanism going will never seem either plausible or interesting. I postulate a moving spirit to 'explain' (which is to say, to live through) things that I already see as actions. Similarly, the divine life which binds the scattered fragments of the phenomenal order into the unifying *Logos* that stems from the One, is not an addendum to 'explain' atheistically perceived phenomena. I postulate, or seek to live in imaginative consciousness of that divinity, because I already (or identically) see the phenomena as the outward body of the God. The point is made by John Compton: 'Scientific analysis of physical nature and of human history has no more need of God as an explanatory factor than the physiologist needs my conscious intent to explain my bodily movements . . . what happens is that the evolution of things is *seen* or *read*, in religious life—as my arm's movement is read in individual life—as a part of an action, as an expression of divine purpose, in addition to its being viewed as a naturalistic process.'[24]

It is here that something *like* agency enters the picture. I have denied that Swinburne's reasons for preferring the hypothesis of a personal agent are compelling. So far as simplicity goes, it seems unnecessary to reckon with a single superbeing to 'explain' what can be explained as well by reference to real, extra-cosmic standards. Either there is no explanation at all for things, no way of eliminating the brute fact of being; or there is an explanation, resting in what must be the case if there is to be any intelligible world at all. In neither case do the determinedly atheistic need to admit an ultimate agency, any more than those who deny personal or quasi-personal agency at the human or animal level have to admit the hypothesis of an extra-physical will and intellect to 'explain' bodily motions. But just as—so most of us will admit—human and animal motions, *qua* behaviour, do embody such a will and intellect, so may the world at large. In attuning ourselves to what might have proved to be a wholly 'impersonal' *Logos*, we find the whole comes to have something like a personal meaning. As Berkeley saw, the whole phenomenal universe is the very speech of God, but of a God eternally hidden from our eyes. So far from being unusual 'it is the

[24] J. J. Compton, 'Science and God's Action in Nature', in *Earth Might Be Fair*, I. G. Barbour (ed.) (New York: Prentice-Hall, 1972), 33–47, 39.

blindness to God's presence that is exceptional: the awareness of God's presence is and ever has been the most persistent specific trait of our species being'.[25]

Implications for Biological Science

Neither Swinburne nor I have spent much effort in these papers to relate a rational theism, whether of his kind or of mine, specifically to biology. It would perhaps be of some interest if I now made that attempt.

First, I reiterate a point I made before. The one overwhelming requirement on any supposedly scientific understanding of our biological aspect is to explain how it could reasonably be expected that creatures like us should have the capacity (which science must assume we have) to see things straight, to see right back into the beginnings. If I were myself a naturalist, believing that only neo-Darwinian pressures have moulded our inheritance, I would conclude that we had no good reason to expect our categories to reach beyond the little confines of what might be immediately useful in pursuing our next meal and avoiding being one. I would conclude that the entire modern scientific synthesis, including evolutionary theory, could be no more than the latest conjuring trick, or fantasy, or instrumentally useful story. Such an account of the theory, unfortunately, means that it could tell me nothing about what I really am. Naturalism begets pragmatism, which begets agnosticism about 'real causes', which destroys the root of naturalism. If real mathematical (and moral and metaphysical) facts could—as naturalists must suppose—have no direct causal impact on us and our evolution, we can have no right to suppose that we have access to them. It is characteristic of this modern age that Papineau, for example, recognizes that such real facts could have an impact on us in so far as they are subsumed in what our predecessors recognized as God, and dismisses the theory in a sentence: 'I take it that such stories are no longer available'.[26] But if they are not, and reality is forever opaque to us, there is an end of the rationalistic attack on religion.

By contrast Swinburne and I at least have theories which make it not unreasonable to think that we could get things right. In so far as you all

[25] E. Kohak, *The Embers and the Stars* (University of Chicago Press, 1984), 185.

[26] D. Papineau, *Reality and Representation* (Oxford: Blackwell, 1987), 160f. Compare similarly unargued and dismissive comments in J. L. Mackie, *Ethics: Inventing Right and Wrong* (Harmondsworth: Pelican, 1976), 48, 230ff.

believe (be honest!) that we can, you would find life more intellectually comfortable on our side of the fence!

Secondly, it is worth pointing out that my appeal to the Platonic Forms is true not only to the insistent urge towards a mathematization of reality (and so to physics) but also to the actual data of biology. The very same forms are represented in the biological realm by organisms with quite different DNA. Much the same DNA enables creatures to embody different Forms. Why is it terrestrial life perennially recreates just the same shapes, and possibilities? Once a Form is even minimally present the world conspires to reveal it more exactly, but what mechanism is there to allow its entry? Well, what do you expect? On Platonic terms the whole world is a moving-image of the *Logos*, as much as ever a reflection of the moon in water: disturb the water and the image dances into fragments for a while, but will reform. More subtly, if the phenomena are shaken loose from one attractor they may solidify again to their old form, or be nudged across some watershed and begin to embody some distinctly different other. The DNA we sort out of the mass of physical events is, as it were, the symbol which invokes the form, the structure which allows the organism to tune in. Some of those to whom I have pointed out the perennial reappearance of the 'same forms', and the irrelevance of 'natural selection' to any explanation of, say, the hexagonal honeycombs of bees (which are consequences of fundamental physical principles, not the 'fittest' variations), reply that these are merely the effects of 'physical laws'. Of course, but such laws must either be grounded in real and necessary connections between extra-cosmic standards such as Plato's Forms, or else be merely helpless generalities of the kind I derided earlier.

Thirdly, if Platonism is correct, in its essentials, it seems possible to make a daring, but still fairly safe, prediction. I do not think that a merely naturalistic account of our origins should give us much confidence that there are any alien intelligences 'out there'. Even if the wild chance of a self-replicating molecule has occurred elsewhere, the tangled tracks of molecular replication are unlikely to be followed twice in the same way. All we terrestrials, from humans to whales to squids to deep-sea annelids to oak trees, lichens and the smallpox virus, are more like each other, biochemically, than any of us are likely to be like the extraterrestrials. Expecting an intelligence even remotely like us, on naturalistic terms, seems optimistic.[27] Even Swinburne's theism does not, I think, make such intelligences expectable: obviously the God could make as many kinds of person as He chose, but He seems, from

[27] These and related issues are discussed in E. Regis Jr. (ed.), *Extraterrestrials: Science and Alien Intelligence* (Cambridge University Press, 1985).

observation of the terrestrial sphere, to have no special need to, nor to make the ones 'like us'. As Haldane remarked, when asked what could be inferred about the Creator from His Creation: 'He has an inordinate affection for beetles'.[28] But if Platonism is correct, the world is heaving itself together to reflect, refract, embody the same *Logos*: just as terrestrial history shows us the same form being achieved again and again, so the universe at large is likely to be populated by children of the same inheritance. In other words: if there are intelligent aliens, then that is additional evidence that Platonism is essentially correct. I did say that it was a daring, but quite safe, hypothesis.

Concluding Unscientific Postscript

The imagined spectacle of a universe of intelligent beings in the image of the *Logos* does of course raise a further problem for believers, and especially for *Christian* Platonists. Most other believers, and even Platonists who are happy to think of themselves as Jews, Muslims or dissenting Christians, need have no great difficulty. Those alien intelligences will seek to awaken to the *Logos* imprinted in their hearts and minds, and find evidences of that order in a life-world suffused with spiritual meaning. Some such tribes, however fallen from Plato's heaven, may still be our superiors: most of those that are may also be less technically and militarily adept than we. Mainstream Christians have the further problem that they have endorsed the doctrine that a certain Galilean hasid was, and is, Himself the *Logos*. If He were only an inspired agent of the *Logos* there would be no problem: so may many be. But if He is the *Logos* then whatever alien intelligence may be it is really in the image of that Christ,[29] and will only see the world straight when it sees it united to that definitely terrestrial and human person. And there I too reach the limit of what I can readily explain.

[28] Cf. G. K. Chesterton, 'God must love common things; He made so many of them'.

[29] See R. Puccetti, *Persons* (London: Macmillan, 1968), T. W. Morris, *The Logic of God Incarnate* (Ithaca: Cornell University Press, 1986), and my *Limits and Renewals* Vol. III (Oxford: Clarendon Press, forthcoming).

Supervenience and Singular Causal Statements*

JAMES WOODWARD

In his recent book, *Causation: A Realistic Approach*,[1] Michael Tooley discusses the following thesis, which he calls the 'thesis of the Humean Supervenience of Causal Relations':

> (T) The truth values of all singular causal statements are logically determined by the truth values of statements of causal laws, together with the truth values of non-causal statements about particulars (p. 182).

(T) represents one version of the 'Humean' idea that there is no more factual content to the claim that two particular events are causally connected than that they occur, instantiate some law or regularity, and perhaps bear some appropriate non-causal (e.g. spatio-temporal) to each other. This is an idea that is tacitly or explicitly assumed in most familiar accounts of singular causal statements. For example (T) is assumed by many probabilistic theories of singular causal statements, by theories which attempt to analyse singular causal statements in terms of necessary and sufficient conditions, and, as I shall argue below, by David Lewis' counterfactual theory.[2]

* I am grateful to the participants in the Royal Institute of Philosophy Conference on Explanation and Its Limits for helpful comments. I am also indebted to Dan Hausman, Richard Otte, and particularly to Paul Humphreys for very helpful correspondence.

[1] M. Tooley, *Causation: A Realist Approach* (Oxford University Press, 1987).

[2] I have in mind here probabilistic theories of causation which attempt to characterize particular causal sequences or actual causal connections between particular events (so-called 'token' causation as in E. Eells and E. Sober, 'Probabilistic Causality and the Question of Transitivity', *Philosophy of Science* **50** (1983): 35–57), in terms of the idea that a cause must raise the probability of its effect in suitably characterized background circumstances. Such theories should be distinguished from theories which attempt to give a probabilistic characterization of generic or population-level causal connections (what Eells and Sober call 'property-causation'). These last theories are not touched by the criticism advanced in this paper. The falsity of (T) undermines both reductionist 'Humean' theories of token-causation which attempt to

As Tooley recognizes, (T) should be distinguished from another central assumption to which Humean theories of causation are committed: that laws of nature are susceptible of some sort of regularity analysis which does not rely on unreduced notions of physical possibility and necessity. Whether or not a regularity analysis of natural laws is correct, it appears to be a distinct question whether (T) is true. (T) should also be distinguished from another claim that is frequently made about the connection between singular causal statements and causal laws: that every true singular causal claim entails the existence of some associated causal law. Here too, this last claim might be true, even though (T) is mistaken.[3]

[3] Reasons of space preclude a detailed discussion of the relationship between (T) and the claim (N) that a (deterministic or probabilistic) law of nature must underlie or otherwise be associated with every true singular causal claim. It is perhaps worth adding, however, that while the truth of (N) is logically compatible with the falsity of (T), my view is that the considerations advanced in this paper against the truth of (T) also help to undermine the plausibility of (N). They do so by making the assumption that (N) is true look gratuitous and unmotivated. First, the belief that singular causal statements are analysable in terms of facts about laws of nature and non-causal facts about particulars has been, historically, an important motive for believing that every true singular causal statement 'must' have a law of nature associated with it. The idea has been that by providing such an analysis, one renders the opaque and mysterious notion of a particular causal connection metaphysically and epistemically acceptable. In denying (T), we deny the possibility of such an analysis, and thus undercut this motive for belief in (N).

Secondly, (and relatedly) if the arguments of this paper are successful, they show that to state the truth conditions for a singular causal claim, it is not enough just to make reference to facts about an associated law and non-causal facts about particulars—reference to an additional fact F, having to do with the obtaining or non-obtaining of particular causal connections, is also required. It then becomes natural to wonder whether the holding of this further fact F is not by itself a sufficient condition for the truth of the singular causal claim in question, even when there is no associated law. Once we recognize that reference to facts about particular causal connections is unavoidable, reference to laws as well in stating truth-conditions for singular causal claims begins to

---

characterize token causation purely in terms of facts about probabilities, and non-reductionist theories of token causation which permit reference to generic causal facts but not singular causal facts in their specification of background circumstances.

Theories which attempt to characterize particular connections in terms of necessary and sufficient conditions include various variants of the idea that a cause is a necessary or non-redundant condition in a set of conditions which are jointly sufficient for some effect as in C. Hempel, *Aspects of Scientific Explanation* (New York: The Free Press, 1965) and J. L. Mackie, 'Causes and Conditions', *American Philosophical Quarterly* **2** (October 1965) 245–64.

Tooley holds that (T) is false. Although I do not find his reasons very persuasive, I think that his conclusion is correct. My aim in this essay is to argue in a more detailed way for the falsity of (T) and to explore the implications of this fact, which I think are extensive, for understanding singular causal claims. As we shall see, (T) bears importantly on issues concerning realism about causation. If (T) is false, then in so far as we wish to hold a realistic theory of the content of singular causal claims (according to which the truth-conditions for such claims are statable without reference to subjective facts about us, or about our beliefs, expectations or epistemic organizing activities), we must admit causal connections between particular events as irreducible and ineliminable features of the world. Moreover, issues about the truth or falsity of (T) are closely bound up with issues about the procedures which we may legitimately use to establish or reject singular causal claims. We can thus use (T) to characterize, in an illuminating way, a variety of different possible positions about the metaphysics and epistemology of singular causal claims.

My general strategy in this paper will be to try to show that an idea which I take to be central to our notion of causation—the idea that causes must exhibit some measure of context-independence or invariance in their operation—is inconsistent with (T). In a bit more detail, my discussion will proceed as follows. In section I, I describe two kinds of cases in which facts about the laws and about the occurrence of particular events do not seem to determine whether these events are causally connected. I will then consider whether one can defend (T) by appealing in addition to a special class of non-causal facts (facts about spatio-temporal relations) and conclude that such a defence is unlikely to succeed. Sections II and III examine and reject two attempts to

---

look superfluous. This does not of course show that it is false that a law is associated with every true singular causal claim, but it does again raise the question of what positive reasons we have for believing that (N) is true. Facts about particular causal connections seem by themselves sufficient to distinguish causal from non-causal sequences, provide support for appropriate counterfactuals, and so forth.

Third, (and again relatedly) the methods we use to establish whether singular causal claims are true or false—and in particular, the use of the eliminative method described below—do not seem to require or to depend upon the truth of (N); there is nothing in those methods which *commits* us to (N). In particular, to establish that c_1 is a possible cause of e and to eliminate other possible candidates for the cause of e, and in this way to establish that c_1 is the cause of e, we do not need to rely on the assumption that there is a law of nature linking c_1 and e. Here too, while this does not show that (N) is false, it again makes the assumption that it 'must' be true look gratuitous and unmotivated.

defend (T) by ruling out on non-empirical grounds the possibility that certain singular causal claims might be true. In sections IV and V, I argue that a strategy which is frequently used to establish singular causal claims, which I call the eliminativist strategy, supports a kind of realism about such claims which is inconsistent with (T). Finally, in section VI, I attempt to allay some epistemological worries that may be raised by my rejection of (T).

I

I begin with a rather abstract and schematic description of two kinds of cases which seem to suggest that (T) is false. Both kinds of case exploit the observation that it seems perfectly possible for there to be a causal law linking some causal factor or type of event C_1 and some effect E, for tokens of C_1 and E to occur on some particular occasion, and yet for this token of C_1 to fail to cause this token of E. The existence of the right sort of law or generalization linking the types of event C_1 and E does not seem to settle whether any two particular occurrences of C_1 and E are causally connected.[4]

Suppose first that there are probablistic laws (D_1, D_2) linking C_1 and E and C_2 and E. It will be important to my subsequent argument that

[4] I claim no originality at all for this general observation, which appears in various forms in many, different writers. Discussions which are at least loosely related occur in J. Kim, 'Causation, Nomic Subsumption, and the Concept of Event', *Journal of Philosophy* **70** (1973) 217–36 and in P. Achinstein, *The Nature of Explanation* (Oxford University Press, 1983) in the context of a criticism of the covering-law model of explanation. More recent and closely related discussion occurs in a series of papers by E. Eells and E. Sober including 'Probabilistic Causality and the Question of Transitivity', E. Sober, 'Causal Factors, Causal Inference, Causal Explanation', *Aristotelian Society Proceedings 1986* and E. Eells, 'Probabilistic Causal Levels', in B. Skyrms and W. Harper, (eds.) *Causation, Chance and Credence* (Dordrecht: Kluwer Academic Publishers, 1988). As noted above, Eells and Sober distinguish between what they call 'property causation' and 'token causation' and give examples having essentially the same structure as my (Ex. 1) below to show that the inference from property causation within a population to token causation between individual occurrences within that population is less straightforward than one might suppose. On their view, probabilistic theories of causation are most plausibly construed as characterizations of property causation rather than token causation. My claim in what follows is the (I believe) different and more radical claim that no characterization of token causation, probabilistic or otherwise, which just makes reference to regularities or generic relations between events and to non-causal facts about particulars will be adequate. I would thus reject the account of token-causation

these laws be understood in a definite way. I shall take the law D_1 linking C_1 and E as specifying the probability P_1 (strictly between 0 and 1) with which C_1 would cause E_1 were C_1 to occur, or, equivalently as specifying that if C_1 occurs and no other causes of E are present and E does not occur spontaneously, then E occurs with fixed probability P_1. The law linking C_1 and E thus has a counterfactual element built into it—it has to do with the probability with which E would occur, were some perhaps (non-actual) condition to obtain. Moreover the law D_1 has to do with the effect of the causal factor C_1 taken in isolation, in circumstances in which no other factors relevant to the probability of E are present. What the law specifies is what difference the occurrence of C_1 by itself makes to the probability of E. Similarly, the law D_2 linking C_2 and E specifies the probability P_2 (also strictly between 0 and 1) with which C_2 would cause E, were C_2 to occur, or equivalently that if C_2 occurs (and no other causes of E are present and E does not occur spontaneously), then E would occur with probability P_2.[5]

Consider now a case (Ex. 1) in which, on some particular occasion tokens or instances of C_1, C_2 and E (which we may represent by the

[5] I take both of these features—that the law D_1 has counterfactual import and that it describes the difference which the factor C_1 taken in isolation makes to the occurrence of E—to be characteristic of laws in general and not just probabilistic laws. Our stipulation that D_1 has these features is thus not *ad hoc*.

The tendency in a great deal of recent philosophical discussion has been to represent or analyse probabilistic laws as claims about conditional probabilities or differences between conditional probabilities. In what follows, I do *not* assume that the probability P_1 will in general coincide with any of the obvious candidates constructed from conditional probabilities such as $P(E/C_1)$, $P(E/C_1)-P(E/\overline{C_1})$ or $P(E/C_1\overline{C_2})-P(E/\overline{C_1}C_2)$. My reasons for resisting this assumption include, among other considerations, the fact that a probabilistic law like D_1 can be true in a situation in which the relevant conditional probabilities are not well defined, whether because the denominators P(C) (or $P(\overline{C})$, etc.)=O or because C or (or $\overline{C}$, etc.) does not have a well defined probability distribution or is not a random variable at all. More generally, I would distinguish sharply between the conditional probability P(E/C) and the probability of E, given a counterfactual condition C. Probabilistic laws have to do with the latter notion, not the former. I should also emphasize that I take the law D_1 to specify the probability P_1, with which C_1 *causes* E. Even putting aside the above problems, this will not, for example, equal $P(E/C_1)-P(E/\overline{C_1})$ if other causes of E are also present and cause E, or if E sometimes occurs spontaneously. For reasons which I lack the space to discuss here, I do not think that the strategy of trying to deal with this difficulty by conditionalizing on the absence of other causes of E will always work.

suggested in Eells' 'Probabilistic Causal Levels'. For reasons which will become obvious below, I am also sceptical of the 'Connecting Principle' between property and token causation proposed by Sober in his 'Causal Factors, Causal Inference, Causal Explanation'.

lower case letters c_1, c_2 and e) all occur.[6] Supose also that we have good reason to believe that C_1 and C_2 occur independently of each other and do not interact with respect to their tendency to produce E. We shall also assume that (Ex. 1) does not involve causal pre-emption: the case is not one in which c_1 in fact causes e, but in which, if c_1 had not caused e, c_2 would have.[7]

Clearly C_1 and C_2 are both (as we might variously call them) possible causes of E, or positive causal factors for E, or risk factors for E. When C_1 is present, the probability of E is higher than it would be in an otherwise similar situation in which C_1 is absent, and similarly for C_2. But equally clearly, this does not seem to settle the distinct question of what the actual causal connections are—which singular causal claims are true—in the specific situation (Ex. 1).

From the point of view of (what I shall call) a naively realistic analysis of singular causal claims, it looks as though there are three distinct possibilities regarding the causation of e: (1) c_1 alone causes e, (2) c_2 alone causes e, (3) c_1 and c_2 both cause e. Apparently, these possibilities compete. It looks as though at most one of them can be true. While I shall not advance anything like a systematic analysis what (1), (2) and (3) mean, I suggest that we may understand each as entailing a counterfactual claim. (1) claims, among other things, that if on this particular occasion, c_1 had not occurred, (but c_2 had occurred), then e would not have occurred. (2) makes the corresponding claim (with c_1 and c_2 interchanged) regarding c_2. (3) represents a case of 'redundant causation', in which there is nothing to break the symmetry between the redundant causes.[8] If (3) is true, then c_1 causes e and c_2 causes e; if c_1 had not occurred, c_2 still would have caused e, and similarly, if c_2 had not occurred, c_1 still would have caused e. If neither c_1 nor c_2 had occurred, e would not have occurred.[9]

[6] I have generally followed the convention of using upper case letters to refer to types of events and lower case letters or locutions like 'particular occurrence of (type) C' to refer to particular event-tokens. However, at a few points at which it seemed awkward, distracting, or intolerably pedantic to explicitly distinguish between type and token, I have run roughshod over the distinction, thinking that it would be clear enough how to sort things out.

[7] One may, if one likes, think of this assumption as implicit in the stipulation that C_1 and C_2 act independently, but it is useful to make the assumption explicit. I have excluded the possibility of pre-emption to avoid certain irrelevant complications and not because I think that such cases create difficulties for the treatment of singular causal claims I shall defend. In fact, I think that cases of pre-emption provide additional reasons for rejecting (T).

[8] Cf. Lewis *Philosophical Papers*, Vol. II (Oxford University Press, 1986), 193 ff.

[9] I should emphasize that my claim is simply that (1), (2) and (3) *entail* the

Let us assume for the moment that the naively realistic view is correct. (I shall explore this asssumption in greater detail below.) Consider (A) the conjunction of the laws linking C_1 and E_1, and C_2 and E_2 and sentences specifying that c_1, c_2 and e have occurred. Plainly (A) is not sufficient to fix which of the alternatives (1), (2) or (3) above is true, since each alternative is consistent with the truth of (A). The correctness of (T) will thus turn on whether there are additional non-causal facts about particulars, which in conjunction with (A) are sufficient to determine which of (1), (2) or (3) is true.

Secondly, consider a case (Ex. 2) in which, as above, there is an indeterministic law linking C_1 and E, but suppose in addition that E sometimes occurs spontaneously, without any cause at all. Suppose that some particular token (c_1) of C_1 occurs and that some particular token (e) of E occurs. Looking at the matter again from a naively realistic point of view it looks as though there are two possibilities, only one of which can be correct: either (4) c_1 caused e or (5) c_1 did not cause e, and instead e occurred spontaneously. Here again, the conjunction of (B) the law linking C_1 and E_1 and the fact that c_1 and e have occurred does not fix which of (4) or (5) is correct. The truth of (T) will turn on whether there are additional non-causal facts which, in conjunction with (B), determine the truth values of (4) and (5).[10]

What might these additional facts be? Undoubtedly the most frequently cited candidates, particularly by those with Humean sympathies, are facts about spatio-temporal relationships. One common suggestion along these lines, sometimes attributed to Hume himself, is that from the point of view of what is actually in the world, there is nothing more to the claim that two events c and e are causally related than that they instantiate some regularity, are spatially contiguous and

[10] For a concrete case, consider an example in which someone has a serious illness. Suppose that he is treated with a drug (c_1) which has probability P_1 of causing recovery (e), but that there is also a non-zero probability of recovery (e) occurring spontaneously.

---

above counterfactuals. I do not claim that the above counterfactuals exhaust the meaning of (1), (2) and (3) or that the truth of these counterfactuals is a sufficient condition for the truth of (1), (2) and (3). In fact, I do not think it is possible to give an analysis of causal claims in terms of a notion of counterfactual dependence which does not presuppose causal notions. Moreover, I also do not claim (in fact I think it is false) that all causal claims entail counterfactuals parallel to those given above. Instead, the form of the counterfactual entailed by a causal claim will depend upon the details of the claim and upon which other causal claims are true in the situation under investigation. For example, if (Ex. 1) were instead a case involving pre-emption or linked overdetermination in which if c_1 had not caused e, c_2 would have, the counterfactual (1) would not hold.

that c temporally precedes e. Thus suppose that in the first example above, (1) c_1 caused e. Then it might be suggested that the truth of (1) supervenes on facts about the spatio-temporal relationships between c_1, c_2 and e, in conjunction with (A) facts about the occurrence of c_1, c_2 and e, and the laws linking them.

I think that there are a number of reasons for doubting that this sort of defence of (T) by means of an appeal to spatio-temporal relationships will work for all singular causal statements. Since the rejection of this defence will be important for my subsequent discussion, I shall explore these reasons in some detail. To begin with, it is true enough that we often use facts about spatio-temporal relationships in conjunction with other kinds of facts, as *evidence* regarding the truth of singular causal claims. But the sorts of generalizations on which we rely seem to have something like the following form:

(S_1) If c causes e, then c bears spatio-temporal relation R to e.

To anticipate a point to be developed in more detail below, spatio-temporal facts are characteristically used in causal inference as part of a general strategy of eliminative induction: one appeals to generalizations of form (S_1) together with the observation that some putative cause c does not bear R to e, to rule out the possibility that c caused e and in this way to build a case for some uneliminated rival causal hypothesis. However, the suggestion we are considering, that we may appeal to spatio-temporal facts to defend the supervenience thesis (T), requires the truth of generalizations having the following form:

(S_2) If c bears spatio-temporal relationship R to e (and the appropriate laws and other non-causal facts about particulars obtain), then c causes e.

It seems unlikely that there are true generalizations of form (S_2) covering every cause–effect pair. To begin with, it is worth noting that standard philosophical claims about spatio-temporal constraints on causal connections generally seem to take the form of generalizations like (S_1) rather than (S_2). For example, the claims that all causes must precede their effects or must be spatio-temporally contiguous to their effects are clearly claims of form (S_1) rather than (S_2)—they claim to state a necessary condition for a causal connection to obtain, not a condition which in conjunction with other conditions is sufficient.[11] Moreover, to explain the evidential role of spatio-temporal facts in supporting causal claims, one needs (at most) only to suppose that

[11] In fact, I believe that both of these claims are false. My point here, however, is that even if true, they are generalizations of form (S_2) rather than form (S_1).

appropriate generalizations of form (S_1) hold, not that generalizations of form (S_2) hold.

A deeper reason why it seems implausible that a generalization of form (S_2) will hold in the case of every causal connection is illustrated by (Ex. 1) and (Ex. 2): the truth of such generalizations would require that spatio-temporal information could always be used to rule out rival singular causal claims which are inconsistent with the correct singular causal claim. It seem unlikely that it will always be possible to use spatio-temporal information in this way. Suppose that (1) in (Ex. 1) is true—c_1 alone caused e. Let R_1 be your candidate for the spatio-temporal relationship between c_1, c_2 and e which, (in conjunction with (A) information about laws and other non-causal facts about particulars) is supposed to insure that (1) is true. For R_1 to do this job, it must be true that, in conjunction with (A), R_1 rules out the possibility that (2) or (3) (or indeed any other alternative causal hypothesis that competes with (1)) is true. If it is possible for c_1, c_2 and e to obtain and for R_1 to obtain and yet for e to be caused by c_2 rather than by c_1, then R_1 will have failed to do the job it was assigned—that of insuring that when c_1, e_1 and R_1 all obtain, it is c_1 alone which causes e.

Now it is plausible to think that *sometimes* we can appeal to spatio-temporal relationships to rule out competing causal hypotheses in just this way. For example, it might be the case that C_1 and C_2 are the only possible causes of E, that tokens of C_1 and C_2 cause tokens of E only when they are spatio-temporally contiguous with tokens of E, and that E never occurs spontaneously. If we are then lucky enough to have a case in which c_1 is spatio-temporally contiguous with e and in which c_2 is not, we have facts that insure that (1) c_1 caused e. Here we in effect rely on a generalization of form (S_1). But it also looks as though there is nothing that *guarantees* that things must work out this way. For example, it might have turned out instead that *both* c_1 and c_2 are spatio-temporally contiguous with e, or that c_2 also bears whatever relationship to e that it needs to bear if it is to cause e. In this sort of case, we cannot appeal to generalizations of form (S_1) in conjunction with information (A) to fix the truth-values of (1), (2) and (3). Instead we require generalizations of form (S_2). More generally, if the spatio-temporal constraints on causal connectedness typically take the form of generalizations like (S_1) rather than generalizations like (S_2), it is a contingent matter, dependent on the details of the relevant generalizations, and on which causal factors happen to be present in the system of interest and what their spatio-temporal relations happen to be, whether we can appeal to spatio-temporal considerations to fix the truth-values of all the competing causal claims regarding that system. To insure that we can always do this, as a defence of the supervenience thesis (T) would require, we need to assume the truth of generalizations of form

(S_2), which commit us to implausibly strong claims about our ability use of spatio-temporal considerations to rule out competing causal hypotheses.

The matter becomes even clearer when we consider (Ex. 2). If, I say, C_1 only causes E when tokens of C_1 are spatio-temporally contiguous to tokens of E and (Ex. 2) is a case in which c_1 is not spatio-temporally contiguous to e, these facts may, together with the relevant laws and other non-causal facts about particulars, fix the truth-values of (4) and (5), insuring that the case is one in which c_1 did not cause e and e instead occurred spontaneously. But to suppose that spatio-temporal considerations can always be used in this way is to suppose that there is some spatio-temporal relationship R_2 such that if c_1 and e occur and c_1 bears R_2 to e, the possibility that e occurs spontaneously is always ruled out. Here again, there seems no justification for making this assumption.

There are also other, more diffuse reasons for doubting that we can always appeal to spatio-temporal facts to insure the truth of the supervenience thesis (T). The very strong kind of connection between the obtaining of a characteristic spatio-temporal relationship and the presence of causation contemplated in either (S_1) or (S_2) seems most plausible in the case of relatively simple microscopic physical systems like colliding billiard balls. The assumption of such a connection does not seem even prima facie plausible in the case of causal relations involving psychological or social phenomena. For example, it seems plausible that increases in the price of oil caused part of the general price inflation of the 1970s, but quite implausible that there is any general spatio-temporal relationship between increases in oil prices and inflation which satisfies either (S_1) or (S_2). Moreover, to anticipate a point to be made in more detail below, I report, as an empirical observation, that when one examines the sorts of arguments and assumptions that are used to support singular causal claims in the various sciences, one rarely if ever finds appeals to anything like (S_2). Facts about spatio-temporal relations are appealed as one kind of evidence bearing on the truth or falsity of causal claims, but there is no suggestion that such facts (in conjunction with information about laws or regularities) can always be relied upon to fix the truth values of singular causal claims. Indeed, discussions of methodological problems of causal inference in the social and behavioural sciences often explicitly deny that this is possible.[12] (This is just what we would expect if the

[12] See, for example, the remarks on the limitations of information about temporal relationships in deciding among conflicting causal claims in T. Cook and T. Campbell, *Quasi-Experimentation* (Boston: Houghton Mifflin, 1979).

spatio-temporal constraints on causation were of form (S_1) rather than form (S_2).)[13]

Moreover, whether or not our world contains any actual examples of causal connnections between events which are not mediated by spatio-temporally continuous intervening processes, it seems hard to deny that causal action at a spatio-temporal distance represents a conceptual possibility.[14] But if this is so, it also seem possible that there may be causally related events for which no generalization of form S_2 is true. If there can be a spatio-temporal gap between c and e even though c and e may be causally connected, then it sems very unlikely that the spatio-temporal relationship between c and e can by itself be such as to preclude the possibility that some alternative intervening cause c_1, independent of c, has caused e.[15]

Finally, there is yet another reason why the defender of the supervenience thesis (T) would be well advised not to dig in his heels and insist

[13] It may be responded that these observations bear only on epistemological and methodological issues relating to causal inference, and not on the metaphysical issue of whether spatio-temporal facts, in conjunction with other non-causal facts about particulars and facts about the laws, are as (T) claims, sufficient to fix the truth of every singular causal statement. The observations show only that generalizations of form (S_2) are not (widely) known for many causes or that they are not used or appealed to in causal inference, not that such generalizations are not true. I concede this point. Still, I think that the above observations help to shift the burden of proof onto those who think that we can always appeal to spatio-temporal facts in defence of (T). What positive reasons are there to believe this claim, given the above methodological and epistemological observations?

[14] It is true that it follows from energy/momentum conservation and the requirement that physical laws must be Lorenz-invariant that energy/momentum transfer must be local and that there can be no causal action at a distance in the context of fundamental physical explanation. See, for example, H. Ohanian, *Gravitation and Spacetime* (New York: W. W. Norton and Co., 1976). But at the level of analysis characteristic of sciences like psychology and economics, the imposition of a similar no-action at a distance requirement seems misguided. Moreover, even in physics, it seems to me that the fact that there is no causal action at a distance is not plausibly viewed as a characterization of what causality *is* or must involve. Rather, this fact is a quite contingent physical fact which stands in need of more fundamental physical explanation. Insofar as there is a common core to the notion of causation, it seems to me to involve the idea of counterfactual dependence emphasized above, rather than the idea of action by contact. See my 'The Causal/Mechanical Model of Explanation', in P. Kitcher and W. Salmon (eds.), *Minnesota Studies in The Philosophy of Science*, Vol. 13 (Mineapolis: The University of Minnesota Press, 1989) for further discussion.

[15] I owe this point to David Ruben.

that there must be facts about spatio-temporal relations and other non-causal facts about particulars which in conjunction with facts about the relevant laws, are sufficient to fix the truth values of (1), (2), (3), and (4) and (5). The defender of (T) has another option available. He can reject what I have been calling the naive realist view and insist that to the extent that (1), (2) and (3) fail to supervene on the laws and on non-causal facts about particulars in (Ex. 1), they do not really represent contrasting empirical possibilities, any one of which (but only one of which) might turn out to be true. And he can take a similar line about (Ex. 2). In the next two sections, I shall consider two different forms which this sort of defence of (T) might take.

II

In a postscript (1986) to his well known paper 'Causation',[16] David Lewis considers a case which is a rather more specific version of the cases described in (Ex. 1) and (Ex. 2), and advances a claim which, if correct, would allow us to reconcile these cases with (T). Lewis envisions a case in which

> c occurs, e has some chance (x) of occurring, and as it happens e does occur; if c had not occurred, e would still have had some chance (y) of occurring, but only a very slight chance since y would have been very much less than x . . . In this case also, I think we should say that e depends causally on c, and that c is a cause of e (p. 176).

Elsewhere he formulates the following general claim, which I shall call (L_1) for future reference:

> (L_1) If distinct events c and e both occur, and if the actual chance of e (at times immediately after c) is sufficiently greater than the counterfactual chances of e without c, this implies outright that c is a cause of e (p. 180).

This principle does not cover all cases like (Ex. 1) and (Ex. 2), since it applies only when the chance of e is 'sufficiently greater' than the chance of e without c. Let us accordingly formulate the following generalization of (L_1), which I shall call (L_2):

> (L_2) If c and e occur, and if the chance of e (at time t immediately after c) is greater than the counterfactual chance of e without c, then c caused e.[17]

[16] In D. Lewis, *Philosophical Papers*, Vol. II, (Oxford University Press), 159–213.

[17] (L_2) may seem less plausible than the narrower claim made in Lewis'

Applying this claim (L_2) to (Ex. 1) and (Ex. 2) leads us to conclude that, in both cases, c_1 caused e. We thus conclude that in (Ex. 2), (4) is true, and (5) is false. Assuming (as I have intended throughout) the absence of interaction effects, it will presumably also be true that in (Ex. 1) the presence of c_2 increases the chance of e occurring over what it would be without c_2. It will thus follow that c_2 also caused e. We thus conclude that in (Ex. 1), (3) is true and that (1) and (2) are false.

If this sort of appeal to (L_2) can be made to work, it gives us back supervenience for both (Ex. 1) and (Ex. 2). What appeared to create difficulties for the supervenience thesis was that the laws and the non-causal facts seemed by themselves insufficient to fix which of the alternatives (1), (2) or (3) or which of the alternatives (4) or (5) was true. The appeal to (L_2) deals with this difficulty by, in effect, ruling out, on quite general (perhaps conceptual or *a priori*) grounds, the possibility that (1) or (2) might be true in (Ex. 1) or that (5) might be true in (Ex. 2).[18] Once these alternatives are ruled out, the laws and non-causal facts about particular values fix the truth values of the remaining alternatives (3) and (4).

But while we can appeal to (L_2) to support thesis (T) in this way, there are several reasons for thinking that (L_2) itself is implausible. The root of the difficulty is that (L_2) appears to be in conflict with the ascription of invariant context–independent causal tendencies to causal factors like C_1 and C_2. Consider an experiment having the structure of (Ex. 1), in which the experimenter can control the presence of C_1 and C_2, and in which C_1 and C_2 act independently. Suppose, as before, that the probability that C_1 will produce E when C_1 alone and no other causes of E are present is P_1, and that the corresponding probability for C_2 is P_2. (We may supose that the experimenter is able to check these

[18] I take it that Lewis intends (L_1) as a conceptual truth about causation.

---

remarks. And this is perhaps why Lewis does not formulate or endorse a principle like (L_2). However, a satisfactory account of causality which might be used to defend (T) obviously must be more general than the account explicitly asserted by Lewis; it must say something consistent with (T) about which causal claims are true in cases (like those represented by (Ex. 1) and (Ex. 2)) in which c raises e's chance of occurring but in which e's chance without c need not be small. Indeed, given that one subscribes to (T) (as I believe Lewis does), it is not easy to see how one could consistently defend the view that c causes e *only* when the chance of e, given c is sufficiently greater than the small chance of e without c, but not in the other kinds of cases covered by (L_2). This restriction looks *ad hoc* and motivated only be a desire to avoid difficulties to which the more general (L_2) may be subject. The fact that (L_1) is vague (how much less is 'very much less'?) in a way in which the notion of causation is, I would argue, not vague, is one reflection of the *ad-hoc* character of this restriction.

claims by producing those situations experimentally.) Suppose also that we have no empirical evidence that C_1 and C_2 interact, or that the presence of one modifies the tendency of the other to produce E. When C_1 and C_2 are both present, the probability of production of E is just what we would expect if these two tendencies were to operate independently: the probability of E is $1-(1-P_1)(1-P_2)=P_1+P_2-P_1P_2$.

According to (L_2), we should regard all cases in which C_1 and C_2, and E occur, as cases in which C_1 causes E. Thus we must suppose that, in the presence of C_2, the characteristic, law-governed tendency of C_1 to cause E is such that it causes E with probability $P_1+P_2-P_1P_2$. However, we also know that when C_1 acts alone, it causes E with probability P_1. It follows that we must suppose that in the presence of C_2, the characteristic tendency of C_1 to produce E is somehow transformed and enhanced from what that tendency is in the absence of C_2. We must suppose this even though there is no (other) evidence of any interaction effect between C_1 and C_2, and even though the probability of occurrence of E, in the presence of C_1 and C_2, is just what we would expect if the probabilistic tendencies of C_1 and C_2 to cause E operated completely independently of each other, and if the presence of C_2 did not in any way modify the probabilistic tendency of C_1 to cause E. Similarly, we must also suppose that the characteristic tendency of C_2 to produce E is enhanced by the presence of C_1.[19]

We might express the difficulty with (L_2) in a more general and abstract way as follows: when we examine (i) the behaviour of C_1 alone, when no other causes of E are present, what we seem to find is that when C_1 occurs, it sometimes causes E, and sometimes does not. Similarly, we find that when (ii) C_2 occurs alone, it sometimes, but not always causes E. When (iii) both C_1 and C_2 are present, the probability of E increases over what it would be if C_1 alone were present. In the absence of special evidence to the contrary, it seems natural to attribute the increase in the probability of E in (iii) over the probability of E in (i)

[19] It is worth emphasizing that this argument and the similar arguments which follow on pp. 224–226 require the understanding of laws and their relation to conditional probabilities sketched on pp. 214–215 and in footnote 5. For example, it is because the law linking C_1 and E gives the probability P_1 with which C_1 would cause E and the law linking C_2 and E gives the probability P_2 with which C_2 would cause E that the probability of production of E in (Ex. 1) is $P_1+P_2-P_1P_2$. If instead P_1 were, say, $P(E/C_1)-P(E/\overline{C_1})$ this result would not follow. More generally, the idea about the link between causation and invariance on which my argument relies requires that the law linking C_1 and E specify the contribution made to E by C_1, acting alone. It is this contribution which should remain invariant when we move to contexts like (Ex. 1) in which other causes of E may be present. I am very grateful to Paul Humphreys for impressing on me the importance of being explicit about this point.

to the causal activity of C_2. We know that C_2 can cause E in situations in which it acts alone and we also know from (i) that C_1 can occur (when C_2 is present) and yet fail to cause E. Given these facts, and that C_1 and C_2 seem to act independently, we have no reason to believe that C_2 causes E when and only when C_1 causes E—i.e. we have no justification for regarding all of the cases in (iii) as cases in which both C_1 and C_2 cause E. It is more reasonable to believe instead that some of the cases in (iii) are cases in which C_1 does not cause E, but in which E is caused by C_2 alone.

We are thus led naturally to what I earlier called the naive realistic view regarding (iii): that sometimes when E occurs in cases of kind (iii), it is caused by C_1 alone, that sometimes E is caused by C_2 alone, and that sometimes it is caused by both C_1 and C_2. We may not be able to tell, on some particular occasion on which C_1, C_2 and E all occur, which of these three alternative hypothesis about the causation of E is the correct one, but if so, this merely reflects a limitation in our knowledge. It does not show that all such cases must be regarded as cases in which both C_1 and C_2 caused E.

By contrast, the increased probability of E in the presence of C_1 and C_2 (over the probability of E in the presence of C_1 alone or C_2 alone) seems much more mysterious if, in accordance with (L_2), we say that all cases in which both C_1 and C_2 are present are cases in which both C_1 and C_2 cause E. First, in making this claim we are, as already noted, in effect supposing that somehow the presence of C_2 enhances the tendency of C_1 to cause E over what the tendency would be when C_1 alone is present. We are also supposing that C_1 has a similar enhancing effect on the tendency of C_2 to cause E. Second, we must assume a remarkable coincidence or co-ordination in the behaviour of C_1 and C_2, even in the absence of any mechanism to account for this coincidence. To make this second point vivid, suppose that the tokens of C_1 and C_2 occur at spacelike separation from each other, or that the experimenter is able through the introduction of shielding or other devices to insure the causal independence of C_1 and C_2: neither is causally linked to the other, and there is no third common cause of both. In this kind of case, given what we know about C_1 and C_2 when each occurs alone, we would naturally expect that the probability that both C_1 and C_2 will cause E is P_1P_2 and that roughly $P_1P_2/P_1+P_2-P_1 . P_2$ of the cases in which C_1, C_2 and E are all present are cases in which C_1 and C_2 have both caused E. However, (L_2) appears to imply that whenever C_1, C_2 and E occur, both C_1 and C_2 act in concert to cause E. On the face of it this looks like a remarkable and implausible coincidence. It seems unsatisfactory to adopt an analysis which requires such a coincidence, when an alternative account (the naive realist account) which requires no such coincidences is readily available.

A similar point holds with respect to (Ex. 2). Suppose that we are able to establish that E sometimes occurs spontaneously, without any cause, and that when C_1 is introduced, E occurs more often than it does in the absence of C_1. According to (L_2), we should regard all cases in which C_1 and E occur as cases in which C_1 has caused E. Here again, this seems arbitrary and unmotivated. In adopting (L_2) we are in effect supposing that, with the introduction of C_1, all of the spontaneous occurrences of E are somehow suppressed, and that all occurrences of E are to be attributed to C_1. Here again, (L_2) commits us to this claim even though we may have no independent evidence that C_1 has this suppressive effect and even though the observed rate of occurrence of E in various contexts in which C_1 is present is just what would be expected if it were true that E has the same constant rate of spontaneous occurrence whether or not C_1 is present.

The present remarks draw our attention to a tension between (L_2) and an idea that plays a rather central role in our ordinary thinking about causal relationships. This is the idea that genuine causes will exhibit certain characteristic *tendencies* or modes of operation in producing their effects and that such tendencies must possess some degree of context-independence: the tendencies must remain approximately the same across a range of different contexts and circumstances.[20] This idea is in turn a reflection of the more general idea that one mark of a causal or nomological relationship is the satisfaction of an invariance condition of some kind: causal or nomological relationships are relationships which will continue to hold in a somewhat stable or systematic way, over a range of possible changes in circumstances. The causes discussed above in (Ex. 1) and (Ex. 2) involve one important example of a tendency having this feature of context-independence or invariance. They are (we supposed) causes which produce their effects with certain fixed probabilities and which continue to act in the same way, both when they act alone, and when other causal agents are present.

[20] A rather similar idea is developed in Nancy Cartwright, *Nature's Capacities and Their Measurement* (Oxford University Press, 1989) which appeared just as I was completing this paper. Cartwright also emphasizes the idea that causal capacities (rather than causal laws) must be stable or context-independent; however her understanding of what such conduct-independence involves seems to be rather different from mine. In particular, Cartwright links the notion of context-independence to the notion of unanimity as it occurs in discussions of probabilistic theories of causation: she requires that a cause must increase the probability of its effect across all homogeneous background contexts. I have argued elsewhere (in Woodward, 'Laws and Causes', *The British Journal for the Philosophy of Science*, forthcoming) that this requirement is too strong and that a factor may have a stable causal tendency even if it does not produce an effect with any fixed probability at all.

However, in using the notion of a characteristic tendency, I do not mean to assume that all genuine causes must produce their effects with such fixed probabilities. Instead, the idea that causes have characteristic tendencies should be understood more broadly: the more general idea is, as suggested above, that genuine causes must exhibit some appropriate sort of stable pattern of behaviour across different background conditions and environments, both in the kinds of effects they produce and in the way they produce those effects. Thus, for example, smoking has a characteristic tendency to cause lung cancer because it produces this outcome by means of a characteristic mode of action across a range of different circumstances, in both humans and laboratory animals, among both men and women, among people with quite different diets and occupations, and so forth. Smoking has this tendency even though it may be the case (for all that is now known) that in many circumstances smoking does not produce lung cancer with any fixed stable probability.

Of course, it is true that many causal tendencies can be interfered with or modified and that the effects produced by a cause depend in part on the background circumstances in which it operates. But I think that it is built into our ordinary thinking about causes that we expect such interference or modification to involve other identifiable causes which themselves possesses certain characteristic tendencies, and that there be limits to how plastic and modifiable the original causal tendency is. Our expectation in other words, is that the behaviour and effects of a genuine cause will not be too unstable and context-dependent. If the effects and behaviour of some putative cause factor seem to vary in all sorts of arbitrary and unpredictable ways as the factor occurs in different background circumstances, and especially if we cannot locate other causal factors to which to attribute this variability, we will be inclined to think that we have not accurately identified a genuine cause at all.

It is this general presumption or expectation about the context-independence of causes that we tacitly appealed to in our discussion of (Ex. 1) and (Ex. 2) and which drives us toward what I have called the naively realistic view. Given that C_1 appears to cause E with a stable probability in a variety of different background circumstances when C_1 occurs alone, and similarly for C_2, it seems arbitrary and unreasonable, in the absence of some specific reason for believing otherwise, to deny that C_1 and C_2 continue to behave similarly when both are present together. To suppose that the frequency with which C_1 causes E abruptly increases when C_2 (or any other cause of E) is present, as (L_2) requires, conflicts with the presumption in favour of ascribing stable, context-independent tendencies to causes, in the absence of some specific reason to believe that other causal factors are present which interfere with such tendencies.

III

Is there any other way of preserving (T) in the face of examples like (Ex. 1) and (Ex. 2)? It might seem as though there is a natural alternative which we so far have neglected: one might hold that, to the extent that the laws and non-causal facts about particulars fail to fix the truth-values of (1), (2) and (3) in (Ex. 1) or (4) and (5) in (Ex. 2), there is simply no fact of the matter about which of these alternative hypotheses is true. This idea could of course be spelled out in a variety of different ways, but the basic thought is that to assume that *any* of these hypotheses is true is to attribute more structure to the world than it could (for metaphysical or conceptual reasons) possess. Inquiring into which of (1), (2) or (3) is true in (Ex. 1) is like inquiring which of several competing interpersonal cardinal comparisons of utility is correct, given the assumptions of standard revealed preference theory, or like inquiring which spatial metric is the true one, on views of spacetime structure according to which the choice of a metric is simply a matter of convention.

According to this idea, the defence of (T) explored above which appeals to (L_2) is mistaken in arguing that (3) is true in (Ex. 1). As our previous discussion shows, (Ex. 1) represents a situation in which facts about the laws and non-causal facts about particulars fail to fix which of (1), (2) and (3) is true. Precisely because of this, it is doubtfully consistent with the underlying idea of (T) to fasten on (3) as the correct description of the causal facts in (Ex. 1), as (L_2) requires. It is thus not surprising that the assumption that (3) is true in (Ex. 1) gets the defender of (T) into difficulties. A better strategy, more consistent with the spirit of (T), is simply to deny that there is any fact of the matter about which of (1), (2) or (3) is true. Given that this is the case it follows immediately that (Ex. 1) poses no threat to (T). Of course, our ordinary thinking about causation is such that we tend to suppose that one and only one of (1), (2) or (3) in (Ex. 1) must be true (even though we are unable to tell which). But this merely shows that our ordinary thinking is suffused with a naive and unwarranted realism. As Hume made clear long ago, not every feature of our everyday concept of causation corresponds to something real in the world.

There are at least two different ways in which this line of argument might be developed. The first holds that inquries about the causes of particular events like e are perfectly legitimate and merely tries to remain non-commital about whether (1), (2) or (3) is true in connection with (Ex. 1). The second, more radical line, proposes that we dispense entirely with singular causal claims and with inquiries about the cause of particular events, and suggests instead that we just stick with talk about initial conditions, and generic relationships described by laws, regularities, and conditional probabilities.

It seems doubtful that the first, more modest line of defence of (T) can be coherently developed in a way that prevents it from collapsing into the appeal to (L_2) explored and rejected above. The adherent of the first line agrees that in (Ex. 1) some occurrence (or occurrences) caused e and that it is unsatisfactory to simply say that e is uncaused or that inquiries into the cause or causes of e are misguided. He thus owes us an account of the legitimate factual content of causal claims, according to which some locution of the form '——caused e' comes out true in (Ex. 1), with the blank in this locution filled out in some definite way. If we go back to the fundamental idea which underlies (T) and which may also inspire the thought that there is no fact of the matter about which of (1), (2) or (3) is true, it is presumably this: there is nothing more to the factual content of a singular causal claim than the claim that (i) the cause and effect occur, (ii) instantiate some law or generalization and (iii) perhaps bear some appropriate non-causal (e.g. spatio-temporal) relationship to each other. There are no further facts about the existence of causal ties between particular occurrences over and above (i), (ii) and (iii). But if we apply this idea to (Ex. 1) we seem naturally led to the conclusion that it is hypothesis (3) [c_1 and c_2 caused e.] which is true in (Ex. 1). Both c_1 and c_2 instantiate laws linking them to e, and (we have been supposing) the non-causal relationships which c_1 and c_2 bear to e are fully consistent with the possibility that c_1 caused e and that c_2 caused e. The first version of the 'no fact of the matter defence' of (T) thus seems to be self-defeating; it collapses into a defence which appeals to (L_2), and according to which there *is* a fact of the matter (namely that (3) is true) about which of (1), (2) or (3) holds in (Ex. 1).

The second, more radical version of this defence of (T) proposes to avoid this difficulty by insisting that once one has specified the laws linking C_1 and E, and C_2 and E, in (Ex. 1) and the fact that c_1, c_2 and e have occurred and bear certain non-causal relationships to each other, one has said everything factual that there is to say about the causal and nomic relationships in (Ex. 1). There simply is no further legitimate question over and above this, about the singular causal explanation of e or about the cause or causes (or lack of these) for e.

I shall say more about this response below. Here I confine myself to observing that (as our discussion so far has brought out) there certainly *appears* to be a natural question about e which is left unanswered, once the laws and generic causal relations and initial conditions in (Ex. 1) have been specified. This is a counterfactual question, having to do with what the particular token event e depends on or with what on this particular occasion made the difference to the occurrence of e: was it c_1 alone, c_2 alone, or both together? As we shall see below, there are many contexts both in science and ordinary life, in which it is just this sort of question that is the focus of inquiry and which we wish to have answered. Furthermore, as we shall also see, there exist investigative strategies and empirical evidence which are often sufficient to answer such questions about particular causal connections: such questions do not seem to be spurious or objectionably metaphysical. Given that this is so, we require some additional motivation for regarding such questions as illegitimate other than the fact that this would provide a defense of (T).

Let me summarize the discussion so far in a slightly different way. Underlying (T) is, as we have said, the idea that there is no more factual content to a singular causal claim than that the cause and effect occur, instantiate a law of nature, and perhaps bear some appropriate non-causal relationship to each other. When we focus on the simplest and most straightforward case, in which c causes e and c is the only possible cause of e which is present, (T) may seem unproblematic. What our discussion has brought out is that (T) encounters difficulties as soon as we move to more complex cases like (Ex. 1) in which several different possible causes of e are present and in which it is not clear which of these (or whether some combination of these) have caused e. On the one hand, it may seem that if (T) is correct, it follows that (3) is the correct description of (Ex. 1), since both c_1 and c_2 do instantiate laws linking them to e. But there are at least two serious difficulties with this suggestion. First, it appears arbitrary, given what we know about how C_1 and C_2 behave in isolation. If C_1 sometimes causes E, and sometimes not, and similarly for C_2 and if C_1 and C_2 act independently of each other, why should we automatically conclude that every case in which C_1, C_2 and E are present is a case in which both C_1 and C_2 cause E? It seems much more plausible that in some cases of the sort represented by (Ex. 1), (1) C_1 alone causes E, in other cases (2) C_2 alone causes E, and in still other cases (3) C_1 and C_2 together cause E. Secondly, (and relatedly) (3) is inconsistent with (1) and (2). If facts about the laws and non-causal facts about particulars fix (3) as the true causal claim in (Ex. 1), this must be because they somehow undermine the possibility that (1) or (2) is true in (Ex. 1). But it is hard to see how facts about the laws and non-causal facts about particulars in (Ex. 1) can do this. These facts do not seem to differentially support or favour (3), as against (1) or (2), as the correct hypothesis regarding (Ex. 1). Instead (1), (2) and (3) seem perfectly symmetrical with respect to (and equally undetermined by) these facts. Given that this is the case, it again seems arbitrary to fasten on (3) as the correct account.

Given these difficulties with (3), it is natural to look for some alternative account of (Ex. 1) which is compatible with (T). But what would this alternative look like? Clearly, it is equally arbitrary to maintain that, given just the laws and non-causal facts in (Ex. 1), it follows that (1) must be true or that (2) must be true. It also seems false, or at least unsatisfactory and misleading, to claim that (6) nothing caused e or that e had no cause in (Ex. 1). Nor, given their inconsistency, can one say, at least without a great deal of further elaboration, that (1), (2) and (3) are merely three alternative ways of describing the causal facts in (Ex. 1). Moreover, taken together (1), (2), (3) and (6) seem, on the face of it, to exhaust the possibilities for the correct causal account of (Ex. 1). It is very unclear what to make of the suggestion that there is some additional alternative, which can be understood as an account of the causal facts in (Ex. 1), and which is somehow independent of, or non-committal with respect to, the truth or falsity of (1), (2), (3) and (6).

The source of the problem we have uncovered seems to be that a singular causal claim, as ordinarily understood, does after all say more than that the events described as cause and effect occur, instantiate a law, and satisfy some additional non-causal description. The fact that (1), (2) and (3) are inconsistent with each other, but that each seems quite consistent with the facts about which laws are instantiated and which non-causal descriptions are satisfied in (Ex. 1) is one reflection of this additional content. The fact that assumptions about how causes operate in isolation and about their characteristic tendencies constrains what we can plausibly claim about how those causes operate when they occur together is another reflection of this additional content. It is because singular causal claims possess such additional content that we encounter such severe dialectical difficulties when we try to characterize the correct singular causal claims in (Ex. 1) in a way that is consistent with (T)—in talking about singular causal connections at all, we seem to inescapably commit ourselves to claims that do not fit with (T). But to give up such talk entirely and to regard questions about the causal connections between particular events, or about the causes of particular outcomes, as somehow illegitimate or as unsuitable subjects for empirical inquiry seems equally unsatisfactory.

IV

My discussion so far has been rather abstract and schematic; it may seem that it has little connection with realistic problems of causal inference in science and in ordinary life. Nothing could be further from the truth. There are many features of our generally accepted methods and practices for making causal inferences which are best understood against the background of a kind of realism about particular causal connections which is inconsistent with (T).

Let me begin with an idea that is at least implicit in my discussion of (Ex. 1) and (Ex. 2). This is that establishing a causal claim regarding some outcome requires that one rule out or eliminate plausible competing causal claims regarding that outcome. The idea is that typically, when one inquires as to whether some particular occurrence c_1 caused some occurrence e, there will be a range of other factors $c_2 \ldots c_n$ which (as it would ordinarily be expressed) might or could have caused e, or which are possible causes of e, or which might be invoked to explain e.[21] Depending upon the details of the case, the claims that one or more of these other factors caused e will compete with the claim that c_1 caused e, in the sense that if any other claims turn out to be true, it cannot also be true that c_1 caused e. Showing that the factor c_1 was the factor that actually caused e on some particular occasion will then be is a matter of providing grounds which make it reasonable or plausible to reject each of these competing causal claims regarding e. Following what I think is a well-established usage, I shall speak of this goal as a matter of 'ruling out' or 'eliminating' each of these competing claims, but with the understanding that the kind and level of evidence and argument required to justify the contention that some causal claim has been ruled out may be a complex and context-relative matter. I do assume, however, that while ruling out a causal claim need not be a matter of producing evidence which is logically inconsistent with the claim, it does require producing evidence or theoretical considerations which render the claim implausible or unlikely or which (on some plausible account of hypothesis-testing or in connection with some relevant set of epistemic or non-epistemic values) makes it reasonable to reject the claim.[22] When our evidence is such that we are unable, in practice or

[21] I offer no general account here of what it is for a factor or type of event to be a possible cause. I will, however, remark that if C is a possible cause of E, it need not be the case that there is a deterministic or probabilistic law linking events of type C to events of type E, but that it must be the case that C exhibits the sort of stable, context-independent tendency or capacity to produce E's discussed in Section II above. That is to say, C's must be capable of causing E's accross a range of different circumstances and background conditions. (We thus must make use of the notion of a particular causal connection or a true singular causal claim to explicate the notion of a possible cause.) We can think of this as (one form of) an invariance or resiliency requirement; invariance is in general a distinguishing mark of causal or nomological relations.

[22] What I mean by rendering a claim implausible or unlikely is illustrated both by the examples considered below, and by the following case, which is adapted from Paul Humphreys, *The Chances of Explanation* (Princeton

even in principle, to carry out this elimination of alternatives, the proper conclusion is not that there is no fact of the matter about what caused e, but rather that we are unable to tell what caused e.[23]

Many philosophers will no doubt respond that this 'eliminativist' requirement on warranted causal inference is far too stringent and that it frequently will be difficult if not impossible to realize in practice. I claim, however, when we examine real life cases involving causal inference we find (a) wide-spread reliance on this eliminative strategy and (b) that the alternative causal possibilities to which the eliminative strategy is applied are individuated in a way that seems to support a kind of realism about singular causal connections which is inconsistent with (T). The eliminativist requirement is a widely acknowledged ideal in our practice concerning causal inference and helps to explain many salient aspects of that practice, such as the widespread tendency to think that controlled experiments are a particularly good way of learning about causal connections. Showing this in detail would of course require a much more lengthy discussion than is possible here, but consider, for purposes of discussion, three brief examples.

(Ex. 3) Under tranditional tort law, a necessary condition for a defendant to be liable for some harm is that his action or omission be shown to have caused the harm. Inquiries into whether this condition is met are thus paradigmatic examples of investigation into particular causal connections. Suppose that manufacturer A has produced some product P_1, exposure to which has been demonstrated to substantially increase the probability of contracting liver cancer, although P_1 by no means always produces such cancer. Suppose that plaintiff B is able to

[23] The eliminative strategy thus fits with a general account of science and inductive inference in which hypotheses are 'accepted' or 'rejected' and are not just assigned degrees of belief between zero and one. I lack the space to defend such an account here, and merely remark that it seems to correspond to scientific practice. The eliminative strategy also reflects the idea that warranted causal inference is *not*, as is sometimes claimed, a matter of inference to the best explanation; it is rather a matter of inference to the *only* acceptable explanation among some range of alternatives.

University Press, forthcoming). According to classical statistical mechanics, a container of water has a very small, but non-zero probability of freezing spontaneously. Suppose that such a container is placed in a refrigeration unit and freezes. There are two singular competing causal hypotheses regarding the freezing: (1) the freezing was caused by the placement within the refrigeration unit; (2) this was one of those very rare occasions in which the freezing occurred spontaneously. In this sort of case, although (2) is a genuine physical possibility, it is known to be extremely unlikely or implausible on the basis of theoretical considerations. As I shall construe the eliminative requirement, we are entitled to regard (2) as ruled out, because of its great implausibility.

show that he has been exposed to P_1 and that he has developed liver cancer. It seems to be generally agreed that in most legal contexts these facts are *not* sufficient to show that exposure to P_1 caused B's cancer. The reason is just what it is suggested by the eliminative strategy. Even if P_1 increases the probability of liver cancer very substantially the possibility remains open that some other possible cause in fact caused B's cancer. It is generally agreed by legal commentators that to show that exposure to P_1 caused B's cancer, evidence must be provided which undermines or undercuts this alternative possibility. That is, evidence must be provided which, given the relevant standards of proof specified by the law, makes it (legally) justifiable to reject the claim that these various alternative factors caused B's cancer.[24] Of course, meeting this standard of proof is in many cases extremely difficult, and the result is that in many cases it is concluded that it is simply not known whether some carcinogenic product caused the plaintiff's injury, and the manufacturer escapes liability. I do not wish to defend this absence of liability, which stikes me as quite unfair, but merely to note that it seems to follow inevitably from the causal condition on liability described above.

(Ex. 4) In his recent book *Nemesis*,[25] the physicist Richard Muller describes the considerations which led Luis Alvarez to put forward his celebrated theory that the extinction of the dinosaurs was caused by the impact of a large extra-terrestrial object such as a comet or an asteroid. The question 'what caused the extinction of the dinosaurs?' is another clear example of the kind of inquiry into the cause or causes of a particular outcome which we discussed in Part I and shows that it is a mistake to suppose (as it is sometimes claimed) that this kind of inquiry plays no role in contemporary physical science.

Alvarez's thinking, at least as described by Muller, appears to be a paradigm of the eliminative strategy. Alverez begins by attempting to

[24] As a general rule, the burden of proof for showing causation rests with the plaintiff but in tort law, unlike the criminal law, the evidence required to establish a causal claim need only show that is 'more likely than not'. (See W. Prosser, *Law of Torts* (St. Paul, Minnesota: West Publishing Company, 1971). While the standards of evidence required to show that it is reasonable to accept or reject a causal claim are thus different from and in some respects less demanding than those often imposed in science, the basic point remains that providing grounds for accepting a causal claim requires providing grounds for rejecting competing hypotheses. A demonstration that a causal factor was present and that the factor was a possible or frequent cause of the effect, or that it raised the probability of the effect is not taken to show that the factor in fact caused the effect.

[25] See R. Muller, *Nemesis* (New York: Weidenfeld and Nicolson, 1988).

show that the impact of an extra-terrestrial body like an asteriod *could* have caused the mass extinction of the dinosaurs—that it is a possible cause. He does this by first determining, from astronomical data, the largest asteroid that might reasonably have been expected to hit the Earth during the period in which dinosaurs existed. He then calculates the kinetic energy that would be released from such a collision and argues by extrapolating from data from nuclear explosions and by looking at the estimated effects of other impact craters, that such a collision would have thrown up enough dust in the atmosphere to block sunlight for a considerable period of time. This shows that such an impact *could* have caused the extinction of dinosaurs (and, indeed, that impacts with catastrophic effects must have occurred, since we know that suitably large extra-terrestrial impacts have occurred), but it does not of course show that the extinction of the dinosaurs was actually due to such an impact. Alvarez attempts to show *this* by finding evidence that rules out other possible causes of this extinction. The complex chain of reasoning that leads finally to the asteroid impact theory involves the proposal of a number of various new possible causes for the extinction and their systematic elimination. A key piece of evidence in this effort was the presence of an unusually high concentration of irridium in a layer of clay which was thought to have been laid down at the same time as the extinction. Alvarez argues that only an extra-terrestrial impact could account for this excess of irridium.

> By the end of the summer, [Alvarez] had concluded that there was only one acceptable explanation for the irridium anomaly. All other origins could be ruled out, either because they were inconsistent with some verified measurement or because they were internally inconsistent. The irridium had come from space . . . (pp. 51–2).

Initially Alvarez thought that the extra-terrestrial source was a nearby supernova. But a supernova explosion would have injected an isotope of plutonium (Pu-244) into the clay layer. Evidence for such plutonium would have supported the supernova hypothesis, precisely because it would conform so completely to the eliminativist ideal. No other possible cause appears to be consistent with the presence of plutonium.

> If [plutonium] could be found in the clay layer, it would prove the theory. Not just verify it, not just strengthen it, but prove it. There is no other conceivable source of Pu-244. Its presence would strengthen the supernova explanation beyond all reasonable doubt (p. 53).[26]

[26] Many philosophers of science, heavily influenced by recent emphasis on the fallible and conjectural character of all scientific knowledge will smile at this talk of 'proving' a scientific theory. In fact, although I lack the space to

When no plutonium was found, the supernova theory was 'dead' (p. 60). Alvarez then considers a number of other alternative causes, looking for data which would 'rule out' (p. 62) or 'eliminate' (p. 67) these possibilities. He eventually settles on the asteroid theory, since it makes a number of additional predictions that seem to be confirmed (it predicts, apparently correctly, the amount of irridium laid down, and that the irridium will be uniform world-wide) and since alternative possible causes do not seem to be able to explain this additional evidence. As Muller puts it, Alvarez adopts the asteroid theory because he regards it as 'unique in its ability to explain [the irridium] enrichment' (p. 69).

(Ex. 5) Suppose that an investigator wishes to determine whether the application of a certain fertilizer will increase plant height. Here, unlike the previous two examples, the typical investigator's interest is not so much in establishing what caused the increase in height in any particular plant, but rather in establishing that certain causal processes or particular causal connections hold in the case of at least some of the plants in the group. That is, when the investigator asks whether the application of fertilizer has increased plant height in this group, he wishes to know whether some or many plants in the group have been caused to increase in height by the application of the fertilizer. Suppose that the investigator proceeds by selecting a group of plants, measuring their average height, applying the fertilizer and then measuring their average height six weeks later. Suppose that there is a significant increase in average height. Is the investigator justified in concluding that the fertilizer caused this increase? According to both common sense and any textbook on experimental design, the answer is 'no'. The difficulty, of course, is that other possible causes of the increase in height have not been ruled out, or as it is commonly put, 'controlled for'. The increase in height may, for example, have been entirely due to normal processes of maturation within the plants or due to an increase in the water or sunlight they received. A properly designed experiment which allows one to reach a definite conclusion about the causal role of the fertilizer must exclude these (and other) possibilities. A classic strategy for accomplishing this is to randomly assign groups of plants to treatment and control groups in such a way that we may reasonably

---

argue for this claim here, I think that the phrase is quite appropriate. Often scientists really can (a) delimit a set of possible theories which (modulo some agreed upon level of idealization and approximation) exhaust the plausible alternatives regarding some phenomena of interest and (b) obtain evidence which rules out all possibilities from this set except for one. When this can be done it is perfectly appropriate to describe the remaining possible theory as proved or established beyond a reasonable doubt.

expect that the control and treatment of groups do not differ systematically with respect to other factors which may be causally relevant.

In general, as Cook and Campbell claim in their well-known text *Quasi-Experimentation*, the whole point of experimental (or non-experimental) design is to 'rule out alternative interpretations' (p. 96) according to which the effect of interest may be due to some other factor besides the putative cause. Causal inferences based on experiments are often more secure than inferences based on passive obervation in large part because experiments are often more effective at accomplishing this end of ruling out alternative causal claims, and certain kinds of non-experimental designs (e.g. longitudinal designs) are superior to others (e.g. cross-sectional designs) for just the same reason.[27]

We must constrast both the causal conclusions reached in each of the above examples and the methodology employed to reach those conclusions with the conclusions and methodology associated with (T) or (L_2). Suppose, as might be easily the case, that in (Ex. 3) above, B is exposed to P_1 and to P_2, both of which have a non-negligible chance of producing cancer and that there are no other relevant non-causal facts besides those stated. It will follow immediately from (L_2) that P_1 (and P_2) caused B's cancer. The competing claim that P_2 alone caused B's cancer will not be treated as a live empirical possibility; there thus will be no basis for the contention that in order to justifiably conclude that P_1 caused B's cancer, we must be able to rule out this competing claim, and that if we are unable to do so, we simply do not know whether P_1 or P_2 alone, or both together caused B's cancer. This is completely at odds with the judgment that is actually reached regarding this sort of case in legal contexts.

Similarly, suppose that Alvarez was unable to eliminate the possibility that a supernova caused the irridium anomaly (we might imagine, counterfactually, that the mechanism of supernova explosions is such that no plutonium is produced). Suppose also that there was independent evidence for both a supernova explosion in the vicinity of the Earth and of an asteroid impact at the time of the extinction of the dinosaurs. In the absence of other relevant non-causal facts, it would apparently follow from (L_2) (or at least be reasonable to believe, on the basis of (L_2)) that a supernova explosion caused the extinction of the

[27] The central role in causal inference of ruling out alternative explanations of an observed association and the differences among research designs in how effectively they accomplish this end are emphasized by many writers in the social and behavioural sciences. In addition to Cook and Campbell, *Quasi-Experimentation*, see, for example, H. Asher, *Causal Modeling* (Beverley Hills: Sage Publications, 1983).

dinosaurs. It would be a mistake to suppose, as Alvarez did, that to support this or any other singular causal claim about the extinction of the dinosaurs it would be necessary to systematically consider and rule out competing singular causal claims.

Again, suppose that in (Ex. 5) exposure to fertilizer does indeed increase the probability that each member of a group of plants will increase in height. According to (T) or (L_2) it follows (or at least it would be reasonable to believe), even if additional possible causes of the increase in height are known to be present and uncontrolled for, that exposure to the fertilizer caused the increase in height (assuming again, for the sake of argument, that there are no non-causal facts about particulars which are inconsistent with this conclusion).

Let me put all of this in a more general way. Each of the examples above has or may have, given what the investigator knows at some stage in his inquiry, the structure of (Ex. 1): an effect e occurs, a possible cause c_1 occurs, (which, let us suppose, raises the probability of e), but the possibility that some other cause c_2 alone may have caused e has not yet been ruled out. In each case the investigator proceeds by attempting to rule out such alternatives; if and only if this can be done, does he regard himself as justified in concluding that c_1 caused e. This method of proceeding makes some sense if a kind of realism about particular causal connections is correct: if the various alternative singular causal claims (1), (2) and (3) concerning e considered in section I represent definite alternative ways the world might be. It then becomes understandable that the point of using the eliminative strategy is to try to obtain evidence which discriminates among three different possibilities. It is also understandable why, if we are unable to do this, the appropriate conclusion is that we simply do not know which of these competing singular causal claims is correct. Finally, if this sort of elimination of competing alternatives is a requirement on warranted causal inference, it becomes unmysterious how (T) could fail to be true: the laws and non-causal facts about particulars that obtain in a given situtation may fail to rule out all but one of the alternative singular causal claims that might be true in that situation. We see how there could be 'something more' at issue when we inquire whether one particular event caused another than whether the events occur and instantiate a law and bear some appropriate non-causal (e.g. spatio-temporal) relationship to each other. The 'something more' has to do with whether or not other possible causes of the effect are operative in the situation. Whether or not such other possible causes are operative is a (causal) fact (indeed, a fact about *singular* causes) which may be underdetermined by the laws and non-causal facts which hold in the situation. It is because the laws and non-causal facts can fail to determine which of the several possible causes which may be present in a situation is actually operative in producing an effect that (T) is false.

By contast, if (T) or (L_2) is correct, use of the eliminative requirement seems to lack motivation, as do many other familiar canons of experimental design. It looks as though, once we have ascertained that a putative cause and effect occur, instantiate a law or generalization of some appropriate sort, and perhaps bear some additional non-causal relation to each other, there is no room for any further question about whether the putative cause and effect are actually causally connected.

V

The interconnections among these various ideas explored above: (L_1), (L_2) and the supervenience thesis (T), realism about particular causal connections, and the eliminative strategy emerge with characteristic clarity in the above-mentioned postscript (1986) to David Lewis' paper 'Causation'. After claiming that in a case like (Ex. 1), we are justified in concluding that c caused e if the increase in the chance of e due to c is 'sufficiently greater' than the chance of e without c, Lewis goes on to remark:

> Some philosophers find this counterintuitive. They would correct me thus. No; if there would have been some residual chance of e even without c, then the raising of probability only makes it *probable* that in this case c is a cause of e. Suppose, for instance, that the actual chance of e, with c, was 88 per cent; but that without c, there would still have been a 3 per cent probability of e. Then most likely (probability 97 per cent) this is a case in which e would not have happened without c; then c is indeed a cause of e. But this just might be (probability 3 per cent) a case in which e would have happened anyway; then c is not a cause of e. We cannot tell for sure which kind of case this is (p. 180).

Lewis rejects this objection:

> The objection presupposes that the case must be of one kind or the other: either e definitely would have occurred without c, or it definitely would not have occurred without c. But I reject the presupposition that there are two different ways the world could be, giving us one definite counterfactual or the other . . .

Lewis' grounds for rejecting this presupposition seem to rest on something very like (T). He writes that, 'the presupposition is that there is some hidden feature which may or may not be present in our actual world and which if present would make true the counterfactual that e would have occurred anyway without c'. But since 'this is supposed to be a matter of chance, in the counterfactual situation as in actuality, whether e occurs', it follows that the 'laws of nature' and 'matters of historical fact' immediately after the occurrence of c do not 'predetermine' e (p. 181). Nor, of course do the chances present in this situation determine whether e will occur. Lewis suggests that there are no other non-mysterious candidates for the required hidden feature:

> So the hidden feature must be something else still: Not a feature of the history of the world, and also not a feature of its chances, or of the laws or conditionals whereby its chances depend on its history. It fails to supervene on those features of the world on which, so far as we know, all else supervenes. To accept any such mysterious extra feature of the world is a serious matter. We need some reason much more weighty than the isolated intuition on which my opponent relies (p. 182).

Let me begin by noting that the numeral values in Lewis' example may make his general conclusions more plausible than they would otherwise seem. On the account I have presented, to conclude that c caused e, we must be able to rule out or render implausible that possibility that e would have occurred without c. If the probability that e would have happened any way is only 3 per cent, we may very well regard ourselves as justified in ruling out this possibility (3 per cent is, after all, well below the 5 per cent significance level standardly used in the social and behavioural sciences) and hence in concluding, just as Lewis does, that c caused e. The fact that the probability of e, given c, is very high also may help to make this conclusion seem plausible.

But if (T) is correct, and Lewis' line of argument warranted, his general conclusions ought to follow even when there is a more substantial chance of e occurring without c (e.g. 20 per cent) and the probability of e, given c is less high (e.g. 60 per cent). But, as (Exs. 3, 4 and 5) above illustrate, on generally accepted methodological requirements regarding causal inference, one is *not* warranted in concluding in this sort of case that c caused e. Instead the conclusion which is generally taken to be warranted is that we do not know whether or not c caused e, just as Lewis' imagined objector claims. I contend that this generally accepted practice is correct, and that Lewis is mistaken in rejecting the eliminative requirement in this sort of case.

Lewis' discussion brings out very clearly a point made in the previous section: that the eliminative requirement will seem to make good sense if one thinks that there are several alternative ways the world might be which are equally consistent with the information given in his example. Once one agrees that the case might be one in which c causes e (where this is understood to imply in the case under discussion, that if c had not occurred, then e would not have occurred) and might be one in which c does not cause e (in which case, e would have occurred, even if c had not occurred) then it also becomes reasonable to insist that one does not know which of these claims is true unless one is able to rule out or render implausible the competing alternative. For reasons that should by now be familiar, I think that Lewis is also correct to conclude that if there are several different ways the world might be, only one of which actually obtains, then which alternative obtains is not fixed by and does not supervene on facts about the laws and chances and non-causal facts about particulars.

Lewis goes on to argue that it is not reasonable to suppose that there really are these two alternative ways the world might be: to suppose otherwise is to suppose that the world contains some mysterious, hidden feature, the presence or absence of which fixes the truth or falsity of the alternative causal claims and counterfactuals described above. This extra feature corresponds of course to the obtaining or non-obtaining of a particular causal connection between c and e, and as Lewis's discussion makes clear, he finds this feature 'mysterious' precisely because he is committed to something very like (T)—because the feature in question fails to supervene on those facts (about laws, initial conditions and chances) on which, he supposes, all else apparently supervenes. The arguments given above seem to me to provide good grounds for rejecting (T) and for accepting the conclusion that much of our reasoning about causation carries with it a commitment to Lewis' non-supervening hidden feature.[28]

I want to conclude this section by commenting briefly on another theme in Lewis' remarks: the connection between (T), the eliminative

[28] In saying this, I do not mean to claim that in every case having the structure of Lewis' example, we are somehow assured on *a priori* grounds that there is a determinate fact of the matter about whether e would or would not have occurred without c. My claim is rather that it is a mistake to hold that merely because supervenience fails, it follows on conceptual grounds that there could not be a determinate fact of the matter regarding which of these counterfactuals is true. In fact, I think that it is very plausible that there are quantum-mechanical cases having the structure of Lewis' example, in which the above counterfactuals lack a definite truth value, and in which nothing corresponding to Lewis' hidden feature can be present. However, I think that this is so for empirical reasons peculiar to quantum-mechanical phenomena (reasons summarized in the various no-hidden-variables theorems) and not merely because of conceptual considerations having to do with the failure of supervenience. My point is thus a conditional one: that if there is to be a determinate fact of the matter about whether c has or has not caused e, then there must also be a determinate fact of the matter about whether e would or would not have occurred without c and the case must be one in which Lewis' hidden feature might be present.

strategy, and the counterfactual implications of causal claims. I have already contended that if (1) c_1 caused e in (Ex. 1), then (given what else is true in (Ex. 1), this implies a counterfactual claim: if c_1 had not occurred, e would not have occurred. Obviously, there is a close connection between this contention and my claim that in order to establish (1) one must satisfy the eliminative requirement. If we can rule out the possibility that any other possible cause produced e and if we can also rule out the possibility that e occurred spontaneously, then we have good reason to think that the above counterfactual is true. If e did not occur spontaneously and no other cause of e was present, then it seems plausible that if c_1, the only remaining possible cause of e, had also not occurred, e would not have occurred.

Now, as is well known, Lewis also holds (to a first approximation) that the claim (1) c_1 caused e entails the above counterfactual (that if c_1 had not occurred, e would not have occurred) when the relationship between c_1 and e is deterministic, or when, although the relationship between c_1 and e is indeterministic, e has no chance of occurring if c_1 had not occurred. But, as the passages quoted above indicate, Lewis thinks that a different 'kind' (p. 175) of 'chancy causation' which has a different counterfactual implication is operative in the sort of case represented by (Ex. 1) or (Ex. 2), in which e has some chance of occurring without c_1. Here (as L_1 requires) Lewis holds that it is true that c_1 caused e, even though the counterfactual claim that if c_1 had not occurred, then e would not have occurred is false. On Lewis' view, the claim that c_1 caused e entails a different counterfactual: that if c_1 had not occurred, then the (counterfactual) chance of e would have been lower than what the chance of e actually is, given the occurrence of c_1. This second 'kind' of causation, with a distinct counterfactual analysis, is clearly required if one is to retain the supervenience thesis in connection with examples like (Ex. 1), but its introduction looks *ad hoc*, and seems to lack independent motivation. Why should a rather different kind of causation between c_1 and e with a sharply different kind of counterfactual entailment come into play when we move from a case in which c_1 causes e and no other possible causes of e are present to a case like (Ex. 1) in which c_1 still causes e but some other possible cause of e is present as well? It seems to me to be preferable to retain a single, unitary account of causation, according to which the claim that (1) c_1 alone caused e represents an empirical possibility in both cases, and in both cases entails that if c_1 had not occurred, then e would not have occurred. The price of such a unitary account is that the supervenience thesis (T) comes out false and that, in (some) real cases having the structure of (Ex. 1), we may not be able to determine whether or not (1) is true. But, I claim, this accurately represents what the facts of the matter are in connection with causal inference.

VI

Although (T) is a metaphysical thesis, it is clear that an important part of its appeal, both historically and at present, is epistemological. Non-causal facts about particulars and facts about laws (particularly if these are in turn susceptible to some sort of regularity analysis) have seemed to many philosophers working within an empiricist framework to be unproblematically epistemically accessible: such facts look like plausible candidates for what can be established by 'observation' or by the application of generally accepted patterns of inductive inference to what can be observed. By contrast, claims about causal connections between particular events which do not supervene on facts about regularities and other non-causal facts about particulars do not look like claims that can be established just on the basis of observation. In denying the supervenience thesis, (T), we may seem in danger of introducing a class of facts which are unknowable in principle, or at least disturbingly far removed from the possibility of ordinary empirical assessment. The fact that, according to the realist account of singular causal claims I have defended, there may be no evidence which would allow us to tell, in a case like (Ex. 1), whether (1), (2) or (3) is true may further underscore or lend credence to this worry.

In what follows, I shall briefly try to address these epistemological concerns. I shall begin (i) by making the evidential assumptions that underlie the realist account more explicit. I shall then (ii) argue that, given these assumptions, singular causal claims are not epistematically inaccessible in any objectionable sense and that (iii) in holding that in some circumstances we may be unable to tell which of several competing singular causal claims is true, the realist account accurately reflects good scientific practice. It is a defect, rather than an advantage, of accounts based on (T) and (L_2) that they attempt to tie the truth conditions for causal claims so closely to what can be observed (or more generally to evidence which is free of causal assumptions). Such accounts make warranted causal inference look easier and more assumption-free than it actually is.

(i) In using the eliminative strategy sketched above, we rely on assumptions (among others) that are (or appear to be) causal in character in order to assess the truth-value of other causal claims. For example, we rely on the assumptions that events $c_1 \ldots c_n$ are 'possible causes' of some event of interest e and that various other events (some of which may occur in the spatio-temporal vincinity of e) are not possible causes of e. We use these causal assumptions together with other kinds of non-causal information (about, for example, spatio-temporal relations) when we attempt to formulate an eliminative argument in support of some particular singular causal claim concerning e.

This way of proceeding fails to satisfy a certain kind of foundationalist ideal which may appeal to those with classical empiricist sensibilities: we are not shown how to determine the truth-values of singular causal claims, given only evidence and other assumptions that are clearly non-causal in character. In testing singular causal claims, we rely on assumptions which are couched in terms of unreduced causal notions (like the notions of a possible cause) and we are not shown how these can be entirely eliminated in favour of assumptions that are unproblematically non-causal.

(ii) Is the non-foundationalist character of the eliminative procedure objectionable? Consider the more general problem of testing theoretical hypotheses in science. The idea, associated with some early positivists and operationalists, that theoretical hypotheses must be fully translatable into or analysable in terms of claims about observables is now widely rejected by philosophers and scientists. According to more recent accounts, such as Clark Glymour's bootstrap model,[29] or, for that matter, many versions of the hypothetico-deductive model, testing theoretical hypotheses typically involves assuming the truth of various other theoretical hypothesis and using these, in conjunction with claims about what is observed (or, at any event, theoretically uncontentious claims about evidence), to deduce claims which can then be compared with the theoretical hypothesis under test. If there are doubts concerning the assumed theoretical hypotheses, it may be possible to test these emprically by using a similar procedure involving the assumption of still other theoretical claims, and so forth. This procedure does not show how the hypothesis under test may be translated into purely observational claims and the claims to which we appeal in arguing that the hypothesis is empirically well supported are not just claims about what we observe, but involve ineliminable reference to other theoretical assumptions. Nonetheless, provided that the whole procedure satisfies certain requirements designed to make it non-question-begging, a theoretical hypothesis that passes such tests may be regarded as strongly supported by empirical evidence.

My claim is that a similar conclusion is warranted in connection with the testing of causal claims: the demand for elimination of causal assumptions in the testing of causal hypothesis is no more reasonable than the demand for the elimination of theoretical assumptions in the testing of theoretical hypotheses. Like the procedures for testing theoretical hypotheses described above, the eliminative strategy and related testing procedures are not in any obvious way circular or question-begging. We do not, for example, assume as a premise the causal conclusion we are attempting to establish and the causal assumptions

[29] C. Glymour, *Theory and Evidence* (Princeton University Press, 1980).

with which we begin are (or so I would argue) clearly empirical claims, which are themselves susceptible of empirical assessment, given the assumption of still other causal claims. Indeed, in many cases, the background causal assumptions on which we rely in attempting to establish some causal claim of interest are at least as epistemically secure as many observational claims. This is so for many garden-variety claims about possible causes. For example, the causal assumptions that ingesting aspirin is a possible cause of the cessation of headaches or that a blow on the head from a heavy object can cause death are not open to serious doubt. I suggest that as long as we can see how, given the truth of other causal claims, observational evidence can be brought to bear on some causal claim of interest, and as long as these other causal claims are themselves testable by means of a similar strategy, the claim in question satisfies whatever requirements of testability and epistemic accessibility it is reasonable to impose.[30]

(iii) Even on an account of singular causal claims which is committed to (T) or (L_2) or more generally to some sort of regularity analysis, there will of course be many cases in which available evidence does not allow one to draw warranted conclusions regarding what the causal connections are. This will be the case if, for example, an investigator is unable to obtain information about relevant regularities or relevant non-causal facts. But the realist account of singular causal claims presented above allows for the possibility of an additional kind of epistemic undetermination. On the realist account, there is nothing which guarantees, even given full information about the regularities which obtain in a given situation and the relevant non-causal facts about particulars, that we will be able to determine whether some singular causal claim of interest is true in that situation. Whether or not we will be able to do this is a contingent matter, which depends on which other causes are present and on whether there exists evidence which allows us to rule out alternatives which compete with the claim of interest. We might express this by saying that on the realist account the causal structure of the situation itself may be such that (because of the presence of other causal factors) it undermines the drawing of definite

[30] Indeed, I suspect that something stronger is true: not only is it *permissible* or *legitimate* to test or support causal claims by the use of other assumptions which are causal in character it is usually (perhaps always) *necessary* to do this. It is typically (perhaps always) unwarranted to draw causal conclusions just on the basis of evidence and assumptions which are non-causal in character. For an argument in support of this claim in connection with the use of statistical techniques like regression analysis to establish causal conclusions, see my 'Understanding Regression', *PSA* 1988 (Philosophy of Science Association, 1988), 255–69.

conclusions about causal connections, even given full information about regularities and other non-causal facts.

Although this feature of the realist account may make some philosophers uneasy, I claim that it merely reflects a quite general fact about causal inference which is widely acknowledged by non-philosophers—namely that different situations or contexts in which causal interactions occur vary enormously in the extent to which they allow one to draw determinate conclusions about (or provide unambiguous evidence concerning) those interactions. Indeed, it is in large measure because of this fact that there is a role for (experimental and non-experimental) *design* in the investigation of causal connections. Many (although by no means all) naturally occurring situations, in which human intervention or design plays no role, are such that so many different possible causal factors are present that the causal structure of the situation itself makes the ruling out of competing causal claims and reaching of definite causal conclusions extremely difficult or impossible. By contrast, in a well-designed experimental or observational study, many of these competing causal claims will be ruled out by the design (the causal structure) of the investigation itself and it will be possible to follow the requirements of the eliminative strategy and to reach definite causal conclusions.

Consider an investigator who wishes to know whether some particular application c_1 of C_1 has caused e, and who applies c_1 in a situation having the structure of (Ex. 1) in which c_2 is also present, and in which there is (or need be) no further evidence that rules out the possibility that c_2 alone caused e. One possible response to the investigator's plight is, in effect, to adjust our concept of causation in such a way that the investigation is assured of reaching a definite conclusion about what the causal connections are, regardless of the situation in which c_1 is applied. This sort of response is reflected in (T) and (L_2), which embody the demand that full information about the relevant regularities and non-causal facts about particulars must allow us to reliably infer what the causal connections are in every possible situation. A second possible response to this sort of case, which seems to be more in accord with good scientific practise, is to recognize that the investigator has set up a badly designed experiment, the defect in the experiment being precisely that it does not allow one to reach definite conclusions about the causal relation between c_1 and e. Rather than adjusting our concept of causation to allow the drawing of definite causal conclusions even in the sort of case represented by (Ex. 1), we instead ought to recognize that what needs changing is the experimental design represented by (Ex. 1) itself. It is a virtue, rather than a defect of the realist account that it allows us to recognize this.

Contrastive Explanation*

PETER LIPTON

1. Introduction

According to a causal model of explanation, we explain phenomena by giving their causes or, where the phenomena are themselves causal regularities, we explain them by giving a mechanism linking cause and effect. If we explain why smoking causes cancer, we do not give the cause of this causal connection, but we do give the causal mechanism that makes it. The claim that to explain is to give a cause is not only natural and plausible, but it also avoids many of the objections to other accounts of explanation, such as the views that to explain is to give a reason to believe the phenomenon occurred, to somehow make the phenomenon familiar, or to give a Deductive-Nomological argument. Unlike the reason for belief account, a causal model makes a clear distinction between understanding why a phenomenon occurs and merely knowing that it does, and the model does so in a way that makes understanding unmysterious and objective. Understanding is not some sort of super-knowledge, but simply more knowledge: knowledge of the phenomenon and knowledge of its causal history. A causal model makes it clear how something can explain without itself being explained, and so avoids the regress of whys, since we can know a phenomenon's cause without knowing the cause of the cause. It also accounts for legitimate self-evidencing explanations, explanations where the phenomenon is an essential part of the evidence that the explanation is correct, so the explanation can not supply a non-circular reason for believing the phenomenon occurred. There is no barrier to knowing a cause through its effects and also knowing that it is their cause. The speed of recession of a star explains its observed red-shift, even though the shift is an essential part of the evidence for its speed of recession. The model also avoids the most serious objection to the familiarity view, which is that some phenomena are familiar yet not understood, since a phenomenon can be perfectly familiar, such as the blueness of the sky or the fact that the same side of the moon always

* I am grateful to Philip Clayton, Trevor Hussey, Philip Pettit, David Ruben, Elliot Sober, Edward Stein, Nick Thompson, Jonathan Vogel, David Weissbord, Alan White, Tim Williamson, and Eddy Zemach for helpful discussions about contrastive explanation.

faces the earth, even if we do not know its cause. Finally, a causal model avoids many of the objections to the Deductive-Nomological model. Ordinary explanations do not have to meet the requirements of the Deductive-Nomological model, because one does not need to give a law to give a cause, and one does not need to know a law to have good reason to believe that a cause is a cause. As for the notorious over-permissiveness of the Deductive-Nomological model, the reason recession explains red-shift but not conversely, is simply that causes explain effects but not conversely, and the reason a conjunction of laws does not explain its conjuncts is that conjunctions do not cause their conjuncts.

The most obvious objection to a causal model of explanation is that there are non-causal explanations. Mathematicians and philosophers give explanations, but mathematical explanations are never causal, and philosophical explanations seldom are. A mathematician may explain why Gödel's Theorem is true, and a philosopher may explain why there can be no inductive justification of induction, but these are not explanations that cite causes. In addition to the mathematical and philosophical cases, there are explanations of the physical world that seem non-causal. Here is a personal favourite. Suppose that some sticks are thrown into the air with a lot of 'spin', so that they separate and tumble about as they fall. Now freeze the scene at a certain point during the sticks' descent. Why are appreciably more of them near the vertical axis than near the horizontal, rather than in more or less equal numbers near each orientation, as one would have expected? The answer, roughly speaking, is that there are many more ways for a stick to be near the horizontal than near the vertical. To see this, consider purely horizontal and vertical orientations for a single stick with a fixed midpoint. There are infinitely many of the former, but only two of the latter. Less roughly, the explanation is that there are two horizontal dimensions but only one vertical one. This is a lovely explanation, but apparently not a causal one, since geometrical facts cannot be causes.

Non-causal explanations show that a causal model of explanation cannot be complete. Nevertheless, a causal model is still a good bet now, because of the backward state of alternate views of explanation, and the overwhelming preponderance of causal explanations among all explanations. Nor is it *ad hoc* to limit our attention to causal explanations. A causal model does not simply pick out a feature that certain explanations happen to have: causal explanations are explanatory *because* they are causal. Like other accounts of explanation, however, causal models face a problem of underdetermination. Most causes do not provide good explanations. This paper attempts a partial solution to this problem.

2. Fact and Foil

Let us focus on the causal explanation of particular events. The problem here is with the notion of explaining an event by giving *the* cause. We may explain an event by giving some information about its causal history (Lewis, 1986), but causal histories are long and wide, and most causal information does not provide a good explanation. The big bang is part of the causal history of every event, but explains only a few. The spark and the oxygen are both part of the causal history that led up to the fire, but only one of them explains it. So what makes one piece of information about the causal history of an event explanatory and another not? The short answer is that the cause that explains depends on our interests. But this does not yield a very informative model of explanation unless we can go some way towards spelling out how explanatory interests determine explanatory causes.

One way to show how we select from among causes is to reveal additional structure in the phenomenon to be explained, structure that points to a particular cause. We can account for the specificity of explanatory answers by revealing the specificity in the explanatory question. Suppose we started by construing a phenomenon to be explained simply as a concrete event, say a particular eclipse. The number of causal factors is enormous. As Hempel has observed, however, we do not explain events, only aspects of events (Hempel, 1965, pp. 421–3). We do not explain the eclipse *tout court*, but only why it lasted as long as it did, or why it was partial, or why it was not visible from a certain place. This reduces the number of causal factors we need consider for any particular phenomenon, since there will be many causes of the eclipse that are not, for example, causes of its duration.

More recently, it has been argued that explanation is 'interest relative', and that we can analyse some of this relativity with a contrastive analysis of the phenomenon to be explained. What gets explained is not simply 'Why this', but 'Why this rather than that' (Garfinkel, 1981, pp. 28–41; van Fraassen, 1980, pp. 126–9). A contrastive phenomenon consists of a *fact* and a *foil*, and the same fact may have several different foils. We may not explain why the leaves turn yellow in November *tout court*, but only, for example, why they turn yellow in November rather than in January, or why they turn yellow in November rather than turning blue.

The contrastive analysis of explanation is extremely natural. We often pose our why-questions in explicitly contrastive form and it is not difficult to come up with examples where different people select different foils, requiring different explanations. When I asked my three year old son why he threw his food on the floor, he told me that he was full. This may explain why he threw it on the floor rather than eating it, but I wanted to know why he threw it rather than leaving it on his plate. Similarly, an explanation of why I went to see *Jumpers* rather than *Candide* will probably not explain why I went to see *Jumpers* rather than staying at home, and an explanation of why Able rather than Baker got the philosophy job may not explain why Able rather than Charles got the job. The proposal that phenomena to be explained have a complex fact–foil structure can be seen as another step along Hempel's path of focusing explanation by adding structure to the why-question. A fact is usually not specific enough: we also need to specify a foil. Since the causes that explain a fact relative to one foil will not generally explain it relative to another, the contrastive question provides a further restriction on explanatory causes.

While the role of contrasts in explanation will not account for all the factors that determine which cause is explanatory, I believe that it does provide the central mechanism. In this essay, I want to show in some detail how contrastive questions help select explanatory causes. My discussion will fall into three parts. First, I will make three general observations about contrastive explanation. Then, I will use these observations to show why contrastive questions resist reduction to non-contrastive form. Finally, I will describe the mechanism of 'causal triangulation' by which the choice of foils in contrastive questions helps to select explanatory causes.

When we ask a contrastive why-question—'Why the fact rather than the foil?'—we presuppose that the fact occurred and that the foil did not. Often we also suppose that the fact and the foil are in some sense incompatible. When we ask why Kate rather than Frank won the Philosophy Department Prize, we suppose that they could not both have won. Similarly, when we asked about leaves, we supposed that if they turn yellow in November, they cannot turn yellow in January, and if they turn yellow in November they cannot also turn blue then. Indeed, it is widely supposed that fact and foil are always incompatible (Garfinkel, 1981, p. 40; Ruben, 1987; Temple, 1988, p. 144). My first observation is that this is false: many contrasts are compatible. We often ask a contrastive question when we do not understand why two apparently similar situations turned out differently. In such a case, far from supposing any incompatibility between fact and foil, we ask the question just because we expected them to turn out the same. By the time we ask the question, we realize that our expectation was disappointed, but this does not normally lead us to believe that the fact precluded the foil, and the explanation for the contrast will usually not show that it did. Consider the much discussed example of syphilis and paresis (cf. Hempel, 1965, pp. 369–70; van Fraassen, 1980, p. 128). Few with syphilis contract paresis, but we can still explain why Jones rather than Smith contracted paresis by pointing out that only Jones had syphilis. In this case, there is no incompatibility. Only Jones contracted paresis, but they both could have: Jones's affliction did not protect Smith. Of course not every pair of compatible propositions would make a sensible contrast but, as we will eventually see, it is not necessary to restrict contrastive questions to incompatible contrasts to distinguish sensible questions from silly ones.

My second and third observations concern the relationship between an explanation of the contrast between a fact and foil and the explanation of the fact alone. I do not have a general account of what it takes to explain a fact on its own. As we will see, this is not necessary to give an account of what it takes to explain a contrast; indeed, this is one of the advantages of a contrastive analysis. Yet, based on our intuitive judgments of what is and what is not an acceptable explanation of a fact alone, the requirements for explaining a fact diverge from the requirements for explaining a contrast. My second observation, then, is that explaining a contrast is sometimes easier than explaining the fact alone (cf. Garfinkel, 1981, p. 30). An explanation of 'P rather than Q' is not always an explanation of P. This is particularly clear in examples of compatible contrasts. Jones's syphilis does not explain why he got paresis, since the vast majority of people who get syphilis do not get paresis, but it does explain why Jones rather than Smith got paresis, since Smith did not have syphilis. The relative ease with which we explain some contrasts also applies to many cases where there is an incompatibility between fact and foil. My preference for contemporary plays may not explain why I went to see *Jumpers* last night, since it does not explain why I went out, but it does explain why I went to see *Jumpers* rather than *Candide*. A particularly striking example of the relative ease with which some contrasts can be explained is the explanation that I chose A rather than B because I did not realize that B was an option. If you ask me why I ordered eggplant rather than sea bass (a 'daily special'), I may give the perfectly good answer that I did not know there were any specials, but this would not be an acceptable answer to the simple question, 'why did you order eggplant?' One reason we can sometimes explain a contrast without explaining the fact alone seems to be that contrastive questions incorporate a presupposition that makes explanation easier. To explain 'P rather than Q' is give a certain type of explanation of P, *given* 'P or Q', and an explanation that succeeds with the presupposition will not generally succeed without it.

My final observation is that explaining a contrast is also sometimes harder than explaining the fact alone. An explanation of P is not always an explanation of 'P rather than Q'. This is obvious in the case of compatible contrasts: you cannot explain why Jones rather than Smith contracted paresis without saying something about Smith. But it also applies to incompatible contrasts. To explain why I went to *Jumpers* rather than *Candide*, it is not enough for me to say that I was in the mood for a philosophical play. To explain why Kate rather than Frank won the prize, it is not enough that she wrote a good essay; it must have been better than Frank's. One reason that explaining a contrast is sometimes harder than explaining the fact alone is that explaining a contrast requires giving causal information that distinguishes the fact from the foil, and information that we accept as an explanation of the fact alone may not do this.

3. Failed Reductions

There have been a number of attempts to reduce contrastive questions to non-contrastive and generally truth-functional form. One motivation for this is to bring contrastive explanations into the fold of the Deductive-Nomological model since, without some reduction, it is not clear what the conclusion of a deductive explanation of 'P rather than Q' ought to be. Armed with our three observations—that contrasts may be compatible, and that explaining a contrast is sometimes easier and sometimes harder than explaining the fact alone—we can show that contrastive questions resist a reduction to non-contrastive form. We have already seen that the contrastive question 'Why P rather than Q?' is not equivalent to the simple question 'Why P?', where two why-questions are explanatorily equivalent just in case any adequate answer to one is an adequate answer to the other. One of the questions may be easier or harder to answer than the other. Still, a proponent of the Deductive-Nomological model of explanation may be tempted to say that, for incompatible contrasts, the question 'Why P rather than Q?' is equivalent to 'Why P?' But it is not plausible to say that a Deductive-Nomological explanation of P is generally necessary to explain 'P rather than Q'. And it is even dubious that a Deductive-Nomological explanation of P is always sufficient to explain 'P rather than Q'. Imagine a typical deductive explanation for the rise of mercury in a thermometer. Such an explanation would explain various contrasts, for example why the mercury rose rather than fell. It may not, however, explain why the mercury rose rather than breaking the glass. A full Deductive-Nomological explanation of the rise will have to include a premise saying that the glass does not break, but it does not need to explain this.

Another natural suggestion is that the contrastive question 'Why P rather than Q?' is equivalent to the conjunctive question 'Why P and not-Q?' On this view, explaining a contrast between fact and foil is tantamount to explaining the conjunction of the fact and the negation of the foil (Temple, 1988). In ordinary language, a contrastive question is often equivalent to its corresponding conjunction, simply because the 'and not' construction is often used contrastively. Instead of asking, 'Why was the prize won by Kate rather than by Frank?', the same question could be posed by asking 'Why was the prize won by Kate and not by Frank?'. But this colloquial equivalence does not seem to capture the point of the conjunctive view. So I suggest that the conjunctive view be taken to entail that explaining a conjunction at least requires explaining each conjunct; than an explanation of 'P and not-Q' must also provide an explanation of P and an explanation of not-Q. Thus, on the conjunctive view, to explain why Kate rather than Frank won the prize at least requires an explanation of why Kate won it and an explanation of why Frank did not. This account of contrastive explanation falls to the observation that explaining a contrast is sometimes easier than explaining the fact alone, since explaining P and explaining not-Q is at least as difficult as explaining P.

The observations that explaining contrasts is sometimes easier and sometimes harder than explaining the fact alone reveal another objection to the conjunctive view, on any model of explanation that is deductively closed. (A model is deductively closed if it entails that an explanation of P will also explain any logical consequence of P.) Consider cases where the fact is logically incompatible with the foil. Here P entails not-Q, so the conjunction 'P and not-Q' is logically equivalent to P alone. Furthermore, all conjunctions whose first conjunct is P and whose second conjunct is logically incompatible with P will be equivalent to each other, since they are all logically equivalent to P. Hence, for a deductively closed model of explanation, explaining 'P and not-Q' is tantamount to explaining P, whatever Q may be, so long as it is incompatible with P. We have seen, however, that explaining 'P rather than Q' is not generally tantamount to explaining P. The conjunction is explanatorily equivalent to P, and the contrast is not, so the conjunction is not equivalent to the contrast.

The failure to represent a contrastive phenomenon by the fact alone or by the conjunction of the fact and the negation of the foil suggests that, if we want a non-contrastive paraphrase, we ought instead to try something logically weaker than the fact. In some cases, it does seem that an explanation of the contrast is really an explanation of a logical consequence of the fact. This is closely related to what Hempel has to say about 'partial explanation' (1965, pp. 415–18). He gives the example of Freud's explanation of a particular slip of the pen that resulted in writing down the wrong date. Freud explains the slip with his theory of wish-fulfillment, but Hempel objects that the explanation does not really show why that particular slip took place, but at best only why there was some wish-fulfilling slip or other. Freud gave a partial explanation of the particular slip, since he gave a full explanation of the weaker claim that there was some slip. Hempel's point fits naturally into contrastive language: Freud did not explain why it was this slip rather than another wish-fulfilling slip, though he did explain why it was this slip rather than no slip at all. And it seems natural to analyse 'Why this slip rather than no slip at all?' as 'Why some slip?'

In general, however, we cannot paraphrase contrastive questions with consequences of their facts. We cannot, for example, say that to explain why the leaves turn yellow in November rather than in January is just to explain why the leaves turn (some colour or other) in November. This attempted paraphrase fails to discriminate between the intended contrastive question and the question, 'Why do the leaves turn in November rather than falling right off?' Similarly, we cannot capture the question, 'Why did Jones rather than Smith get paresis?', by asking about some consequence of Jones's condition, such as why he contracted a disease.

A general problem with finding a paraphrase entailed by the fact P is that, as we have seen, explaining a contrast is sometimes harder than explaining P alone. There are also problems peculiar to the obvious candidates. The disjunction, 'P or Q' will not do: explaining why I went to *Jumpers* rather than *Candide* is not the same as explaining why I went to either. Indeed, this proposal gets things almost backwards: the disjunction is what the contrastive question assumes, not what calls for explanation. This suggests, instead, that the contrast is equivalent to the conditional, 'if P or Q, then P' or, what comes to the same thing if the conditional is truth-functional, to explaining P on the assumption of 'P or Q'. Of all the reductions we have considered, this proposal is the most promising, but I do not think it will do. On a deductive model of explanation it would entail that any explanation of not-Q is also an explanation of the contrast, which is incorrect. We cannot explain why Jones rather than Smith has paresis by explaining why Smith did not get it. It would also wrongly entail that any explanation of P is an explanation of the contrast, since P entails the conditional.

4. Causal Triangulation

By asking a contrastive question, we can achieve a specificity that we do not seem to be able to capture either with a non-contrastive sentence that entails the fact or with one that the fact entails. But how then does a contrastive question specify the sort of information that will provide an adequate answer? It now appears that looking for a non-contrastive reduction of 'P rather than Q' is not a useful way to proceed. The contrastive claim may entail no more than 'P and not-Q' or perhaps better, 'P but not-Q', but explaining the contrast is not the same as explaining these conjuncts. We will do better to leave the analysis of the contrastive question to one side, and instead consider directly what it takes to provide an adequate answer. David Lewis has given an interesting account of contrastive explanation that does not depend on paraphrasing the contrastive question. According to him, we explain why event P occurred rather than event Q by giving information about the causal history of P that would not have applied to the history of Q, if Q had occurred (Lewis, 1986, pp. 229–30). Roughly, we cite a cause of P that would not have been a cause of Q. In Lewis's example, we can explain why he went to Monash rather than to Oxford in 1979 by pointing out that only Monash invited him, because the invitation to Monash was a cause of his trip, and that invitation would not have been a cause of a trip to Oxford, if he had taken one. On the other hand, Lewis's desire to go to places where he has good friends would not explain why he went to Monash rather than Oxford, since he has friends in both places and so the desire would have been part of either causal history.

Lewis's account, however, is too weak: it allows for unexplanatory causes. Suppose that both Oxford and Monash had invited him, but he went to Monash anyway. On Lewis's account, we can still explain this by pointing out that Monash invited him, since that invitation still would not have been a cause of a trip to Oxford. Yet the fact that he received an invitation from Monash clearly does not explain why he went there rather than to Oxford in this case, since Oxford invited him too. Similarly, Jones's syphilis satisfies Lewis's requirement even if Smith has syphilis too, yet in this case it would not explain why Jones rather than Smith contracted paresis.

It might be thought that Lewis's account could be saved by construing the causes more broadly, as types rather than tokens. In the case of the trip to Monash, we might take the cause to be receiving an invitation rather than the particular invitation to Monash he received. If we do this, we can correctly rule out the attempt to explain the trip by appeal to an invitation if Oxford also invited since, in this case, receiving an invitation would also have been a cause of going to Oxford. This, however, will not do, for two reasons. First, it does not capture Lewis's intent: he is interested in particular elements of a particular causal history, not general causal features. Secondly, and more importantly, the suggestion throws out the baby with the bath water. Now we have also ruled out the perfectly good explanation by invitation in some cases where only Monash invites. To see this, suppose that Lewis is the sort of person who only goes where he is invited. In this case, an invitation would have been part of a trip to Oxford, if he had gone there.

To improve on Lewis's account, consider John Stuart Mill's Method of Difference, his version of the controlled experiment (Mill, 1904, bk. III, ch. VIII, sec. 2). Mill's method rests on the principle that a cause must lie among the antecedent differences between a case where the effect occurs and an otherwise similar case where it does not. The difference in effect points back to a difference that locates a cause. Thus we might infer that contracting syphilis is a cause of paresis, since it is one of the ways Smith and Jones differed. The cause that the Method of Difference isolates depends on which control we use. If, instead of Smith, we have Doe, who does not have paresis but did contract syphilis and had it treated, we would be led to say that a cause of paresis is not syphilis, but the failure to treat it. The Method of Difference also applies to incompatible as well as to compatible contrasts. As Mill observes, the method often works particularly well with diachronic (before and after) contrasts, since these give us histories of fact and foil that are largely shared, making it easier to isolate a difference. If we want to determine the cause of a person's death, we naturally ask why he died when he did rather than at another time, and this yields an incompatible contrast, since you can only die once.

The Method of Difference concerns the discovery of causes rather than the explanation of effects, but the similarity to contrastive explanation is striking (cf. Garfinkel, 1981, p. 40). Accordingly, I propose that, for the causal explanations of events, explanatory contrasts select causes by means of what I will call the 'Difference Condition'. *To explain why P rather than Q, we must cite a causal difference between P and not-Q, consisting of a cause of P and the absence of a corresponding event in the history of not-Q.* Instead of pointing to a counterfactual difference, a particular cause of P that would not have been a cause of Q, as Lewis suggests, contrastive questions select as explanatory an actual difference between P and not-Q. Lewis's invitation to Monash does not explain why he went there rather than to Oxford if he was invited to both places because, while there is an invitation in the history of his trip to Monash, there is also an invitation in the history that leads him to forgo Oxford. Similarly, the Difference Condition correctly entails that Jones's syphilis does not explain why he rather than Smith contracted paresis if Smith had syphilis too, and that Kate's submitting an essay does not explain why she rather than Frank won the prize. Consider now some of the examples of successful contrastive explanation. If only Jones had syphilis, that explains why he rather than Smith has paresis, since having syphilis is a condition whose presence was a cause of Jones's paresis and a condition that does not appear in Smith's medical history. Writing the best essay explains why Kate rather than Frank won the prize, since that marks a causal difference between the two of them. Lastly, the fact that *Jumpers* is a contemporary play and *Candide* is not caused me both to go to one and to avoid the other.

The application of the Difference Condition is easiest to see in cases of compatible contrasts, since here the causal histories of P and of not-Q are generally distinct, but the condition does not require this. In cases of choice, for example, the causal histories are usually the same: the causes of my going to *Jumpers* are the same as the causes of my not going to *Candide*. The Difference Condition may nevertheless be satisfied if my belief that *Jumpers* is a contemporary play is a cause of going, and I do not believe that *Candide* is a contemporary play. That is why my preference for contemporary plays explains my choice. The Difference Condition does not require that the same event be present in the history of P but absent in the history of not-Q, a condition that could never be satisfied when the two histories are the same, but only that the cited cause of P find no corresponding event in the history of not-Q, where a corresponding event is something that would bear the same relation to Q as the cause of P bears to P.

One of the merits of the Difference Condition is that it brings out the way the incompatibility of fact and foil, when it obtains, is not sufficient to transform an explanation of the fact into an explanation of the contrast, even if the cause of the fact is also a cause of the foil not obtaining. Perhaps we could explain why Able got the philosophy job by pointing out that Quine wrote him a strong letter of recommendation, but this will only explain why Able rather than Baker got the job if Quine did not also write a similar letter for Baker. If he did, Quine's letter for Able does not alone explain the contrast, even though that letter is a cause of both Able's success and Baker's failure, and the former entails the latter. The letter may be a partial explanation of why Able got the job, but it does not explain why Able rather than Baker got the job. In the case where they both have strong letters from Quine, a good explanation of the contrast will have to find an actual difference, say that Baker's dossier was weaker than Able's in some other respect, or that Able's specialities were more useful to the department. There are some cases of contrastive explanation that do seem to rely on the way the fact precludes the foil, but I think these can be handled by the Difference Condition. For example, suppose we explain why a bomb went off prematurely at noon rather than in the evening by saying that the door hooked up to the trigger was opened at noon (I owe this example to Eddy Zemach). Here it may appear that the Difference Condition is not in play, since the explanation would stand even if the door was also opened in the evening. But the Difference Condition is met, if we take the cause not simply to be the opening of the door, but the opening of the door when it is rigged to an armed bomb.

My goal in this paper is to show how the choice of contrast helps to determine an explanatory cause, not to show why we choose one contrast rather than another. Still, some account of the considerations that govern our choice of why-questions would have to be a part of a full model of our explanatory practices, and it is to the credit of my model of contrastive explanation that it lends itself to this. For example, as I have already observed, not all contrasts make for sensible contrastive questions. It does not make sense, for example, to ask why Lewis went to Monash rather than Baker getting the philosophy job. One might have thought that a sensible contrast must be one where fact and foil are incompatible, but we have seen that this is not necessary, since there are many sensible compatible contrasts. There are also incompatible contrasts that do not yield reasonable contrastive questions, such as why someone died when she did rather than never having been born. The Difference Condition suggests instead that the central requirement for a sensible contrastive question is that the fact and the foil have a largely similar history, against which the differences stand out. When the histories are disparate, we do not know where to begin to answer the question. There are, of course, other considerations that help to determine the contrasts we actually choose. For example, in the case of incompatible contrasts, we often pick as foil the outcome we expected; in the case of compatible contrasts, as I have already mentioned, we often pick as foil a case we expected to turn out the same way as the fact. The condition of a similar history also helps to determine what will count as a corresponding event. If we were to ask why Lewis went to Monash rather than Baker getting the job, it would be difficult to see what in the history of Baker's failure would correspond to Lewis's invitation, but when we ask why Able rather than Baker got the job, the notion of a corresponding event is relatively clear.

5. Further Issues

I will now consider three further issues connected with my analysis of contrastive explanation: the need for further principles for distinguishing explanatory from unexplanatory causes, the prospects for treating all why-questions as contrastive, and a comparison of my analysis with the Deductive-Nomological model. When we ask contrastive why-questions, we choose our foils to point towards the sort of causes that interest us. As we have just seen, when we ask about a surprising event, we often make the foil the thing we expected. Failed expectations are not, however, the only things that prompt us to ask why-questions. If a doctor is interested in the internal etiology of a disease, he will ask why the afflicted have it rather than other people in similar circumstances, even though the shared circumstances may be causally relevant to the disease. Again, if a machine malfunctions, the natural diagnostic contrast is its correct behaviour, since that directs our attention to the causes that we want to change. But the contrasts we construct will almost always leave multiple differences that meet the Difference Condition, and this raises the problem of selecting from among them. A problem of multiple differences also arises for the Method of Difference, in the context of inference rather than explanation. Mill tells us that we may infer that the only antecedent difference between fact and foil marks a cause, but in practice there will almost always be many such differences, not all of which will be causally relevant. Moreover, as Mill seems not to have recognized, his own deterministic assumptions entail that there will always be multiple differences as a matter of principle, since any antecedent difference itself marks an effect that must have a still earlier causal difference. (I owe this point to Trevor Hussey.)

In the case of inference, the central problem is to distinguish those differences that are causally relevant from those that are not. In the case of explanation, on the other hand, all the differences that meet the Difference Condition are, by definition, causally relevant. So all of them may be explanatory: the Difference Condition does not entail that there is only one way to explain a contrast. At the same time, however, some causally relevant differences will not be explanatory in a particular context, so while the Difference Condition may be necessary for the causal contrastive explanations of particular events, it is not generally sufficient. For that we need further principles of causal selection.

The considerations that govern selection from among causally relevant differences are numerous and diverse; the best I can do here is to mention what a few of them are. An obvious pragmatic consideration is that someone who asks a contrastive question may already know about some causal differences, in which case a good explanation will have to tell her something new. If she asks why Kate rather than Frank won the prize, she may assume that it was because Kate wrote the better essay, in which case we will have to tell her more about the differences between the essays that made Kate's better. A second consideration is that, when they are available, we usually prefer explanations where the foil would have occurred if the corresponding cause had occurred. Suppose that only Able had a letter from Quine, but even a strong letter from Quine would not have helped Baker much, since his specialities do not fit the department's needs. Suppose also that, had Baker's specialities been appropriate, he would have gotten the job, even without a letter from Quine. In this case, the difference in specialities is a better explanation than the difference in letters. Note, however, that an explanation that does not meet this condition of counterfactual sufficiency for the occurrence of the foil may be perfectly acceptable, if we do not know of a sufficient difference. The explanation of why Jones rather than Smith contracted paresis is an example of this: even if Smith had syphilis in his medical history, he probably would not have contracted paresis. Moreover, even in cases where a set of known causes does supply a counterfactually sufficient condition, the inquirer may be much more interested in some than in others. The doctor may be particularly interested in causes he can control, the lawyer in causes that are connected with legal liability, and the accused in causes that cannot be held against him.

We also prefer differences where the cause is causally necessary for the fact in the circumstances. Consider a case of overdetermination. Suppose that you ask me why I ordered eggplant rather than beef, when I was in the mood for eggplant and not for beef, and I am a vegetarian. My mood and my convictions are separate causes of my choice, each causally sufficient in the circumstance and neither necessary. In this case, it would be better to give both differences than just one. The Difference Condition could easily be modified to require necessary causes, but I think this would make the condition too strong. One problem would be cases of 'failsafe' overdetermination. Suppose we change the restaurant example so that my vegetarian convictions were not a cause of the particular choice I made: that time, it was simply my mood that was relevant. Nevertheless, even if I had been in the mood for beef, I would have resisted, because of my convictions. In this case, my explanation does not have to include my convictions, even though my mood was not a necessary cause of my choice. (Of course if I knew that you were asking me about my choice because you were planning to invite me to your house for dinner, it would be misleading for me not to mention my convictions, but this goes beyond the conditions for explaining the particular choice I made.) Again, we sometimes do not know whether a cause is necessary for the effect, and in such cases the cause still seems explanatory. But when there are differences that supply a necessary cause, and we know that they do, we prefer them. There are doubtless other pragmatic principles that play a role in determining which differences or combinations of differences yield the best explanation in a particular context. So there is more to contrastive explanation than the Difference Condition describes, but that condition does describe the central mechanism of causal selection.

Since contrastive questions are so common and foils play such an important role in determining explanatory causes, it is natural to wonder whether all why-questions are not at least implicitly contrastive. Often the contrast is so obvious that it is not worth mentioning. If you ask me why I was late for our appointment, the question is why I was late rather than on time, not why I was late rather than not showing up at all. Moreover, in cases where there is no specific contrast, stated or implied, we might construe 'Why P?' as 'Why P rather than not-P?', thus subsuming all causal why-questions under the contrastive analysis. But the Difference Condition seems to misbehave for these 'global' contrasts. It requires that we give a cause of P that finds no corresponding cause in the history of not-Q but, if the foil is simply the negation of the fact, this seems to require that we find a cause of P that finds no corresponding cause of itself, which is impossible, since it is tantamount to the requirement that we find a cause of P that is absent from the history of P.

We can, however, analyse the explanation of P *simpliciter* as the explanation of P rather than not-P. The correct way to construe the Difference Condition as it applies to the limiting case of the contrast, P rather than not-P, is that we must find a difference for events logically or causally incompatible with P, not for a single event, 'not-P'. Thus suppose that we ask why Jones has paresis, with no implied contrast. This would require a difference for foils where he does not have paresis. Saying that he had syphilis differentiates between the fact and the foil of a thoroughly healthy Jones, but this is not enough, since it does not differentiate between the fact and the foil of Jones with syphilis but without paresis. Excluding many incompatible foils will push us towards a sufficient cause of Jones's syphilis, since it is only by giving such a 'full cause' that we can be sure that some bit of it will be missing from the history of all the foils. To explain P rather than not-P, however, we do not need to explain every incompatible contrast. We do not, for example, need to explain why Jones contracted paresis rather than being long dead or never being born. The most we can require is that we exclude all incompatible foils with histories similar to the history of the fact.

One difficulty for this way of avoiding the pathological requirement of finding a cause of P that is absent from the history of P is that there appear to be some facts whose negation also seem to be a single fact (I owe this point to Elliot Sober). Suppose we wish to understand why there are tigers. Here the foil seems simply to be the absence of tigers, and we cannot give a cause of the existence of tigers that is not in the history of tigers. But the existence of tigers is not an event, so this example does not affect my account, which is only meant to apply to the explanation of events. So perhaps the problem does not arise for contrasts whose facts are events and whose foils are either events or sets of events. The Difference Condition will apply to some contrasts that are not explicitly event-contrasts, but not to all of them. Even for P's that are events, however, I am not certain that every apparently non-contrastive question should be analysed in contrastive form, so I am agnostic on the issue of whether all why-questions are contrastive.

Finally, let us compare my analysis of contrastive explanation to the Deductive-Nomological model. First, as we have already noted, a causal model of explanation has the merit of avoiding all the counter-examples to the Deductive-Nomological model where causes are deduced from effects. It also avoids the unhappy consequence of counting almost every explanation we give as a mere sketch, since one can give a cause of P that meets the Difference Condition for various foils without having the laws and singular premises necesssary to deduce P. Many explanations that the deductive model counts as only very partial explanations of P are in fact reasonably complete explanations of P rather than Q. The excessive demands of the deductive model are particularly striking for cases of compatible contrasts, as least if the deductive-nomologist requires that an explanation of P rather than Q provide an explanation of P and an explanation of not-Q. In this case, the model makes explaining the contrast substantially harder than providing a deductive explanation of P, when in fact it is often substantially easier. Our inability to find a non-contrastive reduction of contrastive questions is a symptom the inability of the Deductive-Nomological model to give an accurate account of this common type of explanation.

There are at least two other conspicuous advantages of a causal contrastive model of explanation over the Deductive-Nomological model. One odd feature of the deductive model is that it entails that an explanation cannot be ruined by adding true premises, so long as the additional premises do not render the law superfluous to the deduction, by entailing the conclusion outright. This consequence follows from the elementary logical point that additional premises can never convert a valid argument into an invalid one. In fact, however, irrelevant additions can spoil an explanation. If I say that Jones rather than Smith contracted paresis because only Jones had syphilis and only Smith was a regular church-goer, I have not simply said more than I need to, I have given an incorrect explanation, since going to church is not a prophylactic. By requiring that explanatory information be causally relevant, the causal model avoids this problem. Another related and unhappy feature of the Deductived-Nomological model is that it entails that explanations are virtually deductively closed: an explanation of P will also be an explanation of any logical consequence of P, so long as the consequence is not directly entailed by the initial conditions alone. (For an example of the slight non-closure in the model, notice that a Deductive-Nomological explanation of P will not also be a Deductive-Nomological explanation of the disjunction of P and one of the initial conditions of the explanation.) In practice, however, explanation seems to involve a much stronger form of non-closure. I might explain why all the men in the restaurant are wearing paisley ties by appealing to the fashion of the times for ties to be paisley, but this might not explain why they are all wearing ties, which is because of a rule of the restaurant. (I owe this example to Tim Williamson.) The contrastive model gives a natural account of this sort of non-closure. When we ask about paisley ties, the implied foil is other sorts of tie; but when we ask simply about ties, the foil is not wearing ties. The fashion marks a difference in one case, but not in the other.

A defender of the Deductive-Nomological model may respond to some of these points by arguing that, whatever the merits of a contrastive analysis of lay explanation, the deductive model (perhaps with an additional restriction blocking 'explanations' of causes by effects) gives a better account of scientific explanation. For example, it has been claimed that since scientific explanations, unlike ordinary explanations, do not exhibit the interest relativity of foil variation that a contrastive analysis exploits, a contrastive analysis does not apply to scientific explanation (Worrall, 1984, pp. 76–77). It is, however, a mistake to suppose that all scientific explanations even aspire to Deductive-Nomological status. The explanation of why Jones rather than Smith contracted paresis is presumably scientific, but it is not a deduction *manqué*. Moreover, as the example of the thermometer showed, even a full Deductive-Nomological explanation may exhibit interest relativity. I may explain the fact relative to some foils but not relative to others. A typical Deductive-Nomological explanation of the rise of mercury in a thermometer will simply assume that the glass does not break and so while it will explain, for example, why the mercury rose rather than fell, it will not explain why it rose rather than breaking the thermometer. Quite generally, a Deductive-Nomological explanation of a fact will not explain that fact relative to any foils that are themselves logically inconsistent with one of the premises of the explanation. Again, a Newtonian explanation of the Earth's orbit (ignoring the influence of the other planets) will explain why the Earth has its actual orbit rather than some other orbits, but it will not explain why the Earth does not have any of the other orbits that are compatible with Newton's theory. The explanation must assume information about the Earth's position and velocity at some time that will rule out the other Newtonian orbits, but it will not explain why the Earth does not travel in those paths. To explain this would require quite different information about the early history of the Earth. Similarly, an adaptionist explanation for a species's possession of a certain trait may explain why it has that trait rather than various maladaptive traits, but it may not explain why it had that trait rather than other traits that would perform the same functions equally well. To explain why an animal has one trait rather than another functionally equivalent trait requires instead appeal to the evolutionary history of the species, in so far as it can be explained at all.

With rather more justice, a deductive-nomologist might object that scientific explanations do very often essentially involve laws and theories, and that the contrastive model does not seem to account for this. For even if the fact to be explained carries no restricting contrast, the contrastive model, if it is extended to this case by analysing 'Why P?' as 'Why P rather than not-P?', only requires at most that we cite a condition that is causally sufficient for the fact, not that we actually give any laws. I think, however, that the contrastive model can help to account for the undeniable role of laws in many scientific explanations. To see this, notice that scientists are often and perhaps primarily interested in explaining regularities, rather than particular events (cf. Friedman, 1974, p. 5; though explaining particular events is also important when, for example, scientists test their theories, since observations are of particular events). I think that the Difference Condition applies to many explanations of regularities, but to give a contrastive explanation of a regularity will require citing a law, or at least a generalization, since we here need some general cause (cf. Lewis, 1986, pp. 225–6). To explain, say, why people feel the heat more when the humidity is high, we must find some general causal difference between cases where the humidity is high and cases where it is not, such as the fact that the evaporation of perspiration, upon which our cooling system depends, slows as the humidity rises. So the contrastive model, in an expanded version that applies to general facts as well as to events (a version I do not here provide), should be able to account for the role of laws in scientific explanations as a consequence of the scientific interest in general why-questions. Similarly, although the contrastive model does not require deduction for explanation, it is not mysterious that scientists should often look for explanations that do entail the phenomenon to be explained. This may not have to do with the requirements of explanation *per se*, but rather with the uses to which explanations are put. Scientists often want explanations that can be used for accurate prediction, and this requires deduction. Again, the construction of an explanation is a way to test a theory, and some tests require deduction.

Another way of seeing the compatibility of the scientific emphasis on theory and the contrastive model is by observing that scientists are not just interested in this or that explanation, but in a unified explanatory scheme. Scientists want theories, in part, because they want engines that provide many explanations. The contrastive model does not entail that a theory is necessary for any particular explanation, but a good theory is the best way to provide the many and diverse contrastive explanations that the scientist is after. This also helps to account for the familiar point that scientists are often interested in discovering causal mechanisms. The contrastive model will not require a mechanism to explain why one input into a black box causes one output, but it pushes us to specify more and more of the detailed workings of the box as we try to explain its full behaviour under diverse conditions. So I conclude that the contrastive model of explanation does not fly in the face of scientific practice.

6. Conclusion

The Difference Condition shows how contrastive questions about particular events help to determine an explanatory cause by a kind of causal triangulation. This contrastive model of causal explanation cannot be the whole story about explanation, since not all explanations are causal explanations or explanations of particular events and since, as we have seen, the choice of foil is not the only factor that affects the appropriate choice of cause. It does, however, give a natural account of much of what is going on in many explanations, and it captures some of the merits of competing accounts while avoiding some of their weaknesses. We have just seen this in some detail for the case of the Deductive-Nomological model. It also applies to the familiarity view. When an event surprises us, a natural foil is the outcome we had expected, and meeting the Difference Condition for this contrast will help to show us why our expectation went wrong. The mechanism of causal triangulation also accounts for the way a change in foil can lead to a change in explanatory cause, since a difference for one foil will not in general be a difference for another. It also shows why explaining 'P rather than Q' is sometimes harder and sometimes easier than explaining P alone. It may be harder, because it requires the absence of a corresponding event in the history of not-Q, and this is something that will not generally follow from the presence of the cause of P. Explaining the contrast may be easier, because the cause of P need not be sufficient for P, so long as it is part of a causal difference between P and not-Q. Again, causal triangulation helps to elucidate the interest relativity of explanation. We express some of our interests through our choice of foils and, by construing the phenomenon to be explained as a contrast rather than the fact alone, interest relativity reduces to the important but unsurprising point that different people are interested in explaining different phenomena. The Difference Condition also shows that different interests do not require incompatible explanations to satisfy them, only different but compatible causes. Moreover, my model of contrastive explanation suggests that our choice of foils is often governed by our *inferential* interests. As I argue extensively elsewhere, the structural similarity between the Difference Condition and the Method of Difference enables us to show why the inductive procedure of 'Inference to the Best Explanation' is a reliable way of discovering causes. Because of this similarity, it can be shown that the hypothesis that would provide the best explanation of our contrastive data is also the one that is likeliest to have located an actual cause (Lipton, forthcoming). Finally, the mechanism of causal triangulation accounts for the failure of various attempts to reduce contrastive questions to non-contrastive form. None of these bring out the way a foil serves to select a location on the causal history leading up to the fact. Causal triangulation is the central feature of contrastive explanation that non-contrastive paraphrases suppress.

References

Friedman, M. 1974. 'Explanation and Scientific Understanding', *The Journal of Philosophy* **LXXI**: 1–19.

Garfinkel, A. 1981. *Forms of Explanation* (New Haven: Yale University Press).

Hempel, C. 1965. *Aspects of Scientific Explanation* (New York: The Free Press).

Lewis, D. 1986. 'Causal Explanation', in *Philosophical Papers* Vol. II (New York: Oxford University Press), pp. 214–40.

Lipton, P. forthcoming. *Inference to the Best Explanation* (London: Routledge).

Mill, J. S. 1904. *A System of Logic* (London: Longmans, Green and Co).

Ruben, D.-H. 1987. 'Explaining Contrastive Facts', *Analysis* **47.1**: 35–7.

Temple, D. 1988. 'Discussion: The Contrast Theory of Why-Questions', *Philosophy of Science* **55**: 141–51.

van Fraassen, B. 1980. *The Scientific Image* (Oxford: The Clarendon Press).

Worrall, J. 1988. 'An Unreal Image', *The British Journal for the Philosophy of Science*, **35**: 65–80.

How to Put Questions to Nature

MATTI SINTONEN

1. Introduction

In this paper I propose to examine, and in part revive, a time-honoured perspective to inquiry in general and scientific explanation in particular. The perspective is to view inquiry as a search for answers to questions. If there is anything that deserves to be called a working scientist's view of his or her daily work, it surely is that he or she phrases questions and attempts to find satisfactory answers to them.

The perspective can be given several forms, depending on who or what the supposed answerer is. For Bacon and Kant the answerer was Nature: the idea was to force Nature to give conclusive answers to questions of the inquirer's choice. For some recent writers in the theory of explanation the interlocutor is a fellow inquirer: also scientific explanations are acts in which someone, by uttering an explaining utterance, explains to someone else why something is the case (see Achinstein, 1983 and Tuomela, 1980).

The latter idea is needed in the theory of explanation, because some constraints on acceptable explanations arise from the relationship between questions and answers. But this cannot be the whole story about specifically scientific explanation. If questions to fellow inquirers were sufficient, inquiry would be too easy, as Bacon so clearly saw. He acknowledged the power of old logic to systematize knowledge already available, but criticized it for its sterility in the acquisition of new information. Whatever the merits of Bacon's positive proposal, he surely got this aspect of inquiry essentially right: sooner or later the inquirer must come to terms with Nature, by putting questions to her.

The trouble with Bacon's—and Kant's—idea is that it is a metaphor at best, and a misleading and a powerless one at worst. It is a metaphor because Nature does not have a language, nor a capacity to enter into a genuine dialogue. And it is potentially misleading and powerless, especially in the theory of explanation, because it assumes too much of Nature. Nature, as we shall see, does not understand why-questions.

The main claim of this paper is that this metaphor can be brought down to earth, in a way that avoids taking Nature's answers at face value, and yet explains how even explanation-seeking why- and how-questions are accessible to Her. What is needed, I claim, is a theory notion which is rich enough to show how questions are processed into a form which Nature is able to understand.

2. Reading the Book of Nature

Let us start with Bacon's insight according to which science begins where man begins 'putting nature to the question'. Bacon advocated observation and experiment as the only true means for the interpretation of nature, and indeed these can be construed as situations or set-ups in which the inquirer attempts to force Nature to say either 'Yes!' or 'No!' to a query concerning a descriptive or explanatory hypothesis.

But of course this is a metaphor, because nature does not have beliefs, wants or intentions. Answers, literally construed, are linguistic expressions intended to effect the interlocutor in a certain way, and it is illegitimate animism to think that she says (or tells) anything, or intends by her utterings to get us to know (understand) true answers to our questions. Nature does respond in experimental and observational contexts, but such commerce can only be described in causal and not communicative terms.

The metaphor started to break down already in the hands of Bacon and Kant. Consider the objective of forcing Nature to give unambiguous answers to our questions.[1] Bacon was already familiar with the difficulty, and envisioned the possibility of crucial experiments only in instances 'which exhibit the nature in question naked and standing by itself'.[2] But hypotheses exhibit natures in rather promissory terms. Nature's answers to such questions, in observational and experimental contexts, always are amenable to conflicting interpretations. There are no crucial experiments here, and we must learn to live with Nature's tendency to answer 'Yes and no!'.[3]

[1] Collingwood makes it clear that talk of forcing Nature is not out of place here: 'Bacon was a lawyer and knew what "the question" meant in his profession', viz. torture (see Collingwood, 1940, p. 239).

[2] Bacon (1863), Book II, xxiv. Crucial instances (or crucial experiments) were feasible, he wrote, when 'in the investigation of any nature the understanding is so balanced as to be certain to which of two or more natures the cause of the nature in question should be assigned'. Bacon is here trying to keep conjecture and certainty apart, an insistence that later found its way to the Newtonian ban on hypotheses.

[3] Ruben Abel ascribes this view to Herman Weyl: 'Our science . . . supplies the answers to the questions we have thought of asking. Nature answers only those questions which we put to her (if she answers at all); and when she does answer, it is with a loud No! but with a hardly audible Yes! (as Hermann Weyl puts it)'. See Abel (1982), p. 91.

The metaphor also gives rise to a related worry, for it invites us to look at Nature as an authoritative source: all you need to do is put a question and wait for the answer. But this misplaces the onus of inquiry, as Kant already made clear. He emphasized the ineliminable role of the inquirer in the interpretation of nature, insisting that reason must approach nature, not 'in the character of a pupil who listens to everything that the teacher chooses to say, but of an appointed judge who compels the witnesses to answer questions which he has himself formulated' (Kant, 1933, B xiii–xiv).

Note also that the idea of putting questions to Nature mixes easily with another metaphor, that of interpreting or reading the book of nature. Bacon gave his *New Organon* the subtitle 'true directions concerning the interpretation of Nature', and one of his great admirers, William Whewell, tied the phrase up with the possibility of mastering Nature's language. In interpreting the book of nature, Whewell wrote, an inquirer tries to find laws which are identical with the real laws of nature. 'To trace order and law in that which has been observed, may be considered as interpreting what nature has written down for us, and will commonly prove that we understand her alphabet' (Whewell, 1847, Vol. 2, pp. 64–5).

But this reading amounts to the thought that there is some one language of nature, and one story about the world written in that language.[4] It is identical, or nearly so, with Hilary Putnam's (1978) metaphysical realism, a view in which items of language hook on to their unique counterparts in a ready-made world. It may well be that Putnam and other critics of realism have managed to show that this view indeed is a powerful but ultimately unintelligible metaphor.

But the real challenge is that the metaphor, even when stripped from animism and anthropomorphism, appears to break down completely when we come to questions that interest us here, i.e. explanation-seeking questions. These questions often have the form of a why-question (or a how-question), and take as answers hypotheses which go beyond descriptive accounts to explanatory factors. Such questions do not submit to Nature a restricted, exhaustive and complete, set of alternatives to choose from. There is no way we can put explanation-seeking questions to Nature. To play along with the metaphor one final time, she simply does not understand some why-questions.

[4] The quotation from Whewell continues: 'But to predict what has not been observed, is to attempt ourselves to use the legislative phrases of nature; and when she responds plainly and precisely to that which we thus utter, we cannot but suppose that we have in a great measure made ourselves masters of the meaning and structure of her language'. See also Whewell (1847), Vol. I, 38.

3. The Received View and Question-Answer Sequences

To appreciate the difficulty with why-questions, consider the more recent attempt to employ the question–answer idea, viz. that of Hempel and Oppenheim. Interestingly enough they start with the idea that explanations are answers to explanation-seeking questions, but do not put the idea to use in their analysis. But why is this? One conceivable reason is that writers in the positivist tradition only allowed themselves explicates which were given within a formal language. In the covering law models the laws have slots for singular explanans and explandum facts, and thereby manage to 'subsume' events. In the classic deductive-nomological model (D-N model, for short) a consistent pair or set {T, C} is an explanans for a (singular) E if and only if (1), T is a theory, (2), C is a true singular sentence, (3), {T, C} $\vdash$ E, and (4), there is a class K of basic sentences such that K is compatible with T, K $\vdash$ C, but not K $\vdash$ E. (Hempel and Oppenheim (1948), pp. 277–8).

Such explicates in well-structured languages not only promised to be free of ambiguities, but also ruled out pernicious subjectivism, psychologism, and anthromorphism. D-N first distills from our amorphous notion of an explanation one ingredient, the objective two-placed relation between a set of sentences {T, C} and the sentence E. It then gives a Carnapian explication of the explicandum so refined, in terms of (1)–(4). Inquirers, whether individuals or communities, as well as contexts and other pragmatic items, are thus filtered from both the explicandum and the explicans.

This proposal ran into problems right at the outset, and the consensus now seems to be that it is impossible to capture explanatory arguments, and only them, with a net of the Carnapian design. The details of the downfall of the covering law paradigm cannot be documented here. But a couple of simple examples suggest that the remedy could come from some pragmatic features of discourse, from what counts as an answer to a question.

Just consider condition (4), a syntactic restriction which rules out arguments in which the singular premiss C entails the explanandum E. A complete self-explanation in which C and E are identical is a good example. But what is wrong with it? It is of course an answer in which the inquirer is entertained with information he or she already has. The inquirer knows that E, but not why—and for the latter task new independent information is needed. Another example comes from the need to block arguments with redundant information in the premisses. There were several syntactic candidates for the job. But even if they were successful, it may well be asked what the rationale behind such devices is. One answer volunteers: there is a general communicative requirement to stick to the point and not burden hearers with information that is not needed. What I suggest but cannot prove here is that formal models purport to capture some pragmatic features of questions and answers *by purely syntactic and semantic means* (for discussion, see Sintonen, 1990). Explanations indeed are answers to questions, but their relationship cannot be explicated purely formally, that is, formally in the Hempel and Oppenheim sense.

This result has the advantage that it shows that scientific explanations mirror a familiar dialogue structure, and hence shows that they form a species of common sense-explanations. The drawback is that the dialogue constraints are weak or even trivial, and fail to illuminate the differentia of explanation-seeking questions.

Here we encounter a further reason why D-N cannot make full use of the interrogative idea. The choice of purely syntactic-cum-semantic tools involves, namely, a tacit commitment to the idea that once you have a description of a phenomenon to be explained, you have said all there is to be said about the explanatory task: there is, on that view, a one–one mapping between why-questions and explananda. But this assumption is not always met, because there are explanatory contexts in which there are seemingly innocent, yet revolutionary changes in the very questions that demand attention.

Take as an example the Darwinian Revolution. There were (and still are) two competing models of evolutionary explanation, variational and developmental (see Sober, 1984). In the latter, to be found in linguistics, psychology and elsewhere, development is seen to be constrained by preprogrammed instructions, so that each individual undergoes a series of stages. The environment can speed up or slow down the pace of this natural course, but it cannot make essential alterations in the path itself. However, in variational explanation individuals are required to be relatively stable, and changes in a population are not explained by referring to changes in the individuals. Richard Lewontin has argued that Darwin brought about a revolution in biology by phrasing questions about evolution on the level of a population rather than individual organisms: variation between individuals, rather than development in individuals, was the central aspect of reality *vis-à-vis* evolution (Lewontin, 1974, p. 4).

Another example is from Ernst Mayr (1980) who has discussed the confusion which resulted from mixing two types of causes, proximate and ultimate. The functionalists (experimentalists) and the evolutionists both raised questions like 'Why is this species sexually dimorphic?', but they offered different types of answers, and literally talked past one another. What Darwin did was bring about a respectable way of dealing with ultimate or evolutionary causation, and hence a new set of questions or a new way to understand the old questions.

Sober (1984, p. 155) writes that the Darwinian focus on irreducible population level questions involved a new explanatory model and new propositions that required explanation. The surface question, say, 'Why do members of species X have trait Y?' appeared to mean the same in the mouth of Darwin and Lamarck, but a closer look reveals that this is an illusion: there are different contextual restrictions on potential answers, and hence different explananda. There is, here, more to a question than meets the ear.

Notice that this fact makes ineffective an initially attractive aid to the formal program. Hempel and Oppenheim had no access to a logic of questions, but we do. This suggests a promising move because the logic uses the analysis of the ordinary two-person question-and-answer situations to explicate answerhood (Belnap, 1963, p. 450). It makes use of the syntax of questions and their substantive presuppositions to provide a formal explication of a response (or a list of responses) and cancellation (or matching) procedure.

Such a logic can be conjoined with epistemic logic to study what Hintikka (1976) calls conclusive answers to questions. Thus the wh-question 'Who murdered Olof Palme?' takes as direct answers terms which refer to persons, and any such candidate is conclusive for a questioner if and only if he or she is, after hearing it, in a position to say, truly, 'I know who murdered Olof Palme'. A correct direct answer, say 'The 42-year old', does not automatically fulfill this condition, if the questioner does not know who the 42-year old is. Clearly, an answer is conclusive only if it does not give rise to such further questions as 'But who *is* the 42-year old?'

The analysis has appeal. Although what counts as a satisfactory answer is up to the context, such pragmatic dependence makes no reference to anything idiosyncratic. The analysis is not subjective, psychologistic, or anthropomorphic, and merely continues the Carnapian programme by other means. Just consider Hempel's celebrated question 'Why did the radiator of my car crack?' Its complete direct answers have the form 'The radiator of my car cracked because L_1, L_2, . . ., L_n and C_1, C_2, . . ., C_m', where the variables range over laws and singular facts. Here the question signals a request for laws and singular facts sufficient to entail (or, make probable) the sentence 'The car radiator cracked.' And of course it depends on the background knowledge of the inquirer as to how much needs to be explicitly mentioned. Hence, the erotetic-cum-epistemic logic gives a precise analysis of what can be omitted, for pragmatic reasons, from a complete objective answer.

The trouble is that the recipe does not work, or does not work as a general analysis. The logic catches situations in which the questioner is able to express his or her problem effectively in terms of a set of possible alternatives, and merely needs to know which alternative is the right one. But although some why-questions initiate a search for evidence to determine which one of a set of well-defined potential answers is the actual one, the most interesting and troublesome ones do not. In those cases the relationship between the presuppositions of the questions and the potential answers is opaque, and the inquirer is not in the relatively happy situation of a car mechanic: there is no trouble-shooting manual to consult.

4. The D-Ideal and Explanatory Tasks

There is also a further neglect which arises from the syntactic-cum-semantic mould. The covering law accounts instantiate the deductive ideal (D-ideal, for short), in which the laws contain slots for initial conditions. Such accounts are essentially one-step models, and the intellectual tasks they set have the form: what (true, nomic and empirically contentful) laws and initial conditions are needed to get a deductive (or inductive) argument with the explanandum sentence as the conclusion? The salient assumption is that all explanans items operate on a single level. But a model which so construes the explanatory task fails to take into account the contextually determined mediating steps in the derivation of an explanation.

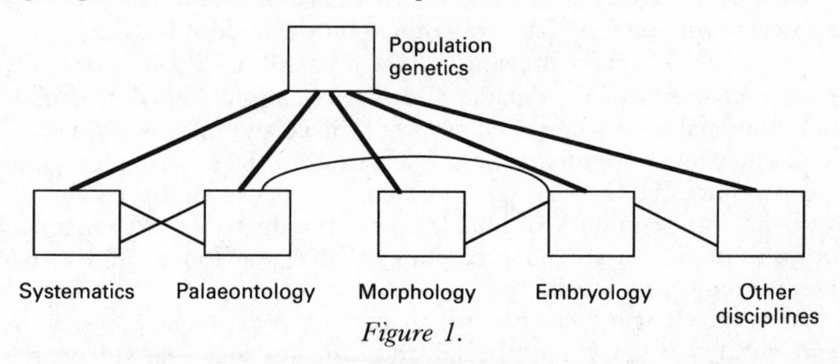

Figure 1.

Take Darwin's *Origin of Species* as an example (Darwin, 1964). The traditional view is that Darwin's theory is deductive in structure, and empirically supported by the various classes of facts deducible from its core (see, e.g., Ghiselin 1984, pp. 65–8, and Ruse, 1979, p. 198). The same goes for the modern synthetic theory, according to Michael Ruse (1973, p. 49). He gives the picture of its structure shown in Figure 1. The rectangles stand for the various disciplines, the thick lines relationships between the fundamental discipline of population genetics and the various subsidiary disciplines, and the thin lines those between the subsidiary disciplines. And, Ruse writes, for several reasons—'the newness of the theory, the fact that many pertinent pieces of information are irretrievably lost, the incredible magnitude and complexity of the problem'—much of the theory is just 'sketched in'. Evolutionists subscribe to the hypothetico-deductive ideal, although they 'are far from having it as a realized actuality'.

But the credibility gap between actuality and the ideal is too wide to warrant further adherence to the ideal. It is better in keeping with Darwin's intent and works to think that his theory was not intended to approximate the D-ideal. And it has also become evident that biologists are not even interested in weaving deductive structures—indeed, as an activity model building is a far cry from the D-ideal.

Just consider some of the anomalies that face the deductivist when he tries to fit Darwinian or modern explanations into the picture. To start with, there is no consensus over the precise nature of the Darwinian deductive core, and even if there were, it could be rightly asked if formulating it was the main task in Darwin's book. Furthermore, the theory was sketchy and gappy. Even after the discovery of the principle of Natural Selection there were outstanding unanswered questions: Darwin lacked an account of the divergence of character, and especially of the hereditary mechanism.

The D-ideal also presupposes that a theory is precise enough to deliver detailed predictions and explanations. The rationale is easy to appreciate: if enough cards are not laid on the table at the outset, the inquirer may cheat by drawing *ad hoc* cards. But Darwin repeatedly confessed ignorance of details, and was unable to pinpoint a single observational consequence which would have annihilated his theory. The same goes for explanations. All variants of the covering law paradigm require that there are laws with precise slots and initial conditions with matching contours to entail (or support inductively) the explanandum. But there were no predictions and explanations, under that description, in the book.

It is no longer possible to claim that evolutionary theory is empirically vacuous, but the advances made since Darwin do not vindicate the D-ideal. Consider how Ruse criticizes Goudge's (1961, p. 16) anti-deductive account. Goudge writes that the theory of evolution makes essential reference to particular historical statements, and these cannot be deduced as logical consequences of a set of postulates. But, Ruse (1973, p. 66) retorts, this fact is of no great moment because physical theories also presuppose reference to particular facts.

This observation by Ruse—that all theories need initial conditions to yield predictions or singular explanations—is helpful but not conclusive. The point is the less obvious one that in biological inquiry the extra information needed to derive explanations is heterogeneous in nature, and pertains to different levels. The slots to be filled by initial conditions are not on the level of the high principles but subtheories or models developed with an eye on mundane applications.

Just consider model building in modern biology. Here there may be general laws (such as the Hardy-Weinberg law in population genetics) which serve as a core of a sort. However, for predictive and explanatory purposes biologists build more specific theoretical models of various levels of generality. And this model building involves the highly non-trivial search for application-specific equations, values for the relevant parameters and constants, etc. Although such models presuppose general laws to be explanatory, it is not until we have the low-level (close to empiria) models that slots for singular explanatory facts emerge.

There is no need to deny that theories have core laws, nor that there are deductive steps in explanation. But it is unrealistic to think that the relevant more or less nomic generalizations operate on one level, and are ready for initial conditions to plug in. And the image in which the inquirer first formulates a theory-component and then starts hunting for initial conditions must go. Biologists often take theories, say, in population biology, to be clusters of models, some more general than others, and related with one another in complex ways. They may serve as alternative models to produce robust theorems (theorems that result from models built on different assumptions); as complementary models for different aspects of a problem; or as hierarchially arranged 'nested' models in which a model provides the parameters of another one at the next higher level (see Levins, 1966). In the latter case especially a clear-cut distinction to laws and initial conditions is misleading.

This brings us to a final shortcoming of formal models, viz., the fact that they make no reference at all to time or progress in inquiry. A focus on deep frozen arguments not only fails to accommodate the fact that application-specific generalizations are needed, but also that some of them surface in the midst of the process of inquiry. To put this idea into question-theoretic parlance: when an inquirer puts an explanation-seeking question to nature, she or he must be prepared to raise a series of small questions concerning auxiliary assumptions and initial conditions.

5. The Structuralist Proposal

Let us sum up the two major weaknesses in attempts to apply the question–answer idea to explanation. First, the relationship between questions and answers cannot be captured by purely syntactic and semantic means. Secondly, there is the extra difficulty with explanation-seeking questions, for the relationship between their presuppositions and their potential answers is opaque. The first weakness highlights the need to insert communicative contraints, as in all dialogue, and the second the need to specify what more there is to an explanation-seeking question.

The first need urges us to bring inquirers back into the picture. And having given the little finger to the devil, I suggest that we go further: scientific theories are tools which inquirers can use to construct answers. Inquirers aim at theories which are rich enough to turn questions which nature does not understand into a form which she does understand. The way to appreciate Bacon's dictum goes through the realization that the weak logic of questions must be supplemented with a resourceful notion of a theory.

To develop this idea I shall outline (in a non-technical mode) what is now known as the structuralist theory notion. It is a variant of the so-called semantic view, but puts more emphasis on the global structure of mature scientific theories, and on the dynamic aspects of theory construction.[5] Interestingly enough, this theory notion grew out of dissatisfaction with the logical positivist view of theories as sets of sentences axiomatized in a language. Actually, then, I shall propose that a satisfactory view of explanation only emerges hand in hand with a satisfactory theory notion, and both are motivated by a concern to meet new metatheoretic demands.

A crucial feature in the structuralist view is that a theory, or rather, theory-element T is given through a definition: one first gives a set of laws or generalizations (the conceptual core K of a theory) and stipulates that a structure is a model of T if and only if it satisfies the laws (generalizations). Apart from the conceptual core a theory also has, associated with it, a range of intended applications I, which typically divides into disparate classes. I is thus a set of classes of applications, whose elements may be of set-theoretical types different from the models of the conceptual core. A theory, in any case, is a dual entity which allow its holders to phrase claims to the effect that particular structures in fact are (expandable to) models of T: it consists of conceptual cutlery, K, and food for thought, I.[6]

[5] The differences between the structuralist and the semantic views are for the present purposes, inessential. Beatty (1980) gives a useful introduction to evolutionary theory from the point of view of the semantic view, and Lloyd (1984) applies it to population genetics. The structuralists have not, thus far, explored biological theories in any detail, but the notion of a theory-net touched upon here seems eminently suitable for the representation of model building based on fundamental biological theories.

[6] The phrase 'expandable to models of T' is needed, because intended applications as such are not models of T: they are, in the structuralistic view, obtained by help of non-theoretically characterized structures and need to be

A single theory-element T does not, however, cover the complexity of and interrelationships between various specific theories in a mature field of inquiry. Happily, theory-elements of the form ⟨K, I⟩, often form larger 'maxi-theories' or theory-nets, finite sequences of theory-elements connected with one another by various types of intertheory relations. A theory-net has a fundamental or basic theory-element and a number of sub-elements introduced to make more detailed claims about some more limited (classes of) applications. Underneath the fundamental core of the theory-net there may then be several branches of more restricted and nested theory-elements, and the result often is a hierarchial tree-structure. We can then distinguish between specific empirical claims made by using sub-cores, and the over-all claim made by help of the core of the net. What the theory-net (or, rather, one who subscribes to it) claims, then is that all of the empirical claims made by help of the various cores hold good.

There are also, in the structuralist arsenal, ways to explicate scientific progress, evolutionary and revolutionary. Theory evolutions are simply sequences in which more refined theory-nets replace already existing ones. Thus normal-scientific progress may consist of the introduction of a more stringent specialized core to deal with an application (theoretical progress), of widening of scope by the addition of new applications (empirical progress), and by finding new evidence for the inclusion of a prospected application among the models of an element (progress in confirmation).[7] Revolutionary change, in turn, takes place when a new basic core is introduced.

To give an example, the fundamental core of classical particle mechanics comprised the three Newtonian laws, but these were immediately specialized to distance-depending forces and then, by adding the gravitation law, into a specialized theory-element for gravitational phenomena. Newton's theory was a branching structure in that the distance-depending theory-element was also specialized to non-gravitational phenomena (such as oscillating systems), and there were still other branches. There was evolution of all three types, until the Newtonian basic core was given up in favour of a new one (see Moulines, 1979).

[7] For more details, see Stegmuller (1979), §§4–6.

appended with theoretical functions to be of the right type. In the interest of readability I have refrained from discussing the internal structure of the conceptual core where theoretically enriched potential models and non-theoretical partial potential models are distinguished. The set of intended applications is a subset of the latter. There is also a difficulty here that I cannot go into: if there are infinitely many ways to expand intended applications into models of a theory-element, the theory claim must be qualified to become falsifiable. For discussion, see e.g. Sneed (1971).

Theories characteristically start their careers as research programmes formed around some core assumptions. But there is always more to a programme than an abstract apparatus and a set of concrete chunks of reality: there are various types of group commitments which provide guidance in attempts to expand the apparatus to new areas of application. One way to codify such aspirations is to think that an inquirer (or community of inquirers) possesses not just conceptual cutlery K and food for thought I, but also paradigmatic exemplars, analogies, explanatory ideals and cognitive values.

Suppose, then, that we view Darwinian evolutionary theory analogously: it had the theory of natural selection as the conceptual core and the various classes of facts (geographical distributions, morphological similarities, palaeontological facts) as intended applications. The various domains of application fed in specific questions, and the basic laws provided some of the tools for construing answers. The paradigmatic exemplars consisted of some exemplary (though sketchy, in modern terms) explanations. There were general metaphysical and ontological commitments, such as Darwin's early commitment to harmonious design and Lyellian gradualism, strengthened by his discoveries of geological changes at coastal Chile and his researches into the formation of coral reefs. There were, furthermore, broad methodological maxims and cognitive values concerning the conduct of inquiry, such as the requirements of objectivity, non-anthropomorphism, 'materialism', etc.[8] And there was of course the real news, too, the peculiar way to read questions through the variational glasses.

A structure with a basic element and subordinate elements does not mimic the D-ideal. Although the basic laws and the classes of intended applications generate a picture very much like Figure 1 above, the structure as a whole is ill at ease with the D-ideal. The relationships between the various theory-elements are not deductive (or only deductive), and the paths from the basic laws to descriptions of facts is too many-tiered to fit the D-ideal. Philip Kitcher (1985, p. 163) nicely illustrates this fact. The theory of evolution by natural selection aims at explaining the properties of various groups of organisms by construing their Darwinian histories, narratives which trace the successive modifications of the groups in terms of a number of factors, such as natural selection (but not only natural selection). But, first, the theory does not itself single out a particular Darwinian history for any explanandum

[8] The role of analogies were of course crucial for in the process of conceiving the basic laws—but we can take it that they as well as the other general metaphysical and methodological commitments guided and restricted possible attempts to expand the theory-net also during the later career of the Darwinian paradigm.

fact: in tracing the ancestral forms of groups of organisms different types of steps are available, and it is a matter of empirical investigation to find out the right one. And secondly, the choice of a particular history does not nail particular observational facts to focus on.

It is more useful to look upon the basic theory-element, the theory of natural selection, as a tool which suggests types of Darwinian histories to deal with different types of applications, but leaves much of the details to the specific contexts. Similarly, the structure of modern evolutionary theory is messier than Ruse's picture indicates. In modern structuralist literature the most important (or at least best studied) mode of concretization is the adding of special laws to the basic laws (specialization), but there are other types of more or less intimate relationships between theory-elements. One element T_1 may be used to test another element T_2; or T_1 may serve to interpret T_2; or indeed T_1 may be reduced to T_2. It is impossible to settle *a priori* the applicability or usefulness of various possible inter-theory relations in a field.[9]

Note that an inquirer may of course face two kinds of explanatory tasks, of unequal importance. First, it is possible that she or he can resort to an already available and well-confirmed pattern of explanation for a domain of phenomena. There already are good candidates for a suitable specialized theory-element which can therefore be used to construe candidate theoretical models for a detailed explanation. Here the inquirer may attempt to derive an explanatory answer to a singular question from the basic theory by finding suitable relations between the basic theory element and the explanandum, using a series of auxiliary questions to find out required information.

Second, an inquirer may face a set of questions without being able to avail of a satisfactory theory-element. There may be descriptive generalizations pertaining to the domain, but these rest unconnected with the existing theory-net. Nevertheless, if the phenomena display sufficient similarity with some exemplars claimed to fall within the range of applications of a theory-element, there is a clearly perceived incentive to find one. In such a case the inquirer must find a pigeonhole for a somewhat fuzzy equivalence class of phenomena, and try to design relevant models.

Using an already existing theory-element to construe a singular explanation is far from trivial, but it does involve more use of familiar tools. Darwin was the first to conceive the outlines of a theory-net with natural selection as the basic core. Even if he did not produce detailed models to deal with the various types of questions, he did outline paths

[9] For a tentative classification of intertheory relationships, see Krüger (1980). See also Balzer, Moulines and Sneed (1986).

from the basic core to sets of intended applications. That is why we can say that Darwin invented gunpowder, his successors more of it.

Darwin, then, brought about a revolution by introducing a new variational basic theory-element to replace the prevailing developmental core. A theory in this sense is not finished when the core is proposed, because there is normal-scientific work to be done along the lines suggested. And clearly, there has been theoretical, empirical and confirmatory progress since Darwin. And arguably, the emergence of the new synthetic core in the 1930s counts as a revolution. The core of population genetics came to replace the Darwinian core, and the previous theory-net moved to live under the new umbrella, in a modified form.

This brings us to the problem of gaps and impressions. The D-ideal focuses on finished products and text-book examples, often written with the benefit of hindsight: an incomplete and imprecise theory is not worth the name at all. However, a look at actual science shows that unfinished business is the norm rather than the exception: theories that catch the scientists' attention and deserve their devotion are never finished. The structuralist theory notion makes graphic sense of this allegedly sorry state, for theories are praised both for precise answers and precise questions. Indeed the built-in tension between the claim made by help of a theory-net and the achievements secured at a time draw attention to the fact that there almost always are missing special laws, and if not these, at least uninvented theoretical models.

6. Explanation: From Single Steps to Strategies

Notice the contrast between the essentially one-step covering law models and the structuralist-interrogative view sketched here. The intellectual task in the former boils down to a search for a set of laws and initial conditions, but it is constitutionally incapable of dealing with the mediating and contextually determined steps. The latter idea codifies a strategy of raising and answering questions on several levels. What it does, then, is explicate the process of inquiry from the fundamental core to the lower-level elements, not just the finished argument on the level of the latter.

The view which emerges is that an inquirer is equipped with a rich theory which in part structures empirical questions by putting constraints on admissible answers. But in the search for precise answers a series of more specific questions are needed to fix contextual assumptions. Take a question in the philosophers' canonical form 'Why Q?', where Q ranges over descriptions of singular events or facts, such as distributions of traits. An answer to such a question has the form 'Q because p', whereas p ranges over sentences or sets of sentences proposed as alternative answers. The semantics of p, we saw, does not usually tell an inquirer much about relevant answers. But here the proposal makes some headway towards a solution. Why- and how-questions contain more than meets the ear, namely certain contextual restrictions on potential answers imposed by the extant theory-net and relevant neighbouring fields. If there already are accepted stories of kindred singular explananda, a good strategy is to try to trace an analogous path from the basic core to Q.

The very question 'Why Q?' arises in part as a result of a perceived similarity or articulated analogy between the states of affairs depicted by Q and Q′. And if Q is similar to Q′ by the theory's lights, as we can safely assume, this recognition triggers the search for a suitable substitution instance for p. Nevertheless it is genuinely up to empirical investigation to find out the additional premisses. And of course there always are several possible paths from the core to the explananda. In any case, the uncovering of new facts by help of small questions is not exclusively a matter of confirming facts already entailed by a theory.

It may be objected that a model like this in which questions generate further questions really does not help to corner Nature, for the inquirer is interested in answers, too. A brief reply goes along the following lines. The strategy provided by a theory indeed amounts to pinpointing more specific types of questions on the levels of the intermediate theory-elements. But there is an important gain: what a theory does is split the initially amorphous and therefore unmanageable (untestable) why-question into a series of more tangible yes–no questions concerning particular Darwinian histories. The availability of previously trodden paths from the basic core K to similar explandum phenomena means that there are questions in which potential answers are restricted to these two: Is this p the right Darwinian history for Q, or is it not? And this question can be further split to a series of experimental and observational questions by help of theoretical models. A query which starts as a loose why-question concerning an area of experience turns (in the 'paradigmatic' phase) into a what-question: what model-specific assumptions are needed to deal with this particular phenomenon? It is a further task to find which-questions which portray specific alternatives and still narrow down admissible substitution instances. There is of course no logic for generating questions—but there is a contextual and non-trivial heuristics. And this opens a channel for a dialogue with Nature, the only one I can conceive of.

The important insight is here, too. I started with the dilemma of the question–answer model, which was that Nature does not understand explanation-seeking why-questions. Now we have got some relief: Nature does not understand why-questions which are not paraphased as which-questions, that is, a series of yes–no-questions. But of course what theories do is precisely chop the unmanageable why-questions to potential wh-questions.

This brings us back to the initial metaphor: How do we put explanation-seeking questions to nature? There is no direct way because nature is bewildered by such questions. But theories understood in a sufficiently rich sense are devices which turn why-questions into a form which nature does understand. They first specify some plausible patterns which single out types of answers. Thus evolutionary theory specifies what happens if no evolutionary forces operate, as well as what comes out when forces do operate. This amounts to specifying a set of alternatives to choose from, that is, which-questions. And each alternative potential answer is a yes–no question: was it in fact natural selection, or mutation or random drift, or whatever? And, with luck and ingenuity, the inquirer may be able to derive further contextual theoretical models with very precise questions where Nature does cooperate by answering 'Indeed, yes!' (or 'No!').

Note, finally, that at this stage of inquiry erotetic logic does apply. When a why-question is sliced to potential answers which can be offered to nature, there is also a way to assess how satisfactory a given answer is: an answer is, recall, satisfactory if and only if it does not leave important unanswered questions. There may of course in evolutionary studies be answers which are not very satisfactory. But this is a difficulty which plagues all historical studies.

References

Abel, Ruben 1982. 'What Is An Explanandum', *Pacific Philosophical Quarterly* **63**.

Achinstein, Peter 1983. *The Nature of Explanation* (Oxford University Press).

Bacon, Francis 1863. *The New Organon. The Works of Francis Bacon*, Vol. VIII, J Spedding, R. L. Ellis and D. D. Heath (eds.) (Boston: Taggard and Thomson).

Balzer, W., Moulines, C. and Sneed, J. D. 1986. 'The Structure of Empirical Science', In Barcan Marcus *et al.* (ed.) *Logic, Methodology and Philosophy of Science* VII (Elsevier Science Publishers B.V.), pp. 291–306.

Beatty, J. 1980. 'What's Wrong with the Received View of Evolutionary Theory?', in *PSA 1980* P. D. Asquith and R. N. Giere (eds.), Vol. 2, 379–426.

Belnap, Nuel 1963. *An Analysis of Questions: Preliminary Report* (Santa Monica: System Development Corporation).

Darwin, Charles 1964. *On the Origin of Species* (Cambridge, Mass: Harvard University Press.

Collingwood, R. G. 1940. *An Essay in Metaphysics* (Oxford: Clarendon Press).

Ghiselin, M. 1984. *The Triumph of the Darwinian Method,* (The University of Chicago Press).
Goudge, T. A. 1961. *The Ascent of Life* (University of Toronto Press).
Hempel, Carl 1965. *Aspects of Scientific Explanation and Other Essays in the Philosophy of Science* (New York: The Free Press).
Hempel, Carl and Oppenheim, Paul 1965. 'Studies in the Logic of Explanation', in C. Hempel (ed.), *Aspects of Scientific Explanation and Other Essays in the Philosophy of Science* (New York: The Free Press), 245–95.
Hintikka, Jaakko 1976. *The Semantics of Questions and the Questions of Semantics*, Acta Philsophica Fennnica, Vol. 28, no. 4.
Kant, Immanuel 1933. *Critique of Pure Reason*, trans Norman Kemp Smith (London and Basingstoke: The Macmillan Press).
Kitcher, Philip 1985. 'Darwin's Achievement', in *Reason and Rationality in Natural Science* Nicholas Rescher (ed.), (Lanham, Maryland: University Press of America), 127–89.
Krüger, Lorentz 1980. 'Intertheoretic Relations as a Tool for the Rational Reconstruction of Scientific Development', *Studies in History and Philosophy of Science* **11.2**, 89–101.
Levins, Richard 1966. 'The Strategy of Model Building in Population Biology', *American Scientist,* **54.4**, 421–31. (As reprinted in E. Sober (ed.), *Conceptual Issues in Evolutionary Biology* (Cambridge, Mass. and London: The MIT Press, 1984), 18–27.)
Lewontin, Richard 1974. 'The Structure of Evolutionary Genetics', ch. 1 of *The Genetic Basis of Evolutionary Change* (New York: Columbia University Press). (As reprinted in E. Sober (ed.), *Conceptual Issues in Evolutionary Biology* (Cambridge, Mass. and London: The MIT Press, 1984), pp. 3–13.)
Lloyd, Elizabeth 1984. 'The Semantic Approach to the Structure of Population Genetics', *Philosophy of Science* **51**, 242–64.
Mayr, Ernst 1980. 'Prologue: Some Thoughts on the History of the Evolutionary Synthesis', in *The Evolutionary Synthesis. Perspectives on the Unification of Biology* E. Mayr and W. Provine (eds.), (Cambridge, Mass.: Harvard University Press), 1–48.
Moulines, Carlos Ulises 1979. 'Theory Nets and the Evolution of Theories: The Example of Newtonian Mechanics', *Synthese* **41**, 417–39.
Putnam, Hilary 1978. *Meaning and the Moral Sciences* (Boston, London and Henley: Routledge & Kegan Paul).
Ruse, Michael 1973. *The Philosophy of Biology* (London: Hutchinson).
Ruse, Michael 1979. *The Darwinian Revolution: Science Red in Tooth and Claw* (The University of Chicago Press).
Sintonen, Matti 1989. 'Explanation: In Search of the Rationale', in *Scientific Explanation*. P. Kitcher and W. C. Salmon (eds.), Minnesota Studies in the Philosophy of Science, Vol. XIII (Minneapolis: University of Minnesota Press), pp. 253–82
Sneed, J. 1971. *The Logical Structure of Mathematical Physics* (Dordrecht: D. Reidel).
Sober, Elliott 1984. *The Nature of Selection* (Cambridge, Mass. and London: Bradford Books, The MIT Press).

Stegmüller, W. 1979. *The Structuralist View of Theories: A Possible Analogue of the Bourbaki Programme in Physical Science* (Berlin, Heidelberg and New York: Springer-Verlag).
Tuomela, Raimo 1980. 'Explaining Explaining', *Erkenntnis* **15**, 211–43.
Whewell, William 1847. *The Philosophy of the Inductive Sciences, Founded upon Their History*, 2nd edn (London: Frank Cass & Co).

Explanation and Scientific Realism

PHILIP GASPER

A few years ago, Bas van Fraassen reminded philosophers of science that there are two central questions that a theory of explanation ought to answer. First, what is a (good) explanation—when has something been explained satisfactorily? Second, why do we value explanations? (van Fraassen, 1977, 1980, ch. 5). For a long time, discussions of explanation concentrated on technical problems connected with the first of these questions, and the second was by and large ignored. But, in fact, I think it is the second question which raises the more fundamental and interesting philosophical issues. I shall offer reasons for thinking that the answer to the first question requires acceptance of the sort of full-blown notion of causation that only a scientific realist can love, and that the answer to the second question requires a realist construal of scientific theories and scientific methodology. My argument will be mainly negative, surveying the problems facing some major alternative accounts of explanation. A full elaboration of the realist perspective will have to await the completion of work in progress.

The Covering-Law Model of Explanation

A satisfactory explanation of an event or phenomena should provide us with understanding of what has been explained. But understanding is a notoriously vague notion. Different inquirers may disagree about what is sufficient for understanding and whether or not understanding has actually been achieved. If the search for explanation has a central role to play in scientific reasoning, then it is important to insure that our concept of explanation is free from this kind of vagueness. We want to be able to tell accurately when the goal of scientific explanation has been reached, and when a new theory has sufficient explanatory resources to be accepted or to be preferred over its competitors.

Now it has seemed to many philosophers that our central, pre-analytic, common-sense view of what an explanation is, does little to help us achieve these aims. On such a view, to explain some phenomena is simply to say how it was caused.[1] But this view of explanation is only

[1] We also recognize other kinds of explanations. Some, such as teleological explanations, functional explanations, and explanations in terms of reasons, I take to be special kinds of causal explanation. Other scientific explanations (e.g., explanations in terms of the geometrical properties of objects) may be non-causal. I will ignore these in this paper.

as satisfactory as the notion of causation that goes along with it, and it is this which has traditionally worried philosophers. Many of them have seen causation as a problematic relation, one which science should abandon or at the very least attempt to clarify. What is it for one thing to be the cause of another? How can we tell when a causal relation exists? In the absence of answers to questions like these, we will not have given a satisfactory account of what it is to explain something.

These worries about causation go back at least as far as Hume. Hume argued in the *Enquiry* that if we examine our idea of causality, we will find that the claim that one event was the cause of another amounts to saying no more than that events like the first one are always (or usually, or typically) followed by events like the second. Hume's account leaves vague what it is for one event to be like another, since for Hume this is simply a psychological matter. However, the vagueness in the account can be eliminated by specifying what features of similarity are relevant in any particular case—in other words by specifying general laws that tell us that events with certain characteristics (or that conditions of a certain kind) are followed with a certain degree of probability (less than or equal to one) by events with certain other characteristics, where the characteristics in question may perhaps be specified quantitatively.

Given this account of causation we are led directly to an account of explanation based on our pre-analytic view. According to that view, to explain something is to specify how it was caused. If we now develop this view in the light of a sophisticated Humean account of causation, we arrive at the following account: an event of a certain kind is explained by citing a general law (or laws) that relates events of that kind to events or conditions of some other kind (and perhaps in addition showing that events or conditions of the latter kind took place or were in effect). This is the classic covering-law model of explanation.

The covering-law model is no mere philosophical abstraction. It has been extremely influential in the natural sciences. The idea of explanation as subsumption under laws was central to Newton's view of science, for example. Indeed, we can see Hume's explicit worries about unobservable causal powers prefigured in Newton's dislike of 'occult powers' and in his famous claim that 'hypotheses non fingo'. The covering-law model has also had an important influence on the conduct of the social sciences. As Richard Miller points out, for example, the model has been one of the motivations behind structuralism in social anthropology. If explanation requires general laws, but no such laws are available couched in terms of familiar social roles such as mother or brother, then such laws must be formulated in terms of increasingly abstract relations, such as 'binary opposition' (Miller, 1983, p. 85).

We arrived at the covering-law model by considering the first question that a theory of explanation must address: what is an explanation and what makes it satisfactory? But the model not only provides an answer to this question, it also neatly answers our second question: why do we value explanation? The answer to this question given by the present account is that explanatory power is evidential. On the covering-law model, a satisfactory explanation turns out to have exactly the same logical structure as the derivation of a prediction on the basis of a scientific theory. If we have an adequate explanation of some occurrence, then, in principle, we could have predicted it before it actually took place (or before we knew that it had taken place). Now, for obvious reasons, accurate prediction is important evidence in favour of a theory. It follows that we should value explanatory theories for exactly the same reasons. A theory which provides us with good explanations is confirmed in the same way as one which yields true predictions.

So the covering-law model of explanation nicely addresses both the questions with which we began. But unfortunately there is a fly in the ointment—in fact there are several flies. One general background problem (which I will simply mention in passing here) is whether it is possible to specify which general statements count as genuine laws (or as lawlike statements) without reference to unobservables or to causal factors. If this is not possible (and a strong case can be made that it is not), then the epistemological motivation for the covering-law model would be completely undercut (see Boyd, 1985).

In addition to this general problem, however, the covering-law model can be confronted with a large number of specific counter-examples. These fall into two classes. First, cases in which an event is not explained, even though its occurrence has been derived from general laws and background conditions. Second, cases in which an event is explained, even though its occurrence has not (or cannot) be derived in the appropriate way. Cases of the first sort show that fitting the covering-law model is not sufficient for something to be an explanation. Cases of the second sort show that fitting the model is not necessary for something to be an explanation.

Included in the first set of cases are the so-called asymmetries of explanation. Many mathematical laws link events in such a way that, given information about either one, information about the other can be derived. Thus, given the laws of optics, the position of the sun and the height of a certain flagpole, we can calculate the length of the shadow that the flagpole will cast. Here, the covering-law model conforms with our intuition that the height of the flagpole explains the length of the shadow. But, given the length of the shadow and the other information, we can equally calculate the height of the flagpole. In terms of the covering-law model, the two cases are parallel—yet it seems that we would not want to say that the length of the shadow explains the height of the flagpole.

We can also find examples of derivation without explanation which involve no asymmetry. For instance, from the laws of biology together with the fact that there are mammals on the earth, we can deduce that there is oxygen in the atmosphere. Clearly, we have not explained the presence of oxygen in this way, but neither does the presence of oxygen explain the existence of mammal life.

One particularly important class of cases of this first kind involves examples in which the derivation fails to provide an explanation due to lack of depth. For example, suppose it were possible, given suitable general laws and knowledge of background conditions, to predict the outbreak of the First World War on the basis of information about the assassination of the Archduke Ferdinand in Sarajevo. Such a prediction would still not constitute a satisfactory explanation of the outbreak of war—the events it cites were at best the immediate triggers of the conflict, not its underlying cause. It is likely that, had the war not broken out in this way, then underlying factors would have ensured that war would have broken out in some other way instead. A satisfactory explanation should surely appeal to these factors, but a derivation which meets the standards of the covering-law model may nevertheless fail to appeal to such underlying tendencies.

Not only are there cases of covering-law derivations which fail to explain, there also appear to be explanations which fail to meet the standards of the covering-law model. For example, an explanation of the absence of a mass-based working-class political organization in the United States (a phenomenon known as 'American exceptionalism') in terms of such factors as constitutional design, geography and natural resources, uneven economic development, racism, ethnic and religious divisions, and state repression, has been defended by Joshua Cohen and Joel Rogers, and seems to be worth taking seriously (see Cohen and Rogers, 1986, part 1). Yet there are no general laws connecting the cited factors with the phenomenon to be explained, and no reason to think that the presence of such factors would always be incompatible with the emergence of class-based politics. Moreover, it would be wrong to see Cohen and Rogers' explanation as merely a sketch waiting to be filled out in a manner which fits the covering-law model. It would surely be bad methodological advice to suggest that political scientists should try to fill the explanation out to fit the model. If Cohen and Rogers' explanation is deficient, it is because it ignores salient factors (e.g. the political weaknesses of the U.S. working-class movement during crucial periods), not because it fails to provide a covering law.

The Pragmatics of Explanation

Problems of the kinds just mentioned seem to many philosophers, including me, to undermine the covering-law model of explanation decisively.[2] But in that case, what do we put in its place? Van Fraassen (1977, 1980, ch. 5) argues that explanation is fundamentally a pragmatic matter. He points out that explaining an event typically involves describing some of the events leading up to it. But which ones? Van Fraassen argues that the choice of events is a pragmatic one, determined by our particular interests. With the help of some technical devices he tries to make the pragmatic dimension of explanation clearer, and to show how the problems facing the old covering-law model can be overcome.

The first of these devices is what van Fraassen calls an explanation's 'contrast class'. He claims that we do not explain the occurrence of an event simpliciter, but only its occurrence in contrast to some specified set of other possible occurrences. Thus we are supposed to ask not simply why the First World War broke out but, rather, why it broke out in August 1914 rather than September 1914, or why there was a war rather than an attempt to set up international institutions designed to solve major power conflicts by negotiations. Van Fraassen makes the plausible claim that paying attention to the contrast class allows a solution to the problem of explanatory depth. Thus the Archduke's assassination is not completely irrelevant to the explanation of the First World War—it simply provides an answer to a less interesting question (why war in August rather than September?) than the one we probably want to ask (why war rather than negotiations?).

Appeal to contrast classes also clarifies cases of explanation without general laws, according to van Fraassen. If we want to know why there is no mass-based labour party in the U.S. in contrast to Britain, then mentioning factors of the sort cited by Cohen and Rogers may be relevant, since at least some of them (geography and natural resources, and uneven economic development) are unique to the U.S. However, if Canada is included in our contrast class, then the factors may cease to be explanatory, since there is a Canadian labour party (the New Democratic Party) with a significant base even though, to one degree or another, the cited factors are all present in Canada as well.

The notion of an explanation's contrast class may have some value in cases such as the ones just discussed. But despite van Fraassen's claims, contrast classes do not, I think, make explanation pragmatic in any important sense. They can be seen as a way of specifying identity conditions for the event to be explained (indeed, when the identity conditions are already clear, it is hard to specify an informative contrast

[2] Not all philosophers are convinced by counter examples of the sort I have given. Railton (1978) is one sophisticated attempt to defend the covering-law model. Railton's views are discussed in Miller (1987), 40–3. Also see Michael Redhead's contribution in this volume.

class for the explanandum), and it is not a surprise that it is up to us to choose what we want explained. On the other hand, as Alan Garfinkel points out, even here there can be good choices and bad ones: 'We can stipulate equivalences at will, but the result will be a good piece of science only if the way we are treating things as inessentially different corresponds to the way nature treats things as inessentially different' (Garfinkel, 1981, p. 32).

But what does van Fraassen have to say about the asymmetries of explanation? Here he appeals to a new notion: the 'relevance relation' of an explanation. The relevance relation specifies the form of the explanation that we are looking for. We might be interested in the events leading up to the event to be explained, or in a standing condition that preceded the event, or in the event's function or purpose, and so on. Van Fraassen claims that we may choose whichever relevance relation we like, and that as a consequence there is no such thing as the cause of an event, independent of our particular interests. And he argues that the asymmetries of explanation can therefore be reversed, so that there may be occasions on which the length of the shadow explains the height of the flagpole—for instance, when the pole was chosen to cast a shadow of a certain length.

But one may doubt whether van Fraassen has shown whether the asymmetries can be reversed. Is it the length of the shadow or, rather, the plan that the shadow should be of a certain length that explains the flagpole's height? Moreover, it is far from clear that there is any independent role for relevance relations to play in the assessment of explanations. Once we have decided what is to be explained (perhaps by specifying a contrast class), we cannot then independently go on to specify the form that the explanation is to have. And in many cases, there may simply be no explanation of a certain form, no matter what contrast class is chosen. Very few occurrences serve any purpose, for example, so a request to explain such an occurrence by specifying its function could not be met.

Van Fraassen's pragmatic account of explanation also encounters problems when faced with the question of the relation between explanatory power and theory choice. Van Fraassen claims that the search for explanatory theories is *ipso facto* a search for empirically adequate theories, but he offers no argument for this claim. He also discusses whether explanatory power is evidential when choosing between two theories, each of which is known to be empirically adequate. Van Fraassen answers this question in the negative, claiming that explanatory power gives us only a pragmatic reason for preferring one theory to the other.

In fact, it is not clear that scientists are very often faced with choices of this second kind; but consider an actual example—the choice between Darwinian theory and creationism in the 1860s. Nineteenth-century creationism was empirically adequate in van Fraassen's sense (at any rate, it would not have been hard to modify it to make it consistent with the known data), so was it rejected merely on pragmatic grounds? On the one hand, creationist 'explanations' hardly seem to fail on pragmatic grounds (e.g. by selecting an inappropriate contrast class), while on the other hand, Darwin's own assessment of his theory seems compelling:

> It can hardly be supposed that a false theory would explain, in so satisfactory a manner as does the theory of natural selection, the several large classes of facts above specified (Darwin, 1859, p. 476; cited in Thagard 1978, p. 77).

Darwin's argument is by no means an unusual one. Darwin himself noted this in defending his reasoning:

> It has recently been objected that this is an unsafe method of arguing; but it is a method used in judging of the common events of life, and has often been used by the greatest natural philosophers (Darwin, 1859, p. 476).

Paul Thagard cites other important instances of scientists taking explanatory power as a guide to at least approximate truth—Lavoisier's argument for the oxygen theory of combustion, Huygen's arguments for the wave theory of light and Fresnel's later arguments for the same theory—and concludes that such examples are 'common in the history of science' (Thagard, 1978, p. 77). Yet if van Fraassen's account of the value of explanation as purely pragmatic were right, Darwin's claim and others like it would have to be rejected. And such claims would have to be rejected even if talk of the (approximate) truth of a theory were replaced by talk of the theory's empirical adequacy.

Explanatory Unification

Philip Kitcher has defended an account of explanation that allows him to accept Darwin's claim. According to Kitcher,

> there are certain context-independent features of arguments which distinguish them for application in response to explanation-seeking why questions, and . . . we can assess theories (including embryonic theories) by their ability to provide us with such arguments. . . . historical appeals to the explanatory power of theories involve recognition of a virtue over and beyond considerations of simplicity and predictive power (1981, p. 170).

The feature that Kitcher discusses is the capacity of a theory to provide a small number of explanatory patterns which can unify a wide variety of apparently disparate phenomena. Kitcher attempts to sketch a rigorous and formal account of explanation as unification, and argues that this conception of explanation helps solve the main problems facing the covering-law model. Consider, for example, the following asymmetry: we can explain the period of a pendulum in terms of the pendulum's length, but not the length in terms of the period. Why is this? According to Kitcher, allowing the latter derivation to count as an explanation would require that we have one pattern of explanation for the length of swinging bodies and another for the length of stationary ones, and this would be less unified than allowing only a single pattern for both sorts of body. To the objection that 'we can construct derivations of the dimensions of bodies from specifications of their dispositional periods, thereby generating an argument pattern which can be applied . . . generally', Kitcher replies that '[t]here are some objects . . . which could not be pendulums, and for which the notion of a dispositional period makes no sense' (1981, p. 180).

Kitcher's account of explanation as unification has a number of promising features, but also some problems. Some of these stem from the fact that Kitcher offers a highly general and formal account of unification that ignores the specific content of the theories he is discussing, and in particular the specific causal processes and powers that they postulate. Suppose, for instance, that we lived in a world in which every object could be a pendulum. In such a world, would Kitcher say that the period of a pendulum *does* explain its length? This would be to ignore the fact that there is a causal mechanism whereby the pendulum's length determines its period, but no mechanism which operates in the opposite direction.

A second problem for Kitcher's account is that, while he recognizes explanatory power as relevant to theory choice, he provides no explanation of the link. Why should the fact that a theory is able to explain disparate phenomena in a unified way give us any (non-pragmatic) reason for accepting it? What is the connection between unification and truth (or between unification and empirical adequacy)?

Realism

A realist account of explanation has the virtue of being able to answer the questions that arise for the other accounts we have examined. On a realist view, scientific theories do more than tell us about observable regularities. They provide us with an inventory of what sorts of entities, mechanisms, processes, etc. (observable and unobservable) exist, and tell us something about the relations between them. (None of this need be expressed mathematically or quantitatively, though it is nice if it can be.) The information is often conveyed by means of paradigm examples. Explanations of particular phenomena are constructed on the basis of a theory's inventory, which often requires great ingenuity.

A realist who rejects sceptical arguments about causality and 'secret powers' has no difficulty in explaining the asymmetries. The flagpole is part of the cause of the shadow—the shadow exists because light waves are absorbed or deflected by the flagpole—but the shadow is not part of the cause of the flagpole. Realists also have helpful things to say about explanation by appeal to insufficient factors and about explanatory depth.

The search for explanations is the search for systematic factors operating in the world. Appeal to insufficient factors (as in the explanation of American exceptionalism) can be satisfactory if these are the only systematic factors operating in a particular case. To take a new example, the evolution of some phenotypic characteristics of members of a biological species may be adequately explained by citing relevant selection pressures, even though such pressures on their own were not sufficient to bring about the characteristic, provided that these were the only systematic factors involved; other factors may be too random or haphazard to be explanatory.

Appeal to sufficient factors (if this is ever possible), on the other hand, can be undermined if these factors themselves have some sort of systematic explanation. In particular, sufficient factors can be undermined by showing that if the factors in question had not operated, others would have brought about the same effect (e.g., appeal to the imperatives of economic expansion might undermine an explanation of the outbreak of the First World War in terms of the belligerency of national leaders). Clearly, just when such systematic factors do or do not exist is a matter not for *a priori* philosophical analysis, but for empirical research. In this sense, the concept of explanation is a theory-dependent one.

What, though, of philosophical worries about the nature of causality? Here, too, the realist response is that such worries cannot be answered by *a priori* analysis, but are matters for empirical research. There is probably no informative analysis in non-causal terms. Causal relations are irreducible features of the universe which we can learn more about through scientific inquiry (cf. Kuhn, 1977). Causality, too, is a theory-dependent notion.

Finally, why is explanatory power desirable? This question is crucial. Recall that scientists do not typically independently identify empirically adequate theories and then compare them with respect to explanatory power. Rather, judgments of explanatory power play a central role in the choice of reliable theories to guide further research. This claim finds confirmation in the fact that explanatory power is routinely taken as a guide to theory acceptance in successful areas of science (cf. Thagard, 1978). Realists insist that we take this practice seriously. Explanatory power (like other theoretical virtues, such as simplicity) is at least a guide to which theories are empirically adequate. In other words, explanatory power is evidential.

The realist construal of this aspect of scientific practice is quite straightforward. If a theory initially introduced to explain one sort of phenomena is found to have the resources to explain other, unconnected phenomena (i.e., if it has the power to yield a variety of explanations), then that is evidence that the theory is an accurate description of the world. On the one hand, the best explanation of the theory's success is that the (observable and unobservable) entities, mechanisms and events which it postulates, actually exist (or closely resemble what actually exists). On the other hand, if an account along these lines were not correct, it would be hard to see the justification for the practice of taking explanatory power as a guide to theory acceptance, or why the practice should contribute to the successful development of scientific research.

If we accept that the goal of science is to arrive at accurate descriptions of both the observable and unobservable parts of the world, and that science uncovers not merely regularities but also causal powers and relations; and if we also accept that explanatory unification is not an end in itself, but is a guide to the accuracy of our theories, then many of the problems plaguing alternative accounts of explanation disappear. This is by no means a knockdown argument in favour of scientific realism, but I believe it does give us reason to take the realist account of explanation very seriously.

References

Boyd, Richard N. 1985. 'Observations, Explanatory Power, and Simplicity: Toward a Non-Humean Account', in *Observation, Experiment, and Hypothesis in Modern Physical Science*, P. Achinstein and O. Hannaway (eds) (Cambridge, MA: M.I.T. Press).

Cohen, Joshua and Rogers, Joel 1985. *Rules of the Game* (Boston: South End Press).

Darwin, Charles 1859. *The Origin of Species*, 6th edn, reprinted 1962 (New York: Collier).

Garfinkel, Alan 1981. *Forms of Explanation* (New Haven: Yale University Press).

Kitcher, Philip 1981. 'Explanatory Unification', *Philosophy of Science* **48**, 507–32. Reprinted in *Theories of Explanation*, J. Pitt (ed.) (Oxford University Press, 1988).

Kuhn, Thomas 1977. 'Concepts of Cause in the Development of Physics', in *The Essential Tension* (University of Chicago Press).

Miller, Richard W. 1983. 'Fact and Method in the Social Sciences', in *Changing Social Science*, D. Sabia and J. Wallulis (eds) (Albany: SUNY Press).
Miller, Richard W. 1987. *Fact and Method* (Princeton University Press).
Railton, Peter 1978. 'A Deductive-Nomological Model of Probabilistic Explanation', *Philosophy of Science*, **45**, 206–26.
Thagard, Paul 1978. 'The Best Explanation: Criteria for Theory Choice', *Journal of Philosophy*, **75**, 76–92.
van Fraassen, Bas 1977. 'The Pragmatics of Explanation', *American Philosophical Quarterly*, **14**, 143–50.
van Fraassen, Bas 1980. *The Scientific Image* (Oxford: Clarendon Press).

How Do Scientific Explanations Explain?

JOYCE KINOSHITA

Introduction

My title question as it stands is ambiguous, and is in want of some initial clarification. Does the question ask how the explanandum is logically related to the explanans? Or does it ask about the details of the dynamics of the explanation speech-act? Or does it ask how the linguistic ambiguities of explanation questions and answers should properly be unpacked? Or does it ask yet some other question?

The ways of studying explanation, like the ways of understanding the world, are many and varied. By this, I mean more than that the phenomenon of explanation can be studied as it arises in the different disciplines of biology, physics, the social sciences, and the like. Rather, I mean that there are varied disciplines of explanation-study itself. For example, the Hempelian tradition has largely focused on the logic of explanation, and others have focused on the linguistic, psychological, social, and epistemological angles of explanation. Thus, it is not appropriate for me to begin by arguing that explanation is a set of logically related statements or a speech-act (just as one does not begin by arguing 'the world' is a sociological or physical phenomenon), but appropriate instead to begin by specifying the explanatory discipline within which I ask my title question.

I ask my title question from within the *epistemological* discipline, the discipline addressed by Michael Friedman (1974), and more recently, by Philip Kitcher (1981). How do explanations—in whatever logical, linguistic, or social form—explain? Despite their differences, both Friedman and Kitcher argue that explanation explains by unifying. Moreover, they both see this unification in terms of some kind of numerical decrease. For example, Friedman (1974) says, 'We don't simply replace one phenomenon with another. We replace one phenomenon with a *more comprehensive* phenomenon and thereby effect a reduction in the total number of accepted phenomena. We thus genuinely increase our understanding of the world.' Kitcher (1981) also concludes, 'By using a few patterns of argument in the derivation of many beliefs we minimize the number of *types* of premises we must take as underived. That is, we reduce, in so far as possible, the number of types of facts we must accept as brute.'

The answer (sketch) I eventually propose to the title question does not specifically follow Friedman's and Kitcher's equation of unity and numerical decrease, but rather takes a slightly different path. Clearly, if there are *logical* (derivational) relations among 'types of facts' then explained (derived) facts are logically unified with others. However important these logical connections are, the focus in this paper is on the *epistemological*. There are two fuzzy notions that seem both intuitively correct and in dire need of further clarification. The first notion is that of explaining by unifying. The second notion is the widespread claim that explanation operates within a context, or background, or web of belief, or theory. I try to answer the title question from an epistemological angle, and in such a way so as to preserve and clarify both of these notions.

This paper has three parts. The first two parts are case studies. Hempelian explanations form the basis of the first case study. Hempel did not specifically address the epistemological version of our title question, but by careful reading, we can extract what he perhaps might have answered. Since Hempel was primarily interested in logical, not epistemological, aspects of explanation, our 'Hempelian extractions' will necessarily be vague on fine points important to this discussion. I leave the vagueness intact, for I use this case study only to display and illustrate a general epistemological concern. I make no interpretive claim that they represent what Hempel actually did, or should have, thought.

Hempel's (hypothetical) answer to our title question is insightful, and can be generalized. The second, very brief, case study uses J. M. E. Moravcsik's interpretation of Aristotelian explanation as a vehicle for generalizing the insights gleaned from the first case study.

Finally, the third part of this paper sketches a general answer to the question, 'How do scientific explanations explain?' I will claim that scientific explanations in general explain by answering *what-questions*. This part of the paper is only programmatic; it attempts to illustrate the plausibility of its proposed approach to the general epistemological question.

1. How Hempelian Explanations Explain

While their logical analyses are similar, Hempelian event explanations and Hempelian law explanations explain in different ways. These two ways that Hempelian explanations explain cut across the D-N and I-S logical categories. Hempelian event explanations operate by 'filling out' or *amplifying* the nature of a specific event-pair; they tell us, for a given event, its nomically related antecedent conditions. By contrast, a Hempelian law explanation operates by *orienting* the regularity within a hierarchy of regularities; it tells us what a regularity is by specifying its place in the nomic hierarchy.

Before clarifying these two different modes of explaining, we need to emphasize some of Hempel's ontological presuppositions. Since Hempel mainly had logical concerns, he discussed 'the explanandum' and 'law statements'. But we want to know how these statements explain the phenomena; we are interested in the world which Hempel presupposes. Thus, instead of concentrating on statements of laws and events, we will be concerned about the events and regularities themselves. Here, we follow Hempel in construing the explanandum phenomena quite loosely; risking some confusion for convenience, we will use the term 'event' to stand for events, states, conditions, occurrences, and sets of these. Regularities are patterns in the occurrence of events.

1.1. How Hempelian Event Explanations Explain

For Hempel, explainable events do not occur independently, but rather come paired up (Hempel, 1965a, p. 362). Hempelian explanations thus presuppose a certain realm of discourse; any realm of discourse in which Hempelian explanations are possible must basically contain various kinds of event-pairs. For example, in D-N cases, 'there are certain empirical regularities, expressed by the laws L1, L2, . . ., Lr, which imply that whenever conditions of the kind indicated by C1, C2, . . ., Ck occur, an event of the kind described in E will take place' (Hempel and Oppenheim, 1948, p. 250). Similarly, I-S explanations also presuppose certain kinds of regularities; Hempel says that probabilistic laws 'assert certain peculiar, namely probabilistic, modes of connection between potentially infinite classes of occurrences' (1965a, p. 377). We might characterize a paradigmatic Hempelian event-pair as follows:

> A Hempelian event-pair is a pair of events (conditions, etc.) c and e, whose kinds C and E are nomologically related.

Given this rough characterization, we can see that there are two important aspects to any such Hempelian pair: (1) the kind C, and (2) the kind E. For Hempel, if an event (of kind E) is explainable, there is an event of kind C connected with it. Event explanation assumes that the kind of event in question is part of some C–E pair. To understand the nature of any Hempelian event-pair thoroughly, we must know its two important aspects, its related kind C and kind E.

In discussing how event explanations explain, Hempel is most emphatic in linking their effectiveness to their ability to provide a certain kind of expectation. That is, such explanations show us that the event's occurrence was to be expected, given the lawlikeness of nature (see, for example, Hempel (1965a), p. 337 and Hempel (1959), p. 304). If we know the antecedent conditions and the laws, we are properly prepared for the event's occurrence. After the event's occurrence, the explanation shows us we could have been thus prepared, with certainty (with D-N explanations), or with very high likelihood (with I-S explanations). The effectiveness of event explanations depends on a sort of hindsight quality; this is especially obvious with I-S explanations, where we oddly speak of being able to expect 'with very high likelihood' an event which presumably has already occurred.

Thus, Hempel construes the why-question of an explanation request as a kind of *what-question*; we explain why by explaining what are the nomically connected antecedent events. In a key passage, Hempel writes, 'the question "*why* does the phenomenon occur?" is construed as meaning "according to what general laws, and by virtue of what antecedent conditions does the phenomenon occur?"' (1948, p. 246). We might call these requests for the missing antecedent half of an event-pair *amplification requests*: such event explanations explain by amplifying, or 'filling out', the specific nature of a particular event-pair. They tell us, for an event of a given kind, what kind of events are nomically connected to it. The characterization of a paradigmatic Hempelian event-pair above displayed two nomically related aspects as important: the kind C and the kind E. A thorough understanding of an event-pair requires a knowledge of both.

From these two important aspects, we can construct the two amplification requests appropriate to any Hempelian event-pair: (1) Given that this event (of an event-pair) is of kind E, what is its nomically connected kind C? (2) Given that this event (of an event-pair) is of kind C, what is its nomically connected kind E? The former kind of amplification request arises in requests for explanation; the latter arises in requests for prediction. The amplification requests show that explaining an event is really a matter of amplifying an event-pair.

To put it another way, it might seem a mystery why citing one thing (a cause) provides understanding of another (an effect). Hempel can account for this mystery, as we saw in the 'key' passage above, by equating the 'why' with a particular kind of 'what'. Citing an Hempelian cause (nomically connected antecedent conditions) explains the explanandum phenomenon precisely because we were never simply asking about the explanandum phenomenon alone; we were really asking for an amplification of an event-pair (not just the explanandum phenomenon) of which we knew only one half. In explaining, we do not say what the event in question is, but rather what its 'connected half' is.

1.2. How Hempelian Regularity Explanations Explain

Event explanations explain by providing a kind of hindsight expectation, and by providing the 'missing connected half'. But regularity explanations cannot explain in the same way; they cannot show how, at some earlier time, we could have properly expected their 'occurrence' (with certainty or otherwise). According to Hempel, we cite regularities to explain other regularities. But the regularities used in explaining are themselves patterns among connected event-pairs, not connected 'regularity pairs' or 'event-regularity pairs'. The general laws cited to explain a regularity do not assert the nomic connection between the explanandum regularity and something else. Regularity explanations do not amplify; they supply no 'missing half'.

Regularity explanations do answer a kind of what-question, for they do tell us what the regularity in question is. They do show the regularity was to be expected, but the kind of expectedness is importantly different. The expectedness here is linked not with occurrence and hindsight, but with 'fit' in a nomic hierarchy. A regularity explanation does not *amplify* the nature of a particular regularity, but rather *orients* the regularity relative to other regularities. Regularity explanations show a regularity to be reasonable or proper by showing it to be a special case of one or more (more comprehensive) regularities. Hempel says, 'when the explanandum is not a particular event, but a uniformity . . ., the explanatory laws exhibit a system of more comprehensive uniformities, of which the given one is but a special case' (1966, p. 54).

Hempel speaks of an explanatory hierarchy which is a *logical* one (1959, p. 300), but it is less clear how the regularities, not the statements, are ordered. The regularities themselves do not all fall into a simple taxonomical tree ordered by genus and species. Some regularities are indeed simply species of others. For example, the regularity captured by the law, 'Wood floats on water' is a species of the regularity captured by the law, 'A solid body with a specific gravity less than that of a given liquid will float in that liquid'. The relata of the genus regularity are more comprehensive than the relata of the species.

But Hempelian regularities are not only ordered on this genus-species analogy, for Hempel maintains that a regularity can be a *manifestation* of several other regularities; the genus-species analogy fails here. For example, the regularity captured by the statement, 'Car radiator crackings are regularly connected with certain (specified) conditions of temperature, radiator content, etc.' is explained by showing it to be a manifestation of several other regularities regarding the behaviour of water at various temperatures near freezing, and so on. Hempel states, 'What scientific explanation, especially theoretical explanation, aims at is . . . an objective kind of insight that is achieved by a systematic unification, by exhibiting the phenomena as *manifestations* of common underlying structures and processes that conform to specific, testable, basic principles' (1966, p. 83, my emphasis, see also pp. 70, 71 and 75). Although there is no substantive logical difference between an explanation requiring only one subsuming law and an explanation requiring many, there are substantive epistemological differences between showing a regularity to be a species of a general regularity and showing a regularity to be a manifestation of several fundamental regularities.

The more fundamental 'underlying' regularities are not just spatiotemporal parts of a manifest regularity. Nor can we say that the event-pairs of the fundamental regularities are always construable as microconstituents of the manifest regularity's event-pairs. This may sometimes be the case, but microreduction is a subset of law explanation, and it cuts across the distinction between genus-species and fundamental-manifest. For those primarily interested in exegesis, much clarification needs to be done here. But for our purposes, we may roughly say that the nomic hierarchy's top levels contain the fundamental regularities, and the low levels contain the increasingly more manifest regularities. In addition, regularities at all levels may have other regularities as species. A regularity explanation *orients* the given regularity within this complex hierarchy by showing it to be a special case (a species or a manifestation) of others.

Thus, a regularity explanation answers a what-question; it explains what sort of entity the regularity in question is by orienting it in relation to others. For his logical project, Hempel did not need to distinguish orientation from amplification, for both are cast in a similar logical form. But if we look at how each explains, we see that there is an important epistemological difference. Amplifications supply 'missing' nomic relata; orientations do not. Although regularity explanations, like event explanations, use general laws in the explanans, these general laws do not assert the nomic connection between the explanandum regularity and something else. Indeed, these explaining laws do assert connections; they assert connections among event-pairs, as all regularities do. But whatever the logical similarities, the explaining laws do not treat the explanandum regularity as a relatum; orientations do not explain in the same way amplifications do. Rather, orientations say what the entity (regularity, in this case) in question is: either a species of some more general regularity, or a manifestation of several more fundamental ones.

Obviously, we separate the amplification of event explanations from the orientation of regularity explanations for our epistemological investigations, even though in practice many explanation requests may be for a combination of both. In some explanatory contexts, we are requesting amplification, we ask for *causes*, and minimal covering laws are all we require. In other contexts, we request orientation; e.g. we may ask for an explanation of thermodynamics' regularities, relative to those of statistical mechanics. In yet other contexts, as perhaps in Hempel's well known car radiator example, we may be asking for both causal amplification and nomic orientation.

With some digging then, we see that Hempel's logical analysis is supplemented by some important insights on how scientific explanations explain. If one accepts kinds of event-pairs as basic, then in order to understand a given event-pair we must know both parts. Event explanations are attempts to supply the requested nomically connected parts; understanding an event is really a matter of understanding it as part of an event-pair. If one accepts Hempel's hierarchical ordering of regularities, then a proper orientation will specify not only that the explanandum entity is either a regularity or a part of one, but (for regularities) will also specify that regularity's relation to others in that hierarchy.

The epistemological difference between orientation and amplification is just as important as their logical similarity. Hempel states, 'all scientific explanation . . . seeks to provide a systematic understanding of empirical phenomena by showing that they fit into a nomic nexus' (1965a, p. 488). But the 'fit into a nomic nexus' is different for events and regularities. For regularities, as we have seen, Hempel assumes a sort of regularity hierarchy; the nexus is a system of regularities with respect to which we can orient a given regularity. But we do not explain events by orienting them into some global, nomic system of 'all events'. Rather, in requesting an event explanation, we assume that the given event is nomically connected to some others, and we request *these*.

Hempel can thus be seen as giving an account of how Hempelian explanations explain. But Hempelian explanations require a Hempelian realm of discourse, one basically consisting of a particular hierarchically ordered class of regularities. We might wish to generalize his account to other realms of discourse. In fact, J. M. E. Moravcsik's interpretation of Aristotelian explanation shows us how we might generalize the Hempelian account. After a brief look at Moravcsik's interpretation then, we will sketch an outline of a general account of how scientific explanations explain.

2. How Aristotelian Explanations Explain

In 'Aristotle on Adequate Explanation' (1974), J. M. E. Moravcsik interprets Aristotle as giving an analysis of explanation. Moravscik points out that substances are the Aristotelian basic entities. For Aristotle, all important facts about the world can be stated in terms of substances and their properties; to ask for an explanation is really to ask about substances. A paradigmatic Aristotelian substance, says Moravcsik, may be characterized as: 'a set of elements with a fixed structure that moves itself toward self-determined goals' (1974, p. 5). Substances are characterized by four important aspects, or *aitiai*: material cause (elements), efficient cause (motion-initiating factor), formal cause (structure), and final cause (functional factor). Moravscik says, 'What gives unity to the list of four is that they jointly provide a complete explanation of the nature of a substance, and thus Aristotle supposes that they must contain the ingredients of a complete explanation of any other phenomena, since these must be in some way dependent on substances' (1974, p. 6).

So, Moravcsik tells us, Aristotle's explanations operate in a realm of discourse composed basically of substances. To explain the nature of a substance is to display its *aitiai*, the important things we need to know about its nature. Moravcsik points this out clearly:

> Thus for x to be an aitia of y is for x to be in a relation to y such that the grasp of the relation enables one to understand some important aspect of y, e.g. how it functions, what its nature is, etc. From this it follows that the aitia-relation is an ontological relation, holding independently of the language and the psychology of any given investigator (1975, p. 624).

Given the characterization of a paradigmatic Aristotelian substance which Moravcsik has provided, the four aitiai and the four amplification what-questions are readily apparent. The amplification what-questions which we could ask about a given substance are: (1) What is its material cause? (2) What is its efficient cause? (3) What is its formal cause? (4) What is its final cause? What we have been calling amplification requests, Moravcsik calls dia ti questions. A substance's 'dia ti' are the factors that are responsible for its nature. Moravcsik states, 'The "dia ti"-phrase requests information about certain types of factors, e.g. how an entity functions, what its constituents are, etc. The "aitiai" are the entities referred to in the answers given to the "dia ti" questions' (1975, p. 624). Aristotle's four aitiai and four amplification what-questions follow directly from his commitment to a particular realm of discourse, one in which Aristotelian substances are basic.

3. How Scientific Explanations Explain

Scientific explanations explain by telling us of an entity what it is. Explanation requires a context, some guidelines for acceptable description. These guidelines are of many sorts, including some sort of prior commitment to general ontology, as well as subject matter. Let us further clarify the distinction between the guidelines of ontological commitment and subject matter.

In explaining, we must accept some prior guidelines of what sorts of things are fundamental objects of explanation, and what sorts of things about them are important to know. I call this background commitment a realm of discourse. A realm of discourse specifies what entities are basic, and what aspects of those entities are important to know.

A realm of discourse is more general than what is often thought of as the 'subject matter' of a field of science. For example, Dudley Shapere states, 'Every inquiry, or at least every inquiry that aims at knowledge, has a subject-matter—that which the inquiry is about, which is the set of objects studied' (1984, p. 320). He calls the set of objects studied the 'domain' of the inquiry. Chemical phenomena, for example, (partly) constitute the domain of chemistry. Shapere sets aside the question of what the 'objects' or 'items' of a domain are; he chooses the 'neutral' term of 'items' to cover objects, processes, events, and so on. The domain thus centres on the subject matter of the items rather than on their ontological kind; it emphasizes, for example, that chemistry's domain may include *chemical* substances, rather than that it includes *substances*. By contrast, a realm of discourse specifies what sort of items—e.g., objects, processes, events—can inhabit a domain, while a given domain specifies which particular items do inhabit it. The realm of discourse also specifies what aspects of the objects, processes, etc. are important to know. In the discussion of Hempel above, we have seen that Hempelian explanations presuppose a realm of discourse in which event-pairs are basic. Hempel's commitment to this realm of discourse is more general than a commitment to chemical 'items' or electromagnetic 'items'. A commitment to Hempelian explanations commits one to treating event-pairs as fundamental in any field or subfield of science. It commits one to treating all items in all domains as redescribable in terms of Hempelian event-pairs. Hempel thus controversially urged historians to 'reform' their explanations in terms of a Hempelian realm of discourse. Moreover, a commitment to a Hempelian realm of discourse is also a commitment to a certain specification of what are the important aspects of these event-pairs. Aristotle, on the other hand, would perhaps have been committed to biological substances, chemical substances, etc., rather than to Hempelian biological regularities, etc.

One might, at this point, ask for a concrete example to illuminate the notion of realm of discourse. What, one might ask, is the realm of discourse presumed by chemistry or physics? Unfortunately, it is not as uncontroversial to specify a realm of discourse as it is to specify a domain. It is not possible to provide a detailed realm of discourse for (e.g.) chemistry or physics in order to clarify and illuminate the general notion of realm of discourse, without taking a specific philosophical stand. To specify a realm of discourse is to specify one's ontological and epistemic commitments, and philosophers may come to (scholarly) blows over these. Obviously, Hempel and Aristotle would not agree on the appropriate realm of discourse for (e.g.) biology, even though they might agree on its appropriate domain.

Thus, while it is quite true that scientific explanations within any given scientific field presuppose a given subject matter, they also presuppose a more general conception of what even counts as a 'domain item' and what we need to know about it. A realm of discourse, organized into basic entities and their aspects, thus sets the epistemic stage upon which our more specific field interests can play. What we count as a basic entity determines the range of appropriate ways of explaining what something is. The set of aitiai, or important aspects, of any basic entity provides a complete explanation of the nature of that entity (for that realm of discourse).

Before examining the topic of what-questions a bit more fully, let us pause to make two clarifications. The first is a lesson taught by Shapere and others who attend carefully to the historical development of science. Realms of discourse, like domains, are rarely born full grown. We may begin with some commitments to what sort of entities are basic and what their important aspects are, but clearly these commitments may evolve. A commitment to substances sharply distinct from events or processes may shift to a commitment to some more hybrid entity. A commitment to regularities may shift from those that are absolute and constant to those 'merely' statistical. Some sort of commitment is a necessary precondition to explanation, but our commitments are fluid; they can change as we learn. And the more radically they change, the more radically the nature of our explanations changes.

The second clarification concerns my rather loose use of the phrase 'ontological commitment'. Scientific explainers must be committed to event-pairs, or substances, or whatever, as epistemically basic. But their commitment may be to a full blown ontological view of what there is, or it may merely be a commitment to a convention or a model of illuminating convenience. The commitment to a realm of discourse is, strictly speaking, less specific than a commitment to a given ontology or a commitment to the truth of some given theory. Bas van Fraassen (1980) promotes an anti-realist stand on scientific activity. He claims that accepting a theory involves not the belief that it is true in all respects, but rather the belief that it is observationally adequate. I mean to steer clear of this realist/anti-realist debate, for it is peripheral to the question of how explanations explain. Explanations require one to hold a realm of discourse, and whether one clasps it to one's heart or holds it at arm's length is a separable topic.

3.1. What-Questions

Scientific explanation is either orienting the thing to be explained in a given realm of discourse, or amplifying a basic entity by providing a specification for the aspect in question. That is, explanation is tied to two different types of what-question, orientation requests and amplification requests. Orientation requests generally take the bland form, 'What is this?' The very blandness of this request displays the need for context. While the domain limits the subject matter of the acceptable explanation, the realm of discourse limits its form. The answer to an orientation request identifies 'this' as a certain basic entity or a certain aspect of a certain basic entity. For example, a Hempelian regularity explanation might identify 'this' as a manifestation of specific fundamental regularities.

It is important to emphasize that the focus here is *not* on the particular syntax of what-questions. That they use the word 'what', rather than the words 'why', 'who', etc., is not the issue here. Rather, the focus is on what such questions *do*: they ask for amplifications or orientations. And what counts as a proper amplification or orientation is determined in part by the realm of discourse accepted.

Answers to orientation requests have two different uses. On the one hand, they may be used to bring completely unexplained phenomena under some acceptable description. For example, early chemists asked 'What is this?', where 'this' was what we now refer to as heat. Their question was not what caused heat, but rather what sort of entity it was; they were not sure if it was a substance, an effect, a process, or an activity. On the other hand, answers to orientation requests may be used to bring previously explained phenomena under a new description. One familiar example of this latter use is reduction. Hempel's law explanations, which are not always reductive, are another example.

Amplification requests seek a particular aspect of a given basic entity. For example, Hempel's event explanations specify, for a given event-pair, its kind of antecedent conditions. Orientation requests seek orientation in the realm of discourse; amplification requests seek amplification of an already oriented entity. An amplification request for Hempelian antecedent conditions clearly presupposes an orientation; it presupposes that the explanandum event is the consequent-half of a Hempelian event-pair. An amplification request for an Aristotelian final cause presupposes that the item in question is a substance with a final cause. A minimal orientation is prior to amplification: In order to ask for an amplification, a 'detailing' of a basic entity, we must already have identified something as a particular basic entity having some particular set of important aspects.

Orientation is epistemically prior to amplification. Temporally, the two sorts of explanation may be closely intertwined. Amplification may proceed from some minimally presumed orientation, and subsequent amplification research may refine the specification of the basic entities. Moreover, success or failure in amplification research may lead one to overthrow the orientation presumption.

Not any question may be appropriately queried. Clearly, the domain exerts obvious limits; certain questions of subatomic phenomena are inappropriate in some biological domains. But the realm of discourse also limits proper questioning. Each realm of discourse has its own set of appropriate amplification requests; the characterization of the basic entities determines which questions are properly allowed. Inappropriate questions are inappropriate relative to a realm of discourse; questions appropriate for an Aristotelian realm of discourse are inappropriate, strictly speaking, for a Hempelian one. To ask for a substance's final cause, within a Hempelian realm of discourse, is to misidentify the thing to be explained. Such mistaken presuppositions will cause a question to be rejected as inappropriate.

There are minimal formal constraints on successful explanation. The constraint of domain must be respected. Moreover, the request for an explanation and the explanation itself must be within the context of the same realm of discourse. If the question presumes a realm of discourse different from that of the answer, then the explainer has not answered that what-question; questioner and answerer are talking past each other. The explanation must also properly answer the what-question. If the request is for orientation, the explanation must pick out the proper orientation. If the request is for amplification, the explanation must correctly tell us what the requested important aspect is.

These minimal formal constraints do not require realms of discourse to be the best current ones available to science. Such a requirement would limit the real explanations to a few utterances among a few scientists at best; utterances could then be dubbed 'explanation' and robbed of their rank as the theoretical vogue changed. Rather, explanations can be formally successful, yet *passé*. We must separate the problem of falsely answering an amplification request (and thereby giving no explanation) from the problem of holding a *passé* realm of discourse. For example, one might have asked a third century Alexandrean alchemist, 'What is this event?' and be told, quite legitimately, that it is a case of a certain substance being brought from a manifest state to an occult state. If the explanation was formally successful then, it is formally successful still, even if one rejects the ontology of substances.

3.2. Acceptability

This brings us to an important point. So far, we have only discussed explanation as that which orients or amplifies within some domain and realm of discourse. We have not discussed which realms of discourse one ought to accept. If explanation tells us what something is, then we will want to know if we have picked the right characterizations, the right set of basic entities to begin with.

There are at least two different things a general epistemological theory of explanation must do. One is to specify how explanations explain; the other is to specify what criteria an explanation must meet in order to be acceptable. These are important, and separable, tasks. It is beyond the scope of this paper to provide a theory of explanatory acceptability; the mainly negative remarks that complete this section are offered only to indicate the complexity required by such a theory, and the need for further work.

The criterion of truth might be one candidate for a constraint on realms of discourse. It might seem that Hempel has combined a theory of acceptability with his model of explanation. He limits the concept of law to *true* laws. According to Hempel, an event explanation explains by showing the event to be that which we ought to have expected, given our knowledge of the true laws of nature. The truth of the laws ensures the acceptability of the explanation.

Hempel explicitly relies on Tarski's formal account of truth. But as Hilary Putnam (1981) points out, Tarski's convention T is philosophically neutral, and hence the truth requirement is itself empty. Putnam writes, 'That science seeks to construct a world picture which is *true* is itself a true statement; an almost empty and formal true statement; the aims of science are given material content only by the criteria of rational acceptability in science. In short . . . *truth is not the bottom line*' (1981, p. 130, his emphasis). To claim that the explanation must be true, or that the entities must be real, is not yet an adequate theory of acceptability.

Much work has already been done towards specifying a theory of acceptability. This work has produced long discussions of many familiar criteria such as testability, universality, simplicity, observational adequacy, consistency, predictive power, fruitfulness, and so on. While this work has been important, even a full list of such criteria, together with exhaustive analyses of each, would not alone be adequate as a theory of acceptability. We must also specify how each is to be weighed, individually and relative to one another. This task may prove complex indeed, if the relative importance of these depends on a science's aims, interests, and stage of development. For example, Ian Hacking (1981) 'worries' that Lakatos' criterion of theoretical and experimental progressiveness is applicable only to the science of the past two centuries. Generalizing Hacking's worry, if our criteria of acceptability and/or our evaluation of what satisfies these criteria are historically indexed, then specifying a general theory of acceptability would then become partly, but integrally, an historical exercise.

Conclusion

In the account sketched above, the notion of context, background etc., is preserved not so much in terms of domain, but rather in terms of realm of discourse. The unifying action of explanation is preserved not so much in terms of a numerical decrease, but rather both in this notion of realm of discourse, and in the notions of amplifications and orientation, which operate within a realm of discourse.

Much more work needs to be done to make this general sketch into a real answer to our title question. Still, perhaps even this minimal account may be illuminating. How do scientific explanations explain? Broadly, scientific explanations explain by answering what-questions; answers to what-questions either orient us by identifying what something is, or they amplify the important aspects of an entity which is basic to a given realm of discourse.

References

Friedman, M. 1974. 'Explanation and Scientific Understanding', *Journal of Philosophy* **71**, 5–19.

Hacking, I. 1981. 'Lakatos's Philosophy of Science', in *Scientific Revolutions* I. Hacking (ed.), (Oxford University Press), 128–43.

Hempel, C. G. 1959. 'The Logic of Functional Analysis', repr. in Hempel (1965b), 297–330.

Hempel, C. G. 1965a. 'Aspects of Scientific Explanation', in Hempel (1965b), 331–496.

Hempel, C. G. 1965b. *Aspects of Scientific Explanation* (New York: The Free Press).

Hempel, C. G. 1966. *Philosophy of Natural Science* (Englewood Cliffs, N.J.: Prentice-Hall).

Hempel, C. G. and Oppenheim, P. 1948. 'Studies in the Logic of Explanation', repr. with Postscript (1964) in Hempel (1965b), 245–95.

Kitcher, P. 1981. 'Explanatory Unification', *Philosophy of Science* **48**, 507–31.

Moravcsik, J. M. E. 1974. 'Aristotle on Adequate Explanation', *Synthese* **28**, 2–17.

Moravcsik, J. M. E. 1975. '*Aitiai* as Generative Factor in Aristotle's Philosophy', *Dialogue* **14**, 622–38.
Putnam, H. 1981. *History, Truth and Reason* (Cambridge University Press).
Shapere, D. 1984. *Reason and the Search for Knowldge* (Dordrecht: D. Reidel).
van Fraassen, B. 1980. *The Scientific Image* (Cambridge: Clarendon Press).

Index

Index

Index

Index